Bistabilities and Nonlinearities
in Laser Diodes

For a complete listing of the *Artech House Optoelectronics Library*,
turn to the back of this book.

Bistabilities and Nonlinearities
in Laser Diodes

Hitoshi Kawaguchi

Artech House
Boston • London

Library of Congress Cataloging-in-Publication Data
Kawaguchi, Hitoshi
Bistabilities and nonlinearities in laser diodes/Hitoshi Kawaguchi
Includes bibliographical references and index.
ISBN 0-89006-671-X
1. Semiconductor lasers. 2. Optical bistability. 3. Nonlinear theories. I. Title.
TA1700.K39   1994   94-7669
621.36'6–dc20   CIP

British Library Cataloguing in Publication Data
Kawaguchi, Hitoshi
Bistabilities and Nonlinearities in Laser Diodes
I. Title
621.366

ISBN 0-89006-671-X

© 1994 ARTECH HOUSE, INC.
685 Canton Street
Norwood, MA 02062

International Standard Book Number: 0-89006-671-X
Library of Congress Catalog Card Number: 94-7669

10 9 8 7 6 5 4 3 2 1

Contents

Preface

Since its invention in 1962, the semiconductor laser has come a long way. It was in 1970 that the successful continuous operation of a heterostructure laser at room temperature was demonstrated. The technologies of index guiding and distributed feedback have improved the characteristics of semiconductor lasers and have made possible practical fiber communication systems. Recently, new optical systems such as photonic switching and optical signal processing have been expected to offer broadband information services. Optical functional devices are the key to opening the door to such advanced optical systems. The first attempt to use semiconductor lasers as functional devices was the proposal of the bistable semiconductor laser in 1964. This came very shortly after injection lasing of semiconductors.

This book was motivated by a desire to describe optical functional devices based on nonlinearities in semiconductor lasers and to discuss the potential applications of these devices for optical signal processing. This book, to my knowledge, is the first single-author publication to cover this wide range of topics.

The book is about a very rapidly expanding field. This fact has made it difficult to decide when to publish. Much of the underlying physics of the behavior of these devices is now understood, even though intense studies on ultrafast nonlinearity mechanisms, semiconductor microstructures, and microcavities will surely result in many new break-throughs. It is hoped that this book captures the beautiful and diverse behavior of bistability and nonlinearities of laser diodes in a manner that will serve both as a basis for further physics and device research and as a ready reference for optical signal processing.

In addition and complementary to the specific topics concerning bistability and nonlinearities in laser diodes, the first part of this book reviews basic laser characteristics, so that even a reader without detailed knowledge of laser diodes may follow the text. In order to understand the book, the reader should have a basic knowledge of electronics, semiconductor physics, and optical communications. It is primarily written for the engineer or scientist working in the field of optoelectronics and its system applications; however, since the book is self-contained, it may serve as a textbook for graduate students.

I would like to express my thanks to my student Kazi Sarwar Abedin for his critical comments and review of the manuscript. This book also includes acknowledged and unacknowledged contributions from Katsuaki Magari, Hiroshi Yasaka, Kenju Otsuka, Ian H. White, John E. Carroll, and Tomoyoshi Irie. The optoelectronics class for graduate students at Yamagata University in 1993 helped ferret out inconsistencies and insisted on clearer presentations.

My wife Kyoko is responsible for the typing, and my assistants Eriko Mikami and Naomi Maruyama for the preparation of figures. To them and to all of the above, my gratitude.

Chapter 1
Introduction

1.1 DISTINCTIVE FEATURES OF FUNCTIONAL SEMICONDUCTOR LASERS

It is amazing that simple two-section semiconductor lasers with the electrode separated into two parts show bistability and self-pulsation as well as mode locking, depending on the operating conditions, with only the application of dc bias voltage. Optical bistability, as the name implies, refers to the situation in which two stable optical output states are associated with a single input state. Self-pulsation is a repetitive strong intensity change in output power at repetition rates on the order of gigahertz. Mode-locked semiconductor lasers generate ultrashort optical pulses in the low picosecond to subpicosecond range. The nonlinearities in semiconductor lasers also cause four-wave mixing (FWM), which will be used as the physical mechanism for constructing optical-wavelength converters.

In the present optical transmission systems, these nonlinear phenomena have not been welcome. The self-pulsation in aged semiconductor lasers sometimes degrades the bit error rate. FWM in traveling-wave laser diode (LD) amplifiers causes the crosstalk in multichannel communications. However, functional devices based on semiconductor laser nonlinearities are expected to become the key devices in future communication and signal processing systems, because if the optical signal can be directly treated without conversion to electrical signals, an extremely flexible communications network can be created. Moreover, it is considered that the speed limitation of electronic devices will be several tens of gigabits per second. For signal processing at higher speeds, the use of all-optical devices will be inevitable. Ian M. Ross presented the following comments in his talk entitled ''Telecommunications in the Era of Photonics'' on the occasion of the IEE's bicentennial observation of the birth of Michael Faraday [1]: ''By the end of the decade, it is reasonable to assume that 20 percent of the components in high performance computers could be photonic and that this could rise to 50 percent by the year 2010. The expected application is in interconnect functions; logic functions are less clear. However, speed is an obvious objective in logic design, and optical technique can achieve switching in the subpicosecond range, while electronics reaches a limit at about 10 picoseconds.''

The most distinctive feature of functional devices based on LD nonlinearities is that they have optical gain. For example, in bistable devices, optical gain results in the advantages of low optical switching power, a high ON-OFF ratio, and a large fan-out. The need to reduce the electrical biasing power is one of the most important problems with such devices. This is realized by reducing the LD threshold current. LDs with threshold currents of less than 1 mA have already been reported by some research groups. For example, strained V-groove lasers with a threshold current as low as 203 μA have been demonstrated at room temperature [2]. It is believed that by using advanced technologies such as quantum-confined effects, LDs with a much lower threshold current (1 μA) will be achieved [3]. Yamanishi and Yamamoto have proposed a semiconductor surface-emitting laser structure in which all spontaneous emission is coupled into a single lasing mode by means of a quantum microcavity, and discrete electron-hole pair emission is made free of absorption by means of a dc-biased quantum dot. They estimated that the threshold current of such an LD can be reduced to below 100 nA [4]. In LDs, there are ultrafast nonlinear mechanisms due to the nonequilibrium carrier distribution in the bands, such as carrier heating or the spectral hole-burning effect. Relaxation is characterized by the intraband relaxation time on the order of 10^{-12} to 10^{-13} sec. This nonlinearity has been suggested as a dominant source of the nonlinear gain, which strongly influences the bandwidth of the direct modulation of LDs. It may be possible to construct functional devices using such ultrafast nonlinearities.

1.2 HISTORICAL PERSPECTIVE

Semiconductor lasers have been developed mainly for optical transmission. Development steps for semiconductor lasers are shown in Table 1.1. The advent of the semiconductor laser dates back to the early 1960s. The feasibility of stimulated emission in semiconductor lasers was considered during this period [5–7], and in 1962 several groups reported the lasing action in semiconductors [8–11]. As early as 1963 it was suggested [12] that semiconductor lasers might be improved if one layer of semiconductor material were sandwiched between two cladding layers of another semiconductor that has a relatively wider bandgap (i.e., a heterostructure laser). It was in 1969 that the successful room-temperature operation of a heterostructure laser was demonstrated [13–15] using the liquid-phase epitaxial (LPE) technique for the growth of GaAs and $Al_xGa_{1-x}As$ layers. However, these lasers operated in the pulsed mode. Further work led in 1970 to heterostructure lasers operating continuously at room temperature [16,17]. Then reliability and control of the transverse mode in optical-fiber communication systems became important. The technology of index guiding has improved the characteristics of semiconductor lasers and has made fiber communication systems practical. In the late 1980s, the ultrahigh-bit-rate

Table 1.1
Development Steps of Laser Diodes

Years	1960	1970	1980	1990
LD Quality	Lasing	CW operation	Single-transverse mode	Single-longitudinal mode	Functions
Physics		Confinement	Index guiding	Wavelength filter	Nonlinearity
Structure	Homojunction	DH	BH	DFB, DBR	BLD, LD Amp.
Application			Fiber comm. CD	High bit rate	Photonic switching
				fiber comm.	Optical signal
				Coherent comm.	processing

fiber communication systems operating above 1 Gbps required dynamic single-longitudinal mode lasing. Recently, new optical systems such as photonic switching and optical signal processing have been expected to offer broadband information services. Optical functional devices are the key to opening the door to such advanced optical systems.

The first attempt to use semiconductor lasers as functional devices was the proposal of the bistable semiconductor laser by Lasher in 1964 [18]. This came very shortly after injection lasing of semiconductors. Nathan et al. [19] soon reported bistable operation in a GaAs laser using Lasher's structure. Basov et al. [20] proposed optical logic using Lasher's double diode. However, after these pioneering works, there was no research on or development of bistable LDs (BLDs) for a decade. LD technology, however, was rapidly developed for use in optical-fiber communications. In 1970, continuous wave (CW)–stimulated emission was demonstrated in GaAs/AlGaAs at room temperature. InGaAsP/InP lasers were also developed as an optical source in the wavelength region of 1.3 to 1.5 μm, in which silica fiber loss is minimal. Based on these advanced technologies, the second stage of bistable semiconductor laser study was opened in 1981 by Kawaguchi and Iwane [21] and Harder et al. [22]. They reported remarkable bistable characteristics in InP/InGaAsP and GaAs/AlGaAs lasers, respectively.

Another important research area of functional semiconductor lasers is the short optical-pulse generation. The techniques fall into two general categories: relaxation oscillation and mode locking. Lee and Roldan [23] and Basov et al. [24] used two-section LDs to produce pulses 10 to 100 ps long under pulsed current excitation. The repetition rate of the pulses was determined by the relaxation time of the system and was not directly connected to the round-trip time of a pulse within the system. Paoli and Ripper [25] generated pulses of fractions of nanoseconds by the so-called *second-order mode-locking*, technique where the natural relaxation frequency lies close to the frequency difference of two adjacent axial-mode frequency separations. Morozov et al. [26] used a two-section LD, one for gain and the other as an absorber in an external resonator. The idea was sound, but ahead of its time, because the diodes had poor mode quality and had to use pulsed operation. Harris [27] used an antireflection-coated diode in an external resonator and modulated the drive current. He observed pulses of 1 to 2 ns duration. Observation of a 0.3-ps substructure in the second harmonic generation (SHG) trace from a pulse-excited diode was reported [28], but the results suffered from poor reproducibility. The first CW mode locking of a semiconductor laser that produced reproducible pulses of 20 ps duration [29] and less [30] were carried out. To suppress the formation of multiple pulses with partial reflection on the LD facet and generate single-peak pulses, two-section lasers were used with an external cavity [31]. Two-section semiconductor lasers were passively mode-locked without an external cavity at a rate above 100 GHz by Vasil'ev and Sergeev [32] and Sanders et al. [33].

Intracavity nearly degenerate four-wave mixing (NDFWM) in a semiconductor laser was reported by Nakajima and Frey in 1985 [34], where by using the lasing light in the LD as counterpropagating pump beams, a light of different frequency with the same order of output power as the incident light appears. NDFWM in a traveling-wave semiconductor laser amplifier has been demonstrated by simultaneously injecting two lights with slightly different frequencies in the same direction. On the other hand, the highly nondegenerate FWM, whose detuning is more than 100 GHz, was theoretically predicted in 1987 by Agrawal [35] and was demonstrated experimentally by Provost and Frey in 1989 [36].

1.3 SYSTEM APPLICATIONS

In recent years, as shown in Table 1.2, various kinds of optical functional devices have been constructed based on the nonlinearities of semiconductor lasers, and many proposals and experiments on application of these devices have been reported in the field of optical-fiber communications and photonic switching.

Absorptive and dispersive bistable devices are obtained using inhomogeneously excited LDs and resonant-type LD amplifiers, respectively. Optical retiming was first demonstrated using a bistable Fabry-Perot amplifier decision gate. An optical time-division switching system was developed using BLDs as optical memories. Sometimes self-pulsing as well as bistability is seen in inhomogeneously excited LDs. The locking-in repetition frequency can be obtained through the injection of the optical pulses into the saturable absorber and can be used for all-optical timing extraction. Inhomogeneously excited LDs also produce ultrashort optical pulses in the low picosecond to subpicosecond range with or without an external cavity. These short optical pulses have very small jitters and are useful sources in the distribution of a clock via an optical fiber.

Optical wavelength-division switching is expected to facilitate a flexible switching network. In optical wavelength-division switching, the wavelength of the signal carried through the optical highway is converted to another wavelength by a wavelength conversion switch. Then one or more signals are selected by a wavelength selection switch to connect with the appropriate channel. The signal is reconverted to a different wavelength and sent into output optical fiber. A tunable wavelength conversion device has been constructed using a tandem-type bistable distributed feedback (DFB) LD. Using a two-electrode DFB LD amplifier, a narrow spectrum selection has been achieved. A frequency conversion based on FWM has been demonstrated. This frequency conversion scheme has the advantage of data transparency. Therefore, any kind of modulation is possible.

The organization of this book is as follows. We start by reviewing the lasing principle, basic device structure, dynamic characteristics, and nonlinear phenomena of semiconductor lasers in Chapter 2. Chapters 3 through 6 cover many aspects of BLDs, such as absorptive BLDs, dispersive BLDs, two-mode BLDs, and waveguiding BLDs, which are the highlights of this book. Chapter 7 presents ultrashort optical-pulse generation and pulse position BLDs. Chapter 8 discusses wavelength conversion using BLDs and FWM. Chapter 9 describes wavelength selection using resonant-type LD amplifiers. Although the physics behind this is linear amplification of the injected light by LD amplifiers rather than nonlinearity, these devices will become important in combination with wavelength converters, and this subject is therefore discussed. Chapter 10 describes

Table 1.2

Major Optical Functional Devices Using Semiconductor Laser Nonlinearity and Their Possible Applications

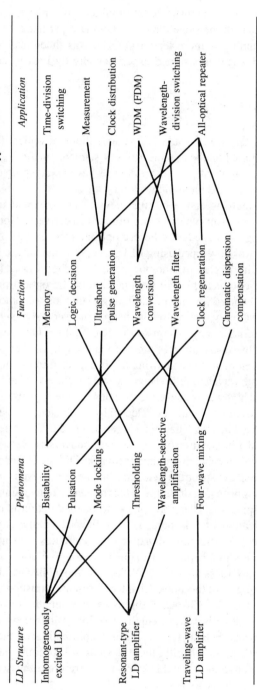

applications of functional semiconductor lasers for photonic switching. Short optical pulses have huge application areas, but here the interest primarily is in photonic switching (Section 10.1). Sections 10.2 through 10.4 cover all-optical repeater and optical exchanging systems using functional semiconductor lasers. Future developments are described in Chapter 11.

REFERENCES

[1] Ross, I. M., "Telecommunications in the Era of Photonics," *Solid State Technology*, Vol. 35, No. 4, April 1992, pp. 36–43.

[2] Tiwari, S., G. D. Pettit, K. R. Milkove, R. J. Davis, J. M. Woodall, and F. Legoues, "203μA Threshold Current Strained V-Groove Laser," *IEDM'92*, pp. 859–862.

[3] Yablonovitch, E., "Inhibited Spontaneous Emission in Solid-State Physics and Electronics," *Phys. Rev. Lett.*, Vol. 58, No. 20, May 1987, pp. 2059–2062.

[4] Yamanishi, M., and Y. Yamamoto, "An Ultimately Low-Threshold Semiconductor Laser With Separate Quantum Confinements of Single Field Mode and Single Electron-Hole Pair," *Japanese J. Appl. Phys.*, Vol. 30, No. 1A, January 1991, pp. L60–L63.

[5] Basov, N. G., O. N. Krokhin, and Yu. M. Popov, "Production of Negative-Temperature States in P-N Junctions of Degenerate Semiconductors," *J. Exptl. Theoret. Phys.* (U.S.S.R.), Vol. 40, June 1961, pp. 1879–1880.

[6] Bernard, M. G. A., and G. Duraffourg, "Laser Conditions in Semiconductors," *Phys. Status Solidi*, Vol. 1, 1961, pp. 699–703.

[7] Dumke, W. P., "Interband Transitions and Maser Action," *Phys. Rev.*, Vol. 127, No. 5, September 1962, pp. 1559–1563.

[8] Hall, N. R., G. E. Fenner, J. D. Kingsley, T. J. Soltys, and R. O. Carlson, "Coherent Light Emission From GaAs Junctions," *Phys. Rev. Lett.*, Vol. 9, No. 9, November 1962, pp. 366–368.

[9] Nathan, M. I., W. P. Dumke, G. Burns, F. H. Dill, Jr., and G. Lasher, "Stimulated Emission of Radiation From GaAs P-N Junctions," *Appl. Phys. Lett.*, Vol. 1, No. 3, November 1962, pp. 62–64.

[10] Quist, T. M., R. H. Rediker, R. J. Keyes, W. E. Krag, B. Lax, A. L. McWhorter, and H. J. Zeiger, "Semiconductor Maser of GaAs," *Appl. Phys. Lett.*, Vol. 1, No. 4, December 1962, pp. 91–92.

[11] Holonyak, N., Jr., and S. F. Bevacqua, "Coherent (Visible) Light Emission From Ga(As$_{1-x}$P$_x$) Junctions," *Appl. Phys. Lett.*, Vol. 1, No. 4, December 1962, pp. 82–83.

[12] Kroemer, H., "A Proposed Class of Hetero-Junction Injection Lasers," *Proc. IEEE*, Vol. 51, 1963, pp. 1782–1783.

[13] Kressel, H., and H. Nelson, "Close-Confinement Gallium Arsenide PN Junction Lasers With Reduced Optical Loss at Room Temperature," *RCA Review*, Vol. 30, March 1969, pp. 106–113.

[14] Hayashi, I., M. B. Panish, and P. W. Foy, "A Low-Threshold Room-Temperature Injection Laser," *IEEE J. Quantum Electronics*, Vol. QE-5, April 1969, pp. 211–212.

[15] Alferov, Zh. I., V. M. Andreev, E. L. Portnoi, and M. K. Trukan, "AlAs-GaAs Heterojunction Injection Lasers With a Low Room-Temperature Threshold, " *Sov. Phys. Semicond.*, Vol. 3, March 1970, pp. 1107–1110.

[16] Hayashi, I., M. B. Panish, P. W. Foy, and S. Sumski, "Junction Lasers Which Operate Continuously at Room Temperature," *Appl. Phys. Lett.*, Vol. 17, No. 3, August 1970, pp. 109–111.

[17] Alferov, Zh. I., V. M. Andreev, D. Z. Garbuzov, Yu. V. Zhilyaev, E. P. Morozov, E. L. Portnoi, and V. G. Trofim, "Investigation of the Influence of the AlAs-GaAs Heterostructure Parameters on the Laser Threshold Current and the Realization of Continuous Emission at Room Temperature," *Soviet Physics-Semiconductors*, Vol. 4, No. 9, March 1971, pp. 1573–1575.

[18] Lasher, G. J., "Analysis of a Proposed Bistable Injection Laser," *Solid-State Electronics*, Vol. 7, 1964, pp. 707–716.

[19] Nathan, M. I., J. C. Marinace, R. F. Rutz, A. E. Michel, and G. J. Lasher, "GaAs Injection Laser With Novel Mode Control and Switching Properties" *J. Appl. Phys.*, Vol. 36, No. 2, February 1965, pp. 473–480.

[20] Basov, N. G., W. H. Culver, and B. Shah, "Application of Lasers to Computers," in *Laser Handbook* 1, F. T. Arecchi and E. O. Schulz-DuBois, eds., Amsterdam, 1972, pp. 1649–1693.

[21] Kawaguchi, H., and G. Iwane, "Bistable Operation in Semiconductor Lasers With Inhomogeneous Excitation," *Electron. Lett.*, Vol. 17, No. 4, February 1981, pp. 167–168.

[22] Harder, Ch., K. Y. Lau, and A. Yariv, "Bistability and Pulsations in CW Semiconductor Lasers With a Controlled Amount of Saturable Absorption," *Appl. Phys. Lett.*, Vol. 39, No. 5, September 1981, pp. 382–384.

[23] Lee, T.-P., and R. H. R. Roldan, "Repetitively Q-Switched Light Pulses From GaAs Injection Lasers With Tandem Double-Section Stripe Geometry," *IEEE J. Quantum Electronics*, Vol. QE-6, No. 6, June 1970, pp. 339–352.

[24] Basov, N. G., V. V. Nitikin, and A. S. Semenov, "Dynamics of Semiconductor Injection Lasers," *Sov. Phys.--Uspekhi*, Vol. 12, No. 2, September/October 1969, pp. 219–240.

[25] Paoli, T. L., and J. E. Ripper, "Direct Modulation of Semiconductor Lasers," *Proc. IEEE*, Vol. 58, No. 10, October 1970, pp. 1457–1465.

[26] Morozov, V. N., V. V. Nitikin, and A. A. Sheronov, "Self-Synchronization of Modes in a GaAs Semiconductor Injection Laser," *JETP Lett.*, Vol. 7, No. 9, 1968, pp. 256–258.

[27] Harris, H. P., "Spiking in Current-Modulated CW GaAs External Cavity Lasers," *J. Appl. Phys.*, Vol. 42, 1971, pp. 892–893.

[28] Bachert, H., P. G. Eliseev, M. A. Manko, V. K. Petrov, and C. M. Tsai, "Interferometric Investigations of the Picosecond Structure and Conditions for the Emission of Ultrashort Pulses From Injection Lasers," *Sov. J. Quantum Electron.*, Vol. 4, No. 9, March 1975, pp. 1102–1105.

[29] Ho, P.-T., L. A. Glasser, E. P. Ippen, and H. A. Haus, "Picosecond Pulse Generation With a CW GaAlAs Laser Diode," *Appl. Phys. Lett.*, Vol. 33, No. 3, August 1978, pp. 241–242.

[30] Glasser, L. A., "C.W. Mode Locking of a GaInAsP Diode Laser," *Electron. Lett.*, Vol. 14, No. 23, November 1978, pp. 725–726.

[31] Derickon, D. J., R. J. Helkey, A. Mar, J. R. Karin, J. E. Bowers, and R. L. Thornton, "Suppresion of Multiple Pulse Formation in External-Cavity Mode-Locked Semiconductor Lasers Using Intrawaveguide Saturable Absorbers," *IEEE Photonics Tech. Lett.*, Vol. 4, No. 4, April 1992, pp. 333–335.

[32] Vasil'ev, P. P., and A. B. Sergeev, "Generation of Bandwidth-Limited 2 ps Pulses With 100 GHz Repetition Rate From Multisegmented Injection Lasers," *Electron. Lett.*, Vol. 25, No. 16, August 1989, pp. 1049–1050.

[33] Sanders, S., L. Eng, J. Paslaski, and A. Yariv, "108 GHz Passive Mode Locking of a Multiple Quantum Well Semiconductor Laser With an Intracavity Absorber," *Appl. Phys. Lett.*, Vol. 56, No. 4, January 1990, pp. 310–311.

[34] Nakajima, H., and R. Frey, "Intracavity Neary Degenerate Four-Wave Mixing in a (GaAl)As Semiconductor Laser," *Appl. Phys. Lett.*, Vol. 47, No. 8, October 1985, pp. 769–771.

[35] Agrawal, G. P., "Highly Nondegenerate Four-Wave Mixing in Semiconductor Lasers Due to Spectral Hole Burning," *Appl. Phys. Lett.*, Vol. 51, No. 5, August 1987, pp. 302–304.

[36] Provost, J. G., and R. Frey, "Cavity-Enhanced Highly Nondegenerate Four-Wave Mixing in GaAlAs Semiconductor Lasers," *Appl. Phys. Lett.*, Vol. 55, No. 6, August 1989, pp. 519–521.

Chapter 2

Basic Concept and Nonlinear Effects of Semiconductor Lasers

This chapter provides some basic background on semiconductor lasers that will be useful for understanding the following chapters. A number of textbooks are recommended for those readers desiring more detail [1–5].

2.1 BASIC CONCEPT OF SEMICONDUCTOR LASERS

The concept of a laser diode is a unique blend of semiconductor device physics and quantum electronics. The basic LD chip consists of two parallel cleaved facets that form an optical cavity (see Figure 2.1). The device has a light-emitting region (active region) sandwiched between p- and n-type regions. The typical cavity length is about 200 to 400 μm.

Early work on semiconductor lasers dealt mostly with homostructures at liquid-nitrogen temperature (77K). The threshold current density J_{th} for these lasers operating at room temperature was very high ($\geq 5 \times 10^4$ A/cm^2). With single heterostructures, the threshold current can be reduced by about a factor of 5 or more. Further reduction of the threshold current is possible by constructing a laser with double heterojunctions, because a further reduction of the thickness of the active layer can be accomplished with these structures without decreasing optical confinement. Because of strong carrier and optical confinement, significantly higher efficiency can be obtained from double heterostructure (DH) lasers.

As is illustrated in Figure 2.2, the DH serves to confine the injected carriers to the p-active (or n-active) layer. Either the N-n-P or the N-p-P structures have the same behavior. In a discussion of heterojunctions, it is convenient to designate the narrower energy gap semiconductor by n or p and the wider energy gap semiconductor by N or P according to the type of majority carrier. The active layer thickness d is greatly reduced

---

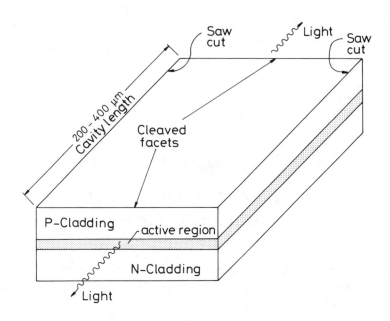

Figure 2.1 Schematic illustration of a typical broad-area laser diode with double heterostructure.

to 0.1 μm or less in GaAs-Ga$_{1-x}$Al$_x$As DH lasers. For the N-p-P DH lasers of Figure 2.2(a), the carrier-confining properties are illustrated in Figure 2.2(b) at high forward bias. Electrons and holes can move freely to the active region under forward bias. However, once there, they cannot cross over to the other side because of the potential barrier resulting from the bandgap difference. This allows for a substantial buildup of the electron and hole populations inside the active region, where they can recombine to produce optical gain. The width of the gain region is determined by the active-layer thickness, typically 0.1 to 0.3 μm. The N- and P-regions have reduced refractive index values, as illustrated in Figure 2.2(c). The heterojunctions thus provide the dielectric waveguide to confine the light to the active layer, as illustrated in Figure 2.2(d). Therefore, the active-layer thickness can be greatly reduced while retaining acceptable values of optical confinement factor in the active layer. Also, the fraction of the propagating mode outside the active layer is in a wider energy gap semiconductor and is not absorbed as in the homostructure case. The use of DHs permitted the reduction of J_{th} to about $1 \pm 0.5 \times 10^3$ A/cm^2 at 300K.

The successful operation of a laser requires that the generated optical field should remain confined in the vicinity of the gain region. In DH lasers, the optical confinement occurs by virtue of a fortunate coincidence. The active layer with a smaller bandgap also has a higher refractive index compared with that of the surrounding cladding layers, as illustrated in Figure 2.2(c). Because of the refractive index difference, the active layer in effect acts as a dielectric waveguide. The physical mechanism behind the confinement is total internal reflection, as illustrated in Figure 2.3. When a ray traveling at an angle θ

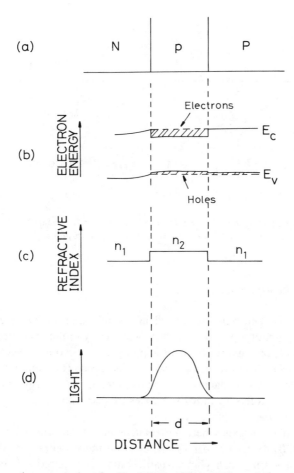

Figure 2.2 (a)Schematic representation of an N-p-P GaAs-AlGaAs DH laser; (b) energy band diagram at high forward bias; (c) refractive index profile; (d) optical field distribution.

(measured from the interface normal) hits the interface, it is reflected back if the angle θ exceeds the critical angle given by

$$\theta = \sin^{-1} \frac{n_1}{n_2} \tag{2.1}$$

where n_1 and n_2 are the refractive indexes of the cladding and active layers, respectively. Thus, rays traveling nearly parallel to the interface are trapped and constitute the waveguide mode. A more detailed discussion of waveguide modes requires the use of Maxwell's equations [1,4].

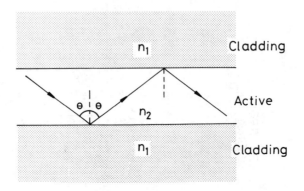

Figure 2.3 Dielectric waveguiding in a DH laser. The relatively higher refractive index $(n_2 > n_1)$ of the active layer allows total internal reflection to occur at the two interfaces for angles such that sin $\theta > n_1/n_2$.

2.1.1 Semiconductor Materials

An excess of electrons and holes is created in a semiconductor by several processes, including the absorption of a photon with sufficient energy to move an electron from the valence to the conduction band (or from an impurity center into the conduction band) or by injection at contacts. The process of photoluminescence (PL) refers to the radiative recombination of electron-hole pairs generated by shining high-energy light on a crystal. Injection luminescence refers to radiative processes occurring in structures containing injecting contacts, such as p-n junctions.

A crucial distinction exists between semiconductors in which the lower energy conduction band (CB) to valence band (VB) electron transition occurs without phonon participation and those in which phonon participation is required [2]. In the simplest conditions of a parabolic band in III-V direct-bandgap material, the energy bands are spherical in shape with the extrema occurring at $\mathbf{k} = 0$, as shown in Figure 2.4(a), and the transition of an electron across the gap at E_g does not involve a change in its momentum. (The small momentum of the emitted or absorbed photon, h/λ, can be neglected for λ in the 1-μm range.) As far as is currently known, direct-bandgap materials are essential for semiconductor laser operation.

In other semiconductors, such as silicon, germanium, and GaP, where the minima in the energy of the conduction band are located at $\mathbf{k} \neq 0$, as shown in Figure 2.4(b), electrons will tend to dwell in that band, whereas holes will be preferentially at $\mathbf{k} = 0$, where their energy is minimal. Hence, when an electron at $\mathbf{k} \neq 0$ recombines with a hole at $\mathbf{k} = 0$, there is a momentum change. Because momentum must be conserved in a band-to-band electron-hole recombination or pair formation process, these *indirect transitions* require the generation or absorption of a phonon with energy E_p with appropriate momentum. The required simultaneous interaction of a photon and a phonon reduces the probabil-

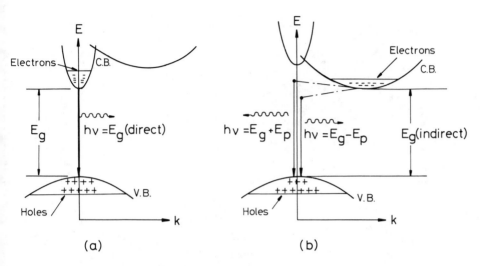

Figure 2.4 Electron-hole recombination in the band-to-band process in (a) a direct and (b) an indirect-bandgap semiconductor. Because of the need for momentum conservation, the energy of the photon emitted in the indirect-bandgap material electron-hole recombination process is either smaller or larger than the bandgap energy by E_p, the energy of the participating phonon.

ity of such an event compared to the direct transition described earlier. Assuming no competing processes for the disappearance of the electron in the conduction band, it will dwell for a relatively long time in the indirect valley (compared to the dwelling time in the direct valley). Unfortunately, competing nonradiative recombination processes involving defects and impurities exist, which allow the electron and hole to recombine in a relatively short time interval in most materials. Therefore, the probability of a radiative band-to-band electron-hole recombination process in indirect-bandgap materials is considerably lower than that in the direct-bandgap materials.

The bandgap and lattice-constant dependence on composition of several important III-V compound semiconductors is shown in Figure 2.5 [6]. The solid lines correspond to direct-gap materials and the dashed lines to indirect-gap materials. Note that for conversion between photon energy E and wavelength λ,

$$E \text{ (eV)} = 1.2398/\lambda \text{ } (\mu m) \tag{2.2}$$

With III-V DH lasers, the performance of the device appears to depend not only on the energy gap E_g and refractive index discontinuities at the heterojunctions that provide carrier and optical confinement, but also on the structural quality of the heterojunction interface and epitaxial layers. A particular problem is the presence of lattice defects resulting from the lattice mismatch between the semiconductors that make up the hetero-junctions. These defects may reduce the efficiency of radiative recombination and reduce the operating life of the device. In III-V systems, the ability to achieve a heterojunction

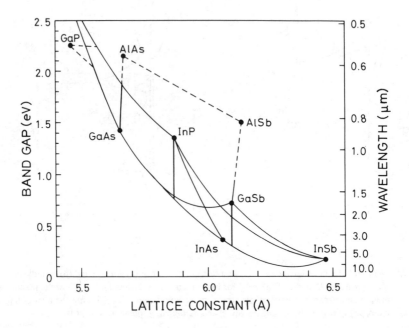

Figure 2.5 Bandgap energy and lattice constant of several binary and ternary compound semiconductors. The heavy lines indicate three important sets of compositions: GaInAsP lattice-matched to InP, GaInAsSb lattice-matched to GaSb, and AlGaAs nearly lattice-matched to GaAs. (From [6].)

of sufficiently high quality to yield useful DH lasers appears to be strongly dependent on the achievement of close lattice-matching between the heterojunction materials.

The successful epitaxial growth of $Ga_{1-x}Al_xAs$ on top of GaAs (and vice versa), which is the main reason for the success of DH lasers, is due to the fact that their lattice constants are the same to within a fraction of a percent over the range $0 \leq x \leq 1$. Note that the line connecting the AlAs ($x = 1$) and the GaAs ($x = 0$) is nearly vertical, which corresponds to a (very nearly) constant lattice constant over this compositional range. The x dependence of the energy gap is approximated by [1]

$$E_g(x < 0.37) = 1.424 + 1.247x \text{ (eV)} \tag{2.3}$$

For $Ga_xIn_{1-x}As_yP_{1-y}$ compositions that are lattice-matched to InP ($y \cong 2.2x$), the bandgap varies as

$$E_g = 1.35 - 0.72y + 0.12y^2 \text{ (eV)} \tag{2.4}$$

Bandgap wavelengths from 0.92 to 1.65 μm are covered by this material system. All GaInAsP compositions lattice-matched to InP are direct-bandgap.

With the advent of strained-layer epitaxy, the chemical matching requirement is relaxed. This allows a much wider selection of semiconductor materials to be used in unison. By employing strained-layer epitaxy, alternating thin layers can be grown such that dislocations caused by lattice mismatch do not form, and instead the atomic spacings of one or both materials shift to accommodate one another. This occurs naturally in very thin layers (e.g., 10 to 100 atoms thick) and can persist in much thicker layers (100 to 1,000 layers) if a low-temperature growth technique is used where dislocations do not have enough energy to form and grow.

2.1.2 Carrier Lifetime

The band-to-band spontaneous recombination rate R_{sp} takes on the simple form [2]

$$R_{sp} = B_r NP \tag{2.5}$$

where $B_r(cm^3/s)$ is a characteristic parameter of the material, N is the density of electrons in the conduction band, and P is the density of holes in the valence band. Under conditions of thermal equilibrium, the hole concentration is P_0 and the electron concentration is N_0, with $N_0 P_0 = N_i^2$. Under nonequilibrium conditions, additional carriers $\Delta N = \Delta P$ are introduced into the material. Therefore, the total recombination rate is

$$R_{sp} = B_r(N_0 + \Delta N)(P_0 + \Delta P) \tag{2.6}$$

The radiative carrier lifetime is defined in terms of the excess carrier pair density ΔN and τ_r as

$$R_{sp}^{exc} = \Delta N / \tau_r \tag{2.7}$$

where R_{sp}^{exc} is the recombination rate of the injected excess carriers. Thus,

$$R_{sp} = R_{sp}^0 + R_{sp}^{exc} \tag{2.8}$$

where R_{sp}^0 is the spontaneous recombination rate in thermal equilibrium. Hence,

$$R_{sp} = B_r[N_0 P_0 + \Delta N(P_0 + N_0) + (\Delta N)^2] \tag{2.9}$$

Since

$$R_{sp}^0 = B_r N_0 P_0 = B_r N_i^2, \qquad R_{sp}^{exc} = R_{sp} - R_{sp}^0$$

$$R_{sp}^{exc} = B_r \Delta N(P_0 + N_0 + \Delta N) \tag{2.10}$$

Using (2.7),

$$\tau_r = [B_r(P_0 + N_0 + \Delta N)]^{-1} \qquad (2.11)$$

We can distinguish two limits for the lifetime depending on the injected carrier density relative to the initial (background) concentration. At high injection levels, where the excess carrier density substantially exceeds the background concentration, $\Delta N > P_0 + N_0$, R_{sp} becomes $R_{sp} = B_r(\Delta N)^2$. τ_r is now a function of the injected carrier density,

$$\tau_r \cong [B_r(\Delta N)]^{-1} \qquad (2.12)$$

This is commonly called the *bimolecular recombination region*, where the lifetime value continually changes as the carrier concentration decays back to its equilibrium value. The term denotes the average carrier decay time starting from a given excess carrier concentration.

At the other extreme is the region in which the injected carrier density is low relative to the background concentration. There, the excess carriers decay exponentially with time with a constant lifetime τ_r determined by the background carrier concentration:

$$\tau_r \cong [B_r(N_0 + P_0)]^{-1} \qquad (2.13)$$

Therefore, in p-type material, the hole concentration ($P_0 \gg N_0$) will determine the lifetime, and in n-type material, the electron concentration ($N_0 \gg P_0$) will be the relevant quantity.

Electrons and holes in a semiconductor can also recombine nonradiatively. *Nonradiative recombination* of an electron-hole pair, as the term implies, is characterized by the absence of an emitted photon in the recombination process. This mechanism includes recombination at defects, surface recombination, and Auger recombination, among others. Although the efficiency decreases, the carrier lifetime of semiconductors is intentionally shortened through nonradiative recombination by impurity doping [7] and proton bombardment [8]. The carrier lifetime can also be shortened by reverse bias [9]. Figure 2.6 shows the lifetimes of photoexcited carriers in H^+ bombarded (200-keV) InP for different damage doses. The lifetime decreases down to 95 fs for a dose of 1×10^{16} cm^{-2} [8].

2.1.3 Optical Gain

The presence of carriers in the active region changes the optical properties of the material, and for sufficiently high carrier densities, gain is possible. The gain is due to a population inversion created by the injection of electrons and holes into the active region.

It is possible to calculate the gain using the theory of solid state [1]. Here we are only interested in some general features of the gain. Figure 2.7(a) illustrates the gain for p-GaAs as a function of the photon energy with the carrier density as parameter [10]. A

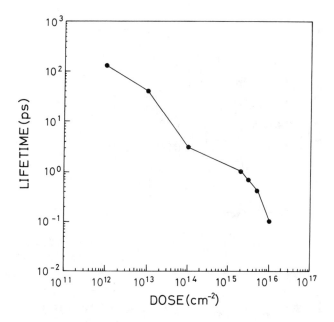

Figure 2.6 Dependence of the carrier lifetimes on the damage dosage is shown for H$^+$ ions with an energy of 200 keV. (From [8]. Reproduced by permission of AIP.)

semiconductor is transparent for photons of low energy, but as soon as the photon energy exceeds the bandgap, there is a very high absorption (or negative gain). As the carrier density increases, a photon energy range with positive gain appears. It turns out that gain is present if the photon energy is larger than the bandgap of the active region but smaller than the separation between the Fermi levels.

There are some important properties of the gain that are specific to semiconductor lasers. The first is that the values of the gain are extremely high, in the range of hundreds of cm^{-1}, which is orders of magnitude greater than in any other type of laser. This is why semiconductor lasers can be made small, typically much less than 1 mm long. The second remarkable feature is that this gain curve is extremely wide, in the range of tens of nanometers. The reason for this is that the optical transition is between a pair of energy bands instead of between well-defined states. There is one more interesting fact. If we look at the gain for a fixed photon energy, we can see that the gain increases with carrier density N, and a very useful approximation is that it increases linearly:

$$g = a(N - N_0) \tag{2.14}$$

where a is the gain constant and N_0 is the carrier density required to achieve transparency (corresponding to the onset of population inversion). This approximation is used extensively.

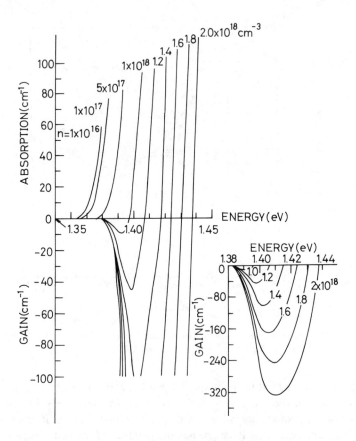

Figure 2.7(a) Gain as a function of photon energy for different values of the minority carrier density for GaAs with $p = 1.2 \times 10^{18}$ cm^{-3}. ($^©$1976 IEEE. From [4].)

If we consider the carrier-density-dependent gain more precisely, it depends on gain medium, doping, carrier concentration, and so forth. For example, the gain-versus-carrier-density curves for undoped and n-type GaAs and GaAlAs active layers usually exhibit a super-linear dependence with increasing carrier density [11]. Hence, the gain can be written as [12]

$$g(N) = aN^2 + bN + c \tag{2.15}$$

where a, b, and c are constant. In contrast to this behavior, the gain-versus-carrier-density relation for the p-type active layer usually appears to be slightly sublinear with increasing minority carrier density.

To simplify the calculation for two-section LDs, the nonlinear gain or loss-versus-carrier-density relation is sometimes approximated with linear functions with different

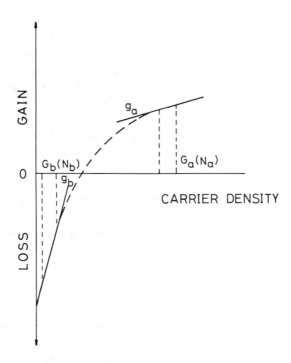

Figure 2.7(b) Nonlinear gain (loss) versus carrier density relation. It is approximated with two tangential lines
with different slopes in gain and loss regions, respectively.

slopes in respective regions, as illustrated in Figure 2.7(b). That is, they are approximated
in respective regions as

$$G_a(N_a) = g_a(N_a - N_{0a}) \tag{2.16}$$

$$G_b(N_b) = g_b(N_b - N_{0b}) \tag{2.17}$$

where g_a is the differential gain ($g_a = dG_a(N_a)/dN_a$), and N_{0a} is the carrier density at which
$G_a(N_a)$ vanishes. This assumption will be valid as long as the carrier density variations
in respective regions are small compared with the mean values.

2.1.4 Stripe-Geometry Lasers

A variety of LDs are available with either symmetric or asymmetric multiple-heterojunction
topology to provide separate optical and carrier confinements in the transverse direction.
For lateral confinement in the junction plane, either a stripe or the buried geometry shown

in Figure 2.8 is commonly used. The simplest configuration is that shown in Figure 2.8(a), where the optical gain region is confined by a stripe electrode with a width of typically 5 to 30 μm. In this case, the P-AlGaAs layer of the heterostructure with a relatively low doping concentration of 5×10^{17} cm^{-3} has been used to reduce the current spreading in a thickness direction. The top GaAs layer is necessary to provide a good electrical contact with the electrode, which is formed by first opening a narrow stripe in the SiO$_2$ layer and then diffusing a shallow depth of zinc through this window before metallic deposition. Alternative methods, which have also been used to provide the stripe geometry, consist of either etching the top GaAs layer to form a shallow mesa or ion implantation to reduce the carrier mobility in the region outside the electrode.

In gain-guided stripe-geometry lasers, the unpumped regions on both sides of the stripe are very lossy with values of absorption coefficient in the range of 200 to 400 cm^{-1}. The photons generated in the region with gain values below the threshold are absorbed by the medium and those that propagate along the stripe contribute to the maximum

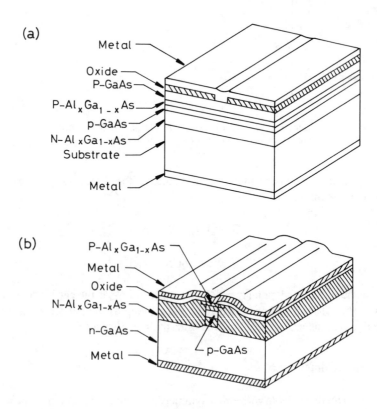

Figure 2.8 Topology of semiconductor heterojunction lasers: (a) stripe-geometry DH laser; (b) buried heterostructure laser.

optical power. The lateral optical modes are determined by both the dielectric constant and the gain (loss) coefficient of the medium. For stripe widths greater than the carrier diffusion length (>10 μm), a dip in the carrier density distribution near the center of the stripe often occurs. Such a change in the spatial gain profile often results in a shift in the gain-guided laser beam and a corresponding "kink" in the light versus current. These undesirable characteristics of wide-stripe gain-guided lasers can be avoided by using very narrow (about 3 to 5 μm) stripe contacts. With narrow stripes, carrier diffusion in the active layer region can smooth out the dip that otherwise develops in wider stripe geometry. The problem associated with very narrow stripes is the rapid increase in threshold current density with decreasing stripe width. Another problem is the increase in astigmatism in the output of very-narrow-stripe gain-guided lasers.

A more sophisticated structure is made by forming a buried p-GaAs active layer in a stripe configuration completely surrounded by AlGaAs, as shown in Figure 2.8(b). Although the fabrication of this device is complicated, it basically involves preferential etching and LPE growth around the mesa. Such a structure can provide very stable output power in a single transverse mode when the width of the active layer in the injection plane is made as narrow as 2 to 3 μm.

The most important feature of the buried heterostructure (BH) laser is that the active GaAs region is surrounded on all sides by the lower index GaAlAs, so that electromagnetically the structure is that of a rectangular dielectric waveguide. The transverse dimensions of the active region and the index discontinuities are so chosen that only the lowest order transverse mode can propagate in the laser waveguide. Another important feature of this laser is the confinement of the injected carriers at the boundaries of the active region due to the energy band discontinuity at a GaAs/GaAlAs interface. These act as potential barriers inhibiting carrier escape out of the active region.

2.1.5 Threshold Condition and Longitudinal Modes [1]

Laser oscillation is usually formed by the use of parallel reflecting surfaces for a medium with gain to form a Fabry-Perot etalon or interferometer. The oscillation can be obtained by considering the plane-wave reflection between parallel partially reflecting surfaces, as illustrated in Figure 2.9(a). This model follows the description of laser oscillation given by Yariv [13].

The plane wave with the complex propagation constant Γ is incident on the left cavity mirror as shown in Figure 2.9(a). Γ is expressed as follows.

$$\Gamma = \gamma + j\beta = j(\bar{n} - j\bar{k})k_0 \tag{2.18}$$

Here, $\bar{n}$ is the refractive index, $k_0 = 2\pi/\lambda_0$, and $\bar{k} = \alpha\lambda_0/4\pi$. The cavity length is L, the ratio of transmitted to incident fields at the left mirror is taken as t_1, and the ratio of transmitted to incident fields at the right mirror is taken as t_2. The ratio of reflected to

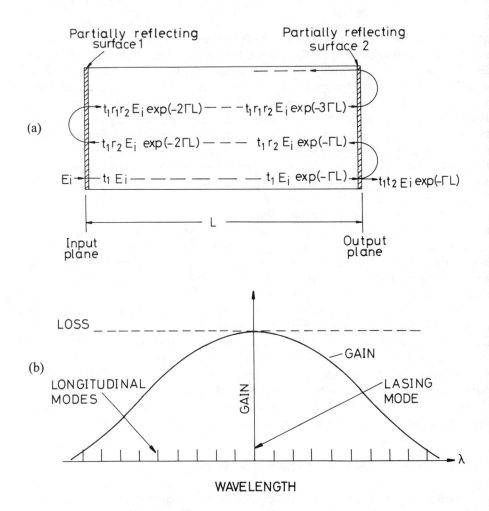

Figure 2.9 (a) Representation of the oscillation condition for a medium with gain and parallel reflecting surfaces. (From [13]: *Optical Electronics*, Fourth Edition, by Amnon Yariv, ©1991 by Saunders College Publishing, reproduced by permission of the publisher.) (b) Schematic illustration of the gain profile and longitudinal modes of an LD with a Fabry-Perot cavity. For the lasing mode in the vicinity of the gain peak, the threshold is reached when gain equals loss.

incident fields within the optical cavity is $r_1 \exp(j\theta_1)$ at the left mirror and $r_2 \exp(j\theta_2)$ at the right mirror. These complex field reflectances are related to the power reflectances $R_1 = r_1 r_1^*$ and $R_2 = r_2 r_2^*$. For a low-loss medium, the phase shifts θ_1 and θ_2 are small and are generally neglected so that $r_1 = \sqrt{R_1}$ and $r_2 = \sqrt{R_2}$.

Without the time dependence, the plane-wave electric field is expressed as $E_i \exp(-\Gamma z)$ so that the field amplitude of the wave traveling in the +z direction is $t_1 E_i$ inside the left boundary and $t_1E_i \exp(-\Gamma L)$ just inside the right boundary. The first portion of the field transmitted at the right boundary is $t_1t_2E_i \exp(-\Gamma L)$ and the reflected field is $t_1r_2E_i \exp(-\Gamma L)$. The next portion of the wave transmitted at the right boundary becomes $t_1t_2r_1r_2 \exp(-3\Gamma L)$ and so on. Addition of these transmitted fields gives

$$E_t = t_1t_2E_i \exp(-\Gamma L)[1 + r_1r_2 \exp(-2\Gamma L)$$
$$+ r_1^2r_2^2 \exp(-4\Gamma L) + \ldots] \tag{2.19}$$

The sum is a geometric progression permitting (2.19) to be written as

$$E_t = E_i \left[\frac{t_1t_2 \exp(-\Gamma L)}{1 - r_1r_2 \exp(-2\Gamma L)} \right] \tag{2.20}$$

When the denominator of (2.20) goes to zero, the condition of a finite transmitted wave E_t with zero E_i is obtained, which is the condition for oscillation. Therefore, the oscillation condition is reached when

$$r_1r_2 \exp(-2\Gamma L) = 1 \tag{2.21}$$

From (2.18), (2.21) becomes

$$r_1r_2 \exp[(g - \alpha_i) L] \exp[-2j(2\pi n/\lambda_0)L] = 1 \tag{2.22}$$

Here, α is the absorption coefficient, which is defined as the fraction of the power lost per incremental length:

$$\alpha = - \frac{dI/I}{dz} \tag{2.23}$$

In (2.22), the absorption term has been written as the difference between the gain and all of the losses α_i.

The condition for oscillation given by (2.22) represents a wave making a round-trip of $2L$ inside the cavity to the starting plane with the same amplitude and phase, within a multiple of 2π. The amplitude requirement for oscillation is

$$r_1r_2 \exp[(g - \alpha_i)L] = 1 \tag{2.24}$$

or

$$g = \alpha_i + (1/L) \ln(1/r_1r_2) \tag{2.25}$$

The phase condition is

$$4\pi nL/\lambda_0 = 2m\pi \qquad (2.26)$$

with $m = 1, 2, 3, \ldots$, and becomes

$$m(\lambda_0/n) = 2L \qquad (2.27)$$

Equation (2.25) is generally given with the power reflectances as

$$g = \alpha_i + (1/2L)\ \ln(1/R_1R_2) \qquad (2.28)$$

or for $R_1 = R_2 = R$, the gain requirement for laser oscillation becomes

$$g = \alpha_i + (1/L)\ \ln(1/R) \qquad (2.29)$$

Equation (2.27) shows that the laser tends to oscillate at a frequency that coincides with that of a longitudinal mode supported by the Fabry-Perot cavity. Which one and how many of them reach the threshold depends on details of the gain spectrum, such as the gain bandwidth and the gain-broadening mechanism (whether homogeneous or inhomogeneous). In the case of homogeneous broadening, only one longitudinal mode, whose frequency nearly coincides with the gain-peak frequency, reaches the threshold, and the laser maintains the single-longitudinal-mode operation even in the above threshold regime (see Figure 2.9(b)).

If an LD is modulated at high data rates, the laser spectrum may broaden due to the transient spectral phenomena, even in the homogeneous broadening case.

2.1.6 Distributed Resonators

An elegant approach to single-frequency operation is the integration of wavelength selectivity directly into the LD structure using a distributed-Bragg grating [6]. In a DFB laser, the grating region is built into a pumped part of the gain region. In a distributed-Bragg-reflector (DBR) laser, an unpumped Bragg grating coupled to a low-loss waveguide is used to replace the usual cleaved mirror on one or both ends of the resonator. Figure 2.10 illustrates schematically the DFB, phase-shifted DFB, and DBR structures. In all cases, the grating is shown in a passive waveguide layer adjacent to the active gain region. The grating can be produced with a periodic variation of either gain or refractive index, or both, along the structure.

The operation of lasers with distributed resonators can be understood in terms of the distributed-Bragg refractive index grating reflector and has been treated in considerable detail [14,15]. In effect, a refractive index grating is a region of a periodically varying

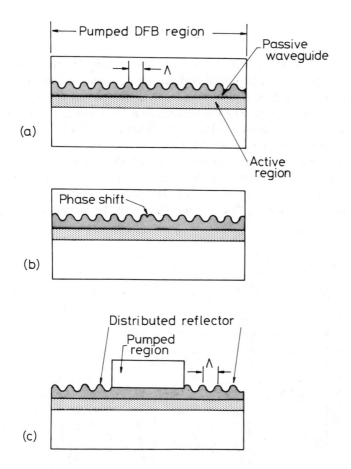

Figure 2.10 (a) DFB LD; (b) phase-shifted DFB LD; (c) DBR LD.

refractive index that serves to couple two counterpropagating traveling waves. The coupling is a maximum for wavelengths close to the Bragg wavelength λ_B, which is related to the grating spatial period Λ by

$$\lambda_B = 2n_e\Lambda/l \tag{2.30}$$

where n_e is the effective refractive index of the mode and l is an integer representing the order of the grating. First-order gratings provide the strongest coupling, but second-order gratings are sometimes used because they are easier to fabricate with their larger spatial period.

In a DFB laser without facet or other extraneous reflections and with an ideal grating, longitudinal modes are spaced symmetrically around λ_B at wavelengths given by

$$\lambda = \lambda_B \pm [(m + 1/2)\lambda_B^2/2nL_e] \tag{2.31}$$

where m is the mode index and L_e is the effective grating length. These modes exist outside of the transmission stop band centered on λ_B. There are two equivalent lowest order modes ($m = 0$) and oscillation on at least two frequencies is expected.

We explain the reason why the conventional DFB LD does not oscillate at the Bragg wavelength [16]. Let us consider the division of a DFB LD, with the first-order sinusoidal modulation in the equivalent refractive index n_e of the waveguide at the center, in two sections, the right and left sections (Figure 2.11(a)). Forward and backward waves, F_r and B_r, or F_l and B_l in Figure 2.11(b), make standing waves in each section. At the Bragg wavelength ($\lambda_B = 2n_e\Lambda$), all nodes of the standing wave are at points where the equivalent refractive index increases most rapidly in the direction of propagation of

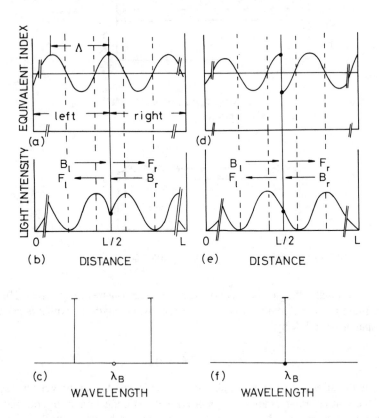

Figure 2.11 Comparison between conventional and phase-shifted DFB LDs. (a) and (d) are equivalent refractive index distributions, (b) and (e) are distributions of light intensity around centers of corrugations, and (c) and (f) are resonance spectra. L is the cavity length and λ_B the Bragg wavelength. (From [16]. Reproduced by permission of IEE.)

forward waves (Figure 2.11(b)). Standing waves in the right and left sections do not connect smoothly at the center (Figure 2.11(b)), and therefore resonance does not occur at the Bragg wavelength (Figure 2.11(c)). Because of grating imperfections and end-facet reflections, actual DFB lasers are not perfectly symmetric and single-frequency operation on one of the modes is often the result. Single-frequency operation with increased power output can also be achieved by intentionally using DFB lasers with cleaved end facets, one coated for high reflectivity and the other (output) coated for low reflectivity.

On the other hand, an introduction of a phase shift into the center of the corrugation by π as shown in Figure 2.11(d) results in the smooth connection of the standing wave (Figure 2.11(e)) and therefore resonance at the Bragg wavelength (Figure 2.11(f)). The half-pitch shift of the first-order corrugation grating corresponds to the $\lambda/4$-shift of a field in the waveguide. Therefore, this is usually called $\lambda/4$-*shifted DFB LD*.

Although it is equivalent with the above consideration, another way of explanation is possible in order to understand why the DFB LD does not oscillate at the Bragg wavelength [17]. We consider a simple rectangular grating (Figure 2.12(a)). The signs of the interface reflections alternate, and the phase shift for each half period is π. All the reflected waves will therefore add in phase. However, if we look at the total round-trip phase for each wave, we find that it is $(2p + 1)\pi$, p being an integer, and not a multiple of 2π as required. Consequently, the wavelength must deviate from λ_B. If an extra phase

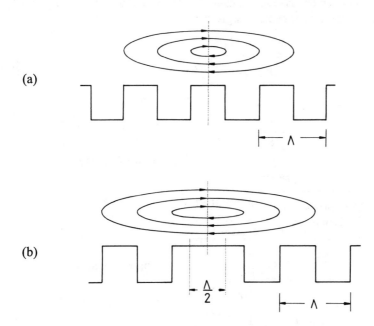

Figure 2.12 Rectangular grating (a) without and (b) with a $\pi/2$ phase shift in the center, showing the individual interface reflections. (From [17].)

shift is inserted into the center of the periodic structure and the length of the phase shift section is $\Lambda/2 = \lambda_0/4n_e$ (phase shift $\pi/2$), corresponding to an extra round-trip phase shift of π, all the reflected waves will add in phase and total round-trip phase will be a multiple of 2π (Figure 2.12(b)).

2.1.7 Rate Equations

The behavior of an LD is illustrated in Figure 2.13 [17]. A current I flows into the active region and supplies charge carriers to the active region. These carriers either recombine spontaneously or give rise to gain for the stimulated recombination. Consequently, there exists a photon density in the active region. Some of the photons are lost, and some of them are emitted from the laser; this gives the optical output power.

Based on this diagram, and considering the balance between the current, the carrier density, and the photon density, we can write down the two rate equations that govern the time dependence of the carrier density and the photon density. The first equation deals with the carriers. Carriers are supplied by the current I to the active volume V. Some of the carriers recombine spontaneously with a lifetime τ_s. Other carriers recombine by stimulated recombination described by a gain factor $g(N)$ and the photon density n_p. The time dependence of the carrier density is given by (q being the unit charge)

$$\frac{\mathrm{d}N}{\mathrm{d}t} = \frac{I}{qV} - g(N)n_p v_g - \frac{N}{\tau_s} \tag{2.32}$$

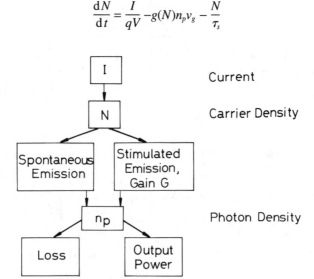

Figure 2.13 Rate equation simulation of an LD. (From [17].)

The gain g is the net gain per unit length and v_g is the group velocity of light in the laser. The time dependence of the photon density is described by the second rate equation:

$$\frac{dn_p}{dt} = g(N)n_p v_g - \frac{n_p}{\tau_p} + \beta_{sp}\frac{N}{\tau_s} \qquad (2.33)$$

The first term on the right-hand side is the stimulated recombination term. We get one photon for each stimulated recombination, so this term is the same as the last term in the carrier density rate equation, but the sign is now positive. Some of the photons are lost, and this is described by a photon lifetime τ_p. Finally, we have a term that is due to the spontaneous emission: a fraction β_{sp} of the spontaneous recombination events happens to supply a photon into the lasing mode. The rate equations (2.32) and (2.33) constitute a set of two coupled nonlinear differential equations.

The rate equations can be used to understand most of the features of light-output-versus-current characteristics. In the case of CW operation at a constant current I, the time derivatives in 2.32 and 2.33 can be set to zero. The solution takes a particularly simple form if spontaneous emission is neglected by setting $\beta_{sp} = 0$. For currents such that $g(N)v_g\tau_p < 1$, $n_p = 0$ and $N = \tau_s I/qV$. The threshold is reached at a current for which $g(n)v_g\tau_p = 1$. The carrier population is then clamped to the threshold value n_{th}. If we use the linear gain function $g(N) = a(N - N_0)$, the threshold carrier density becomes $N_{th} = N_0 + 1/(av_g\tau_p)$. The threshold current is given by

$$I_{th} = \frac{qVN_{th}}{\tau_s} = \frac{qV}{\tau_s}\left(N_0 + \frac{1}{av_g\tau_p}\right) \qquad (2.34)$$

For $I > I_{th}$, the photon number n_p increases with I as

$$n_p = \frac{\tau_p}{qV}(I - I_{th}) \qquad (2.35)$$

The resulting light-power-versus-current and carrier-density-versus-current characteristics are schematically shown in Figure 2.14. Without the spontaneous emission term in (2.32) and (2.33), an absolutely abrupt transition at the threshold would occur; however, due to the spontaneous emission term, one obtains an amplified spontaneous emission (ASE) even below the threshold, yielding a gradual transition around the threshold, which depends on the amount of spontaneous emission in the lasing modes.

We now linearize the rate equations, meaning that we restrict the investigation to small-signal modulation, where the current consists of a constant part and a small modulation term with modulation frequency ω_m:

$$I = \bar{I} + \Delta I \exp(j\omega_m t) \qquad (2.36)$$

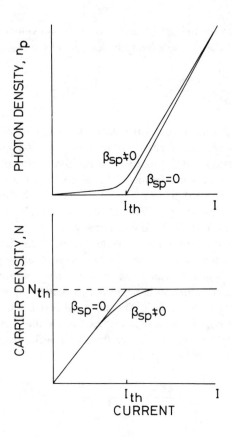

Figure 2.14 Schematic view of the light-output-power-versus-current and carrier-density-versus-current characteristics of an LD.

For small-signal modulation, we can write both the carrier density and the photon density as a constant part and a time-dependent part:

$$N = \overline{N} + \Delta N \exp(j\omega_m t) \tag{2.37}$$

$$n_p = \overline{n}_p + \Delta n_p \exp(j\omega_m t) \tag{2.38}$$

For operation above the lasing threshold, we can use the equations (2.34) and (2.35) as the steady-state solution. We now insert (2.36) to (2.38) into the rate equations, use (2.34) and (2.35), eliminate the steady-state solution, and neglect higher order terms. The result is

$$\frac{d(\Delta N)}{dt} = \frac{\Delta I}{qV} - av_g\{(\overline{N} - N_0)\Delta n_p + \overline{n}_p\,\Delta N\} - \frac{\Delta N}{\tau_s} \tag{2.39}$$

$$\frac{d(\Delta n_p)}{dt} = av_g\{(\overline{N} - N_0)\,\Delta n_p + \overline{n}_p\,\Delta N\} - \frac{\Delta n_p}{\tau_p} \tag{2.40}$$

The normalized transfer function becomes as follows.

$$H(\omega_m) \equiv \frac{\Delta n_p(\omega_m)}{\Delta n_p(0)}$$

$$= \frac{1}{1 - \left(\dfrac{\omega_m}{\omega_r}\right)^2 + j(\omega_m\tau_s)\left\{\dfrac{\tau_p}{\tau_s} + \dfrac{1}{(\omega_r\tau_s)^2}\right\}} \tag{2.41}$$

where

$$\omega_r = 2\pi f_r = \frac{1}{\sqrt{\tau_s\,\tau_p}\,\sqrt{1 - \dfrac{N_0}{N_{th}}}}\sqrt{\frac{\overline{I}}{I_{th}} - 1} \tag{2.42}$$

Since $H(0) = 1$, this gives the dc response we expected. The transfer function $H(\omega_m)$ is schematically shown in Figure 2.15. A resonant peak at $f_m = f_r$ appears. f_r is usually called the *relaxation oscillation frequency* (or relaxation resonance frequency).

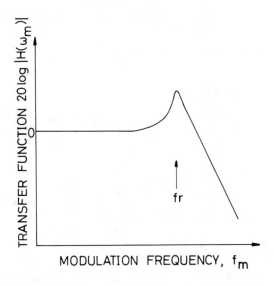

Figure 2.15 Amplitude modulation transfer function as a function of the modulation frequency.

2.2 SEMICONDUCTOR QUANTUM WELLS

2.2.1 Quantum Size Effects

A quantum-well (QW) structure is a finite segment of a superlattice, which is a one-dimensional periodic structure consisting of alternating ultrathin layers of two different materials (see Figure 2.16(a,b)). if characteristic dimensions such as the superlattice period and layer thickness (the QW layer thickness L_z and the barrier layer thickness L_B) in the semiconductor nanometer structure are reduced to less than the electron mean free path, the entire electron system enters into a quantum regime (quantum size effect).

In a QW structure, the confinement of electrons and holes in one dimension causes a quantization in the allowed energy levels and the formation of an nth subband of energy [18]:

$$E_{ne} = \frac{(n\pi\hbar)^2}{2\,m_e L_z^2}$$

(2.43)

GaAs–AlAs Quantum–Well Structure

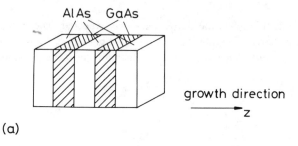

growth direction
z

(a)

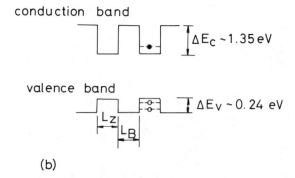

(b)

Figure 2.16(a,b) GaAs-AlAs QW structure: (a) layer structure: z- direction is the growth direction; (b) energy band structure for conduction and valence band.

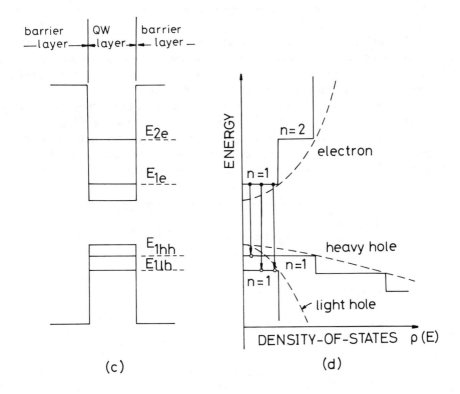

Figure 2.16(c,d) GaAs-AlAs QW structure: (c) quantum energy levels in a QW: E_{1e} and E_{2e} are $n = 1$ and $n = 2$ electron levels, and E_{1hh} and E_{1lh} are $n = 1$ heavy-hole and light-hole levels, respectively; (d) density of states as a function of energy for a QW structure (solid line) and for a bulk crystal (broken line). (From [35].)

for the conduction band, where L_z is the thickness of the QW, m_e is the effective mass of electrons, and $\hbar$ is Planck's constant (h) divided by 2π. A similar expression holds for the valence band. The density of states changes from the parabolic dependence to a steplike structure (see Figure 2.16(a)):

$$\rho_c(E) = \sum_{n=1}^{\infty} \frac{m_e}{\pi\hbar^2} H(E - E_{ne}) \qquad (2.44)$$

where H is the Heaviside function. $H(E - E_{ne}) = 0$ for $E - E_{ne} < 0$ and $H(E - E_{ne}) = 1$ for $E - E_{ne} > 0$.

Since the density of states is constant over a band of energies, rather than gradually increasing from zero density, there is a group of electrons of nearly the same energy available to recombine with a group of holes with nearly the same energy, and gain can be larger than in DHs. Calculations of the gain for multiple quantum well (MQW) are shown in Figure 2.17 [19]. The maximum gain is much larger than that for conventional DHs, although the exact level depends on the assumed value of the intraband relaxation time. It is this effect that reduces the threshold, increases the resonance frequency and decreases the linewidth enhancement factor in MQW lasers.

In the QW structure, a transverse electric (TE) field–polarized optical wave whose electric vector lies in the plane of the QW layers can couple both with the electron to heavy-hole transition and with the electron to light-hole transition, but a transverse magnetic (TM) field–polarized optical wave whose electric vector is perpendicular to the plane of the QW layers is allowed to couple only with the latter transition.

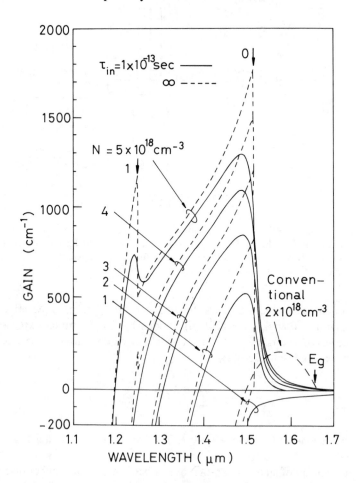

Figure 2.17 Linear gain of GaInAs/InP MQW lasers with intraband relaxation included. The result for a conventional DH LD is shown with a long dashed curve. (©1984 IEEE. From [19].)

The polarization-dependent matrix elements including conduction-to-light-hole (lh) band transitions as well as the conduction-to-heavy-hole (hh) band transitions were calculated based on the $k \cdot p$ perturbation theory by Yamanishi and Suemune [20].

In a zinc blende crystal with a direct gap such as GaAs, base functions in a unit cell for small electron wave numbers k are given as follows [21]. For a conduction band,

$$| S > \cdot |\downarrow' > \text{ or } | \uparrow' > \tag{2.45}$$

For a heavy-hole band,

$$\frac{1}{\sqrt{2}}\{(\cos \theta_0 \cdot \cos \phi_0 \mp i \sin \phi_0) \mid X > + (\cos \theta_0 \cdot \sin \phi_0 \pm i \cos \phi_0) \mid Y >$$

$$- (\sin \theta_0) \mid Z >\} \cdot \mid \downarrow' > \text{ or } \mid \uparrow' >, \tag{2.46}$$

and for light-hole band,

$$\frac{1}{\sqrt{6}} \{(\cos \theta_0 \cdot \cos \phi_0 \mp i \sin \phi_0) \mid X > + (\cos \theta_0 \cdot \sin \phi_0 \pm i \cos \phi_0) \mid Y >$$

$$- (\sin \theta_0) \mid Z >\} \mid \downarrow' > \text{ or } \mid \uparrow' > - \sqrt{\frac{2}{3}}\{(\sin \theta_0 \cdot \cos \phi_0) \mid X >$$

$$+ (\sin \theta_0 \cdot \sin \phi_0) \mid Y > + (\cos \theta_0) \mid Z >\} \mid \uparrow' > \text{ or } \mid \downarrow' >, \tag{2.47}$$

where $\mid S >$ is an s-like isotropic function and $\mid X >$, $\mid Y >$, and $\mid Z >$ are p-like functions. θ_0 and ϕ_0 represent the angles of electron wave vector $\mathbf{k}$ under consideration in the polar coordinate of which the polar axis is the quantization axis. Also, $\mid \uparrow' >$ and $\mid \downarrow' >$ are the spin functions.

Let us consider a simple single-QW structure as shown in Figure 2.18(a) to avoid unessential complications, postulating that electron and hole envelope functions become zero at the wells. In this case, normal components k_z of the wave vector of the electrons or holes in a subband are given by a certain value $(\pi/L_z)n$ (n: integer). In Figure 2.18(a), θ and ϕ are defined as the angles of the electron wave vector $\mathbf{k}$ in the polar coordinate of which the polar axis is the z-axis normal to the well plane. In general, θ and ϕ are different from θ_0 and ϕ_0, respectively. Physical phenomena under consideration are isotropic in a plane parallel to the well plane (x-y plane in Figure 2.18(a)), so that the squared momentum matrix elements may be averaged over the angle ϕ (not ϕ_0) when the k conservation law in the QW plane is assumed. As a result, from (2.45) to (2.47), with the aid of symmetry consideration for the p-like functions, the squared momentum matrix elements averaged over the angle ϕ are given in forms that do not include explicitly θ_0 and ϕ_0 for conduction-to-heavy-hole transitions:

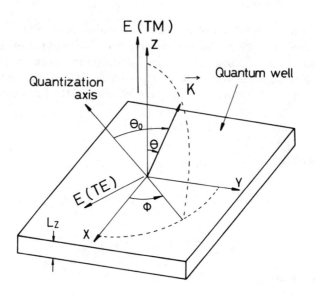

Figure 2.18(a) Geometry of the QW structure; for the TM modes, only the dominant electric field is drawn.

$$<M^2>_{hh,TE} = \frac{3}{4}M^2(1 + \cos^2\theta)$$

$$= \frac{3}{4}M^2\left(1 + \frac{k_z^2}{k^2}\right) = \frac{3}{4}M^2\left(1 + \frac{E_{z,n}}{E_n}\right) \qquad \text{for TE modes } (E \perp z), \qquad (2.48)$$

$$<M^2>_{hh,TM} = \frac{3}{2}M^2\sin^2\theta$$

$$= \frac{3}{2}M^2\left(1 - \frac{k_z^2}{k^2}\right) = \frac{3}{2}M^2\left(1 - \frac{E_{z,n}}{E_n}\right) \qquad \text{for TM modes } (E \text{ // } z) \qquad (2.49)$$

Here, $E_{z,n}$ and E_n are the quantized level and the total energy of subband n, respectively. At the edge of the subband, $E_n = E_{z,n}$. Therefore,

$$<M^2>_{hh,TE} = \frac{3}{2}M^2 \qquad (2.50)$$

$$<M^2>_{hh,TM} = 0 \qquad (2.51)$$

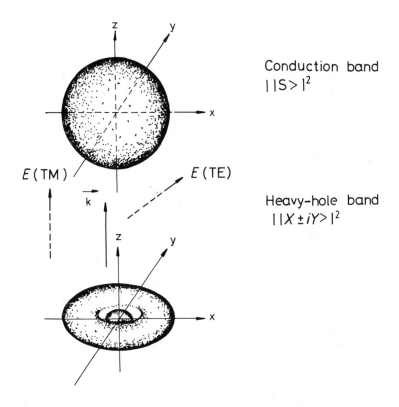

Figure 2.18(b) Squared base function of conduction band and heavy-hole band.

This means that the TM modes do not couple to the conduction-to-heavy-hole transition. This strong polarization dependence of the TM modes is explained as follows. The squared base function for the heavy hole has a doughnutlike shape and the "plane" of the doughnut is perpendicular to **k** as shown in Figure 2.18(b). Therefore, the conduction-to-heavy-hole transition strongly couples to the TE modes, but does not couple to the TM modes.

For the conduction-to-light-hole transition, the squared momentum matrix elements can be obtained as follows.

$$<M^2>_{\text{lh,TE}} = \frac{1}{4}M^2(1 + \cos^2\theta) + M^2\sin^2\theta$$

$$= \frac{1}{4}M^2\left(1 + \frac{E_{z,n}}{E_n}\right) + M^2\left(1 - \frac{E_{z,n}}{E_n}\right) \qquad \text{for TE modes } (E \perp z), \qquad (2.52)$$

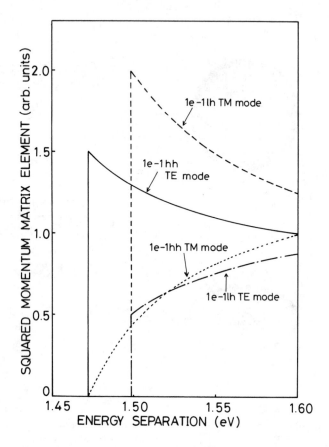

Figure 2.18(c) Estimated squared momentum matrix elements for the TE and TM modes as a function of energy separation between states. 1e-1hh and 1e-1lh denote the transitions between the lowest subbands in the conduction and heavy-hole bands and between the lowest subbands in the conduction and light-hole bands, respectively. (From [20]. Reproduced by permission of JJAP.)

$$<M^2>_{lh,TM} = \frac{1}{2}M^2\sin^2\theta + 2M^2\cos^2\theta$$

$$= \frac{1}{2}M^2\left(1 - \frac{E_{z,n}}{E_n}\right) + 2M^2\frac{E_{z,n}}{E_n} \quad \text{for TM modes } (E \, /\!/ \, z). \quad (2.53)$$

At the edge of subband, these values become as follows.

$$<M^2>_{lh,TE} = \frac{1}{2}M^2 \quad (2.54)$$

$$<M^2>_{\text{lh,TM}} = 2M^2 \tag{2.55}$$

Therefore, the conduction-to-light-hole transition couples four times stronger to TM modes than to TE modes.

For a bulk crystal, the above elements are averaged over the angle θ so that they approach a single value M^2. It should be noted that the obtained results on the matrix elements in the QWs are independent of the direction of the quantization axis for the p-like functions. In other words, the polarization dependence of the present momentum matrix elements is not caused directly by atomic orientation. Rather, it is done by only the directionality of the electron wave vector $\mathbf{k}$ in the QW structures.

We shall consider a single-QW structure as a specific example, where a GaAs active layer with a thickness of 10 nm is sandwiched between $Ga_{0.4}Al_{0.6}As$ barriers. Figure 2.18(c) shows the energy dependence of the squared momentum matrix elements estimated for this QW structure [20]. The elements are polarization- and energy-dependent, different from those in bulk crystals.

2.2.2 Exciton States in Quantum Wells

Exciton is a key concept in the physics of the optical properties of semiconductors. The technology of layered semiconductor growth, which is now able to grow structures at will with atomic monolayer precision, has been developed. Among the many uses of this growth technology is the ability to generate structures that can confine excitons in dimensions less than the usual Bohr radius of the exciton. Although one might expect at first sight that this gross interference with the exciton would destroy it, in fact, the opposite is true. The resulting excitons are actually so robust that clear excitonic absorption resonances can be seen at room temperature, a fact that actually makes them easier to study than bulk excitons.

In order to discuss the absorption spectra, we need to know how electron-hole (e-h) pair states are constructed from these single-particle states [22]. QWs have properties that are intermediate between two-dimensional (2D) and three-dimensional (3D) material; it is therefore instructive first to compare excitonic effects in pure 2D and 3D before discussing the influence of the finite well thickness. The theory of excitons in 3D semiconductors is extremely well documented [23]. It turns out that within the effective mass approximation, it can be easily extended to Wannier excitons with a pure 2D motion and $1/r$ e-h attraction [24].

Let us consider first the ideal case of a semiconductor of bandgap E_g with two isotropic valence and conduction bands with masses $m_v = -m_h$ and $m_c = m_e$, respectively. We shall use as units of energy and length the 3D Rydberg and Bohr radius, respectively, $R_0 = e^4\mu/2\epsilon_0^2\hbar^2$ and $a_0 = \epsilon_0\hbar^2/\mu e^2$, where $\mu^{-1} = m_e^{-1} + m_h^{-1}$ is the reduced e-h mass and ϵ_0 is the dielectric constant. The solution of the 2D Schrödinger equation gives a series of bound states with energies $E_n^{2D} = E_g - R_0/(n - 1/2)^2$ as compared to the 3D result $E_n^{3D} = E_g - R_0/n^2$. The wave function of the 1s state in real space is

$$U_{1s}^{2D}(r) = (2/\pi)^{1/2} \frac{2}{a_0} \exp\left(-\frac{2r}{a_0}\right) \tag{2.56a}$$

and in momentum space

$$U_{1s}^{2D}(k) = \frac{(2\pi)^{1/2} a_0}{[1 + (a_0 k/2)^2]^{3/2}} \tag{2.56b}$$

Let us recall for comparison that in 3D, these functions are

$$U_{1s}^{3D}(r) = \frac{1}{(\pi)^{1/2} a_0^{3/2}} \exp\left(-\frac{r}{a_0}\right) \tag{2.57a}$$

and

$$U_{1s}^{3D}(k) = \frac{(8\pi)^{1/2} a_0^{3/2}}{[1 + (a_0 k)^2]^2} \tag{2.57b}$$

In 2D, the 1s exciton binding energy and Bohr radius are $R_{2D} = 4R_0$ and $a_0/2$, respectively, and the maximum radial charge density occurs at $a_{2D} = a_0/4$. Furthermore, the probability of finding the electron and hole in the same unit cell varies like $|U_n^{2D}(r = 0)|^2 = (2n - 1)^{-3}$ $|U_{1s}^{2D}(r = 0)|^2$, whereas in 3D, $|U_n^{3D}(r = 0)|^2 = (n)^{-3} |U_{1s}^{3D}(r = 0)|^2$, which shows that the e-h correlation in excited bound states decreases more rapidly in 2D than in 3D. Much the same can be said for the scattering states (i.e., the unbound e-h pair states). In 2D, one finds

$$|U_k^{2D}(r = 0)|^2 = \frac{2}{[1 + \exp(-2\pi/a_0 k)]} \tag{2.58a}$$

where $a_0 k = [(E - E_g)/R_0]^{1/2}$ is the dimensionless e-h relative momentum. In 3D,

$$|U_k^{3D}(r = 0)|^2 = \frac{2\pi/a_0 k}{[1 - \exp(-2\pi/a_0 k)]} \tag{2.58b}$$

In real semiconductor QWs that have a finite thickness and depth, one faces many complications. The motion normal to and in the plane of the layer are coupled by the Coulomb interaction. Because of the penetration of the wave functions into the barrier medium, the differences of the valence conduction band effective masses inside and outside the well have to be accounted for. Finally, in III-V compounds, the hole motion in the plane is complex because the degeneracy of the light- and heavy-hole bands at the

Γ point is lifted, giving rise to two series of subbands. Away from $k = 0$, these subbands couple and acquire a highly nonparabolic dispersion so that a constant effective mass can no longer be defined for the holes.

For a trial wave function inseparable in $z_{e,h}$ and $(x,y)_{e,h}$ and infinitely deep wells, a smooth variation of the 1s exciton binding energy from $4R_0$ to R_0 is found for L_z/a_0 varying from 0 to ∞, in good agreement with physical intuition. Unfortunately, this situation does not correspond to real QWs because, in the limit of very thin wells, the wave function penetrates more and more in the barrier medium as a result of the finite confining potential, and the exciton becomes less and less confined. As L_z decreases to 0, the exciton tends toward the 3D exciton of this material and the binding energy tends toward the corresponding Rydberg. Thus, the enhancement of the binding energy shows an optimum $<4R_0$ for narrow but finite QWs.

Figure 2.19 shows an optical absorption spectrum at 300K for a GaAs-AlAs QW structure (solid line) [25]. At the step-edge energies for $n = 1, 2, 3, ...$ electrons to heavy- and light-hole transitions, a double peak structure appears. This double peak structure does not change in a temperature range from a cryogenic temperature, where the peaks have been assigned to excitons, to room temperature, so it is referred to as a *room-temperature exciton*. This is in marked contrast to bulk GaAs, where excitonic absorption peak appears predominantly at lower temperatures. At room temperature, a vanishing trace appears as a shoulder at the absorption edge only for high-purity epitaxial GaAs films (broken line).

2.2.3 Quantum-Confined Stark Effect

The effects of fields perpendicular to the layers of QW are qualitatively different from electroabsorption in bulk materials [26]. The absorption edge shows a clear shift to lower energies while still remaining abrupt, with clearly resolved excitonic peaks at very high fields. We shall approach the explanation of this behavior in two stages. First, we consider the electroabsorption neglecting the Coulomb interaction, just as was done for the Franz-Keldysh (FK) effect. In fact, in this case, the resulting QW electroabsorption can be described as a quantum-confined Franz-Keldysh effect (QCFK), even though the behavior for thin wells is quite unlike the FK effect [27]. This approach is useful in understanding the QW absorption and its relation to the bulk behavior, but it cannot explain the other remarkable phenomenon in perpendicular field QW electroabsorption, namely that the exciton peaks remain resolved at very high fields (e.g., 100 times the classical ionization field). For this, we must include the Coulomb interaction in the quantum-confined Stark effect (QCSE) model [28,29]. The importance of the QCFK is largely conceptual, so we shall consider the extreme case of an infinitely deep QW. Then all the states are bound, the wave functions are all identically zero at the walls of the well, and the problem can be solved exactly with Airy functions as the solutions. Figure 2.20 shows the resulting wave functions for a particular illustrative case.

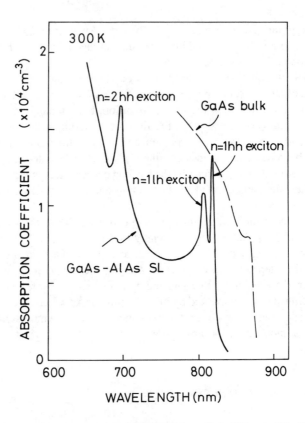

Figure 2.19 Optical absorption spectrum measured at 300K for a GaAs (8.3 nm)–AlAs (9.3 nm) superlattice (solid line). The optical absorption spectrum for high-purity bulk GaAs is also shown for comparison (broken line). (From [25]. Reproduced by permission of JJAP.)

When excitonic effects are included in bulk electroabsorption, the excitonic absorption resonances broaden rapidly with field with comparatively little Stark shift. In Figure 2.21, we show spectra of QW electroabsorption [30,31] for an electric field perpendicular to the layers. Not only are there large shifts as expected from the discussion of the QCFK above, but the exciton peaks also remain well resolved up to very high fields. The spectra in Figure 2.21 were taken in a waveguide sample so that the two distinct polarizations of the optical electric vector in this system, namely parallel and perpendicular to the layers, could both be used. As we have discussed, in the perpendicular polarization, the heavy-hole-to-conduction transitions are forbidden; this has nothing to do with the applied static electric field, although it is apparently not destroyed by it.

2.2.4 Strained Quantum Wells

The combination of strain and quantum confinement gives rise to novel electronic properties as a result of the strain-induced modification of the band structure. This could be

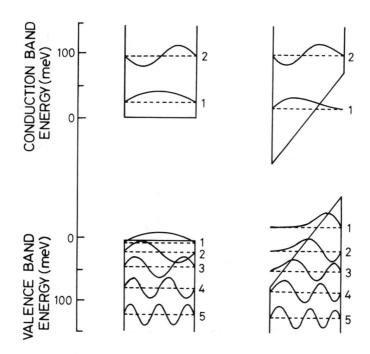

Figure 2.20 Valence and conduction energy levels and normalized wave functions in a 15-nm GaAs-like QW at 0 and 10^5 V/cm, plotted together with the net confining potential, including the effect of field. (From [27]. Reproduced by permission of APS.)

used either to improve the performance of strained-layer $In_xGa_{1-x}As$ MQW lasers over existing 1.5-μm-wavelength MQW lasers or to exploit new properties that cannot be obtained from lattice-matched MQW structures.

High-quality heterostructures can be grown from lattice-mismatched materials as long as the lattice mismatch can be accommodated by elastic tetragonal deformation of the unit cell. The loss of the cubic symmetry can result in significant distortion of the band structure, particularly in regions of degeneracy, such as the top of the valence band. The deformation of the $In_xGa_{1-x}As$ unit cells, with indium fractions $x > 0.53$, $x = 0.53$, and $x < 0.53$, grown coherently on (001) InP and the strain-induced band structure modification are shown in Figure 2.22 [32]. The figure of the strain-induced modification of the band structure only shows the effects of strain and does not include the band gap dependency on its material composition. For $In_xGa_{1-x}As$ grown under biaxial compression (i.e., with $x > 0.53$), the strain-induced modification of the band structure can be summarized as follows: the bandgap is increased, the light-hole–heavy-hole (lh-hh) degeneracy

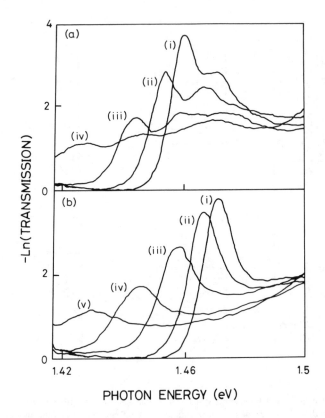

Figure 2.21 Absorption spectra of a waveguide containing 9.4-nm GaAs QWs as a function of electric field applied perpendicularly to the layers: (a) incident optical polarization parallel to the plane of the layers for fields of (i) 1.6×10^4 V/cm, (ii) 1×10^5 V/cm, (iii) 1.3×10^5 V/cm, and (iv) 1.8×10^5 V/cm; (b) incident optical polarization perpendicular to the plane of the layers for fields of (i) 1.6×10^4 V/cm, (ii) 1×10^5 V/cm, (iii) 1.4×10^5 V/cm, (iv) 1.8×10^5 V/cm, and (v) 2.2×10^5 V/cm. (Reprinted with permission of David A. B. Miller. From [30].)

near the zone center is lifted, and this energy splitting of the valence subbands is enhanced by the application of QWs, the hh effective mass in the inplane direction (k_x, k_y) is reduced near the zone center, and the conduction band discontinuity is increased. The reduced inplane hh mass lowers the carrier density needed for population inversion; that is, a reduction of the threshold current is expected. This, together with the increased conduction band discontinuity which reduces the hetero-barrier carrier leakage, and the increased bandgap and energy separation of the valence subbands are expected to result in a reduction of the amount of Auger recombination and intervalence band absorption, respectively. The latter are considered to be the major loss mechanisms in 1.5-μm-wavelength MQW lasers, and reduction of them would result in a significantly improved characteristic

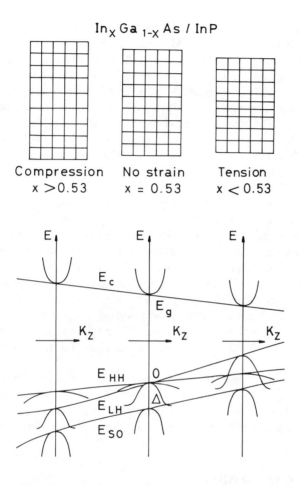

Figure 2.22 Lattice deformation and strain-induced band structure modification of bulk In$_x$Ga$_{1-x}$As/InP. (From [32].)

temperature T_0, differential external efficiency η, and differential gain. The combination of reduced threshold current, enhanced T_0, and η give excellent prospects for high-power operation.

For strained-layer In$_x$Ga$_{1-x}$As MQW structures grown under sufficient biaxial tension (i.e., with $x < 0.53$), the energy levels of hh and lh are reversed, as shown in Figure 2.22, and hence TM-polarized emission is expected because the TM mode couples exclusively to the lh states. This cannot be obtained in lattice-matched or biaxially compressed MQW structures.

Traveling-wave LD amplifiers (TWAs) have an inherent problem in that the signal gain between the TE and TM modes is different even though conventional bulk materials exhibit the polarization-insensitive gain. This is mainly due to the mode confinement factor difference between the TE and TM modes. Various approaches to eliminating this problem have been attempted: introduction of a thick active layer, a large optical cavity (LOC) structure, a hybrid configuration using two TWAs, and so on.

Introduction of strain into the MQW active region of TWAs offers a novel approach to realizing complete polarization insensitivity. The effectiveness of the approach results from the fact that the TM mode gain is enhanced. Magari et al. [33] have confirmed this through the realization of InGaAs TWAs having tensile strain in the barrier layers of the active region. In this device, the signal gain of the TM mode completely agrees with that of the TE mode at a specific amplifier driving current.

The active-region structure is shown in Figure 2.23(a). The lattice constant of 10 $In_{1-x}Ga_xAs$ ($x = 0.47$) wells was matched to that of the InP substrates. The thickness of the wells is 5 nm. The 11 $In_{1-y}Ga_yAs$ ($y = 0.72$) barriers were mismatched to the InP substrate and have an internal strain of -1.7%. The barrier thickness is also 5 nm. Although applying biaxial tension to the wells is a straightforward approach to enhancing TM mode gain, in practice it is difficult to achieve, because the well composition cannot be adjusted to a wavelength of 1.55 μm. A novel approach was adopted to incorporate tensile strain in the barriers to enhance the TM mode gain. This active region was sandwiched between two InGaAsP waveguide layers with a bandgap wavelength of 1.28 μm. The thicknesses of the upper and lower InGaAsP layers were 100 and 50 nm, respectively. The device length and stripe width were 300 and 1.5 μm, respectively. The facets were antireflection (AR) coated.

The measured signal gain characteristics are shown as a function of the amplifier driving current in Figure 2.23(b). The signal gain was calibrated using a coupling loss of 4.5 dB per facet. At an amplifier driving current of 70 mA, the signal gain of the TM mode is equal to that of the TE mode. Under this condition, the signal gain was 7.5 dB.

2.3 NONLINEAR PHENOMENA

Nonlinear optical phenomena in LDs are mainly induced by three types of mechanisms. Both gain/loss change and refractive index change play significant roles in optical functional devices. If we begin with intensity-dependent refraction, we can simply express linear effects in intensity I with

$$n(I) = n_0 + n_2 I \tag{2.59}$$

where the nonlinear refractive index n_2 can conveniently be measured in square centimeters per watt.

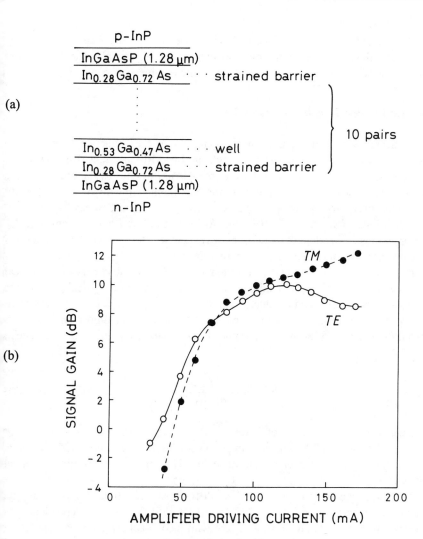

Figure 2.23 (a) Strained MQW amplifier structures; (b) signal gain versus driving current under signal injection at a wavelength of 1.495 μm. (©1990 IEEE. From [33].)

2.3.1 Thermal Effects

The bandgap energy decreases and the carrier energy distribution becomes broad due to the change in the Fermi-Dirac distribution with increasing temperature. Therefore, the gain decreases and the refractive index increases around the wavelength of the maximum gain with increasing temperature. The nonlinear refractive index coefficient n_2 is large, typically 5×10^{-5} [cm^2/W]. However, thermal effects are not considered to be particularly useful in practical devices because of their slow response time (typically 1 to 10 μs), and are usually observed as adverse effects of other, faster nonlinear mechanisms.

2.3.2 Carrier Effects: Interband Transition

The second mechanism is the change in the gain/loss or the refractive index due to variations in the number of free carriers through the band-filling effect (Burstein-Moss shift), whose relaxation is characterized by the free-carrier lifetime on the order 10^{-9} sec due to band-to-band transition. The variation of the maximum gain with the carrier density is approximately linear, and can be written as $g = a(N - N_0)$. Typical values for InGaAsP LDs are in the range of 1.2 to 2.5×10^{-16} cm^2 for a, and 0.9 to 1.5×10^{18} cm^{-3} for N_0, depending on the laser wavelength and doping levels. When the measured wavelength is fixed, both a and N_0 become greater than these values.

Semiconductors that are not carrier-injected act as saturable absorbers for an optical input with wavelength close to the bandgap. In contrast, saturation of optical gain occurs when the optical input to a semiconductor active medium is increased. The refractive index (n) also depends on electron concentration (N). The ratio of dn/dN and dg/dN is conventionally expressed in terms of a parameter b (denoted by α in some references), defined as

$$b = -4\frac{\pi}{\lambda}\frac{\mathrm{d}n/\mathrm{d}N}{\mathrm{d}g/\mathrm{d}N} \tag{2.60}$$

where λ is the wavelength. Values of b are typically found in the ranges of 4 to 7 for 1.3 to 1.6-μm InGaAsP and 2.5 to 4 for 0.85-μm GaAs lasers, respectively [34]. To date, such mechanisms based on carrier effects have mainly been used in nonlinear optical functional devices.

Figure 2.24(a) shows an experimental result for b as a function of photon energy, near the bandgap, measured on 1.5-μm InGaAsP LDs [35]. The strong dependence on photon energy close to the bandgap is consistent with theoretical calculations. The lasing photon energy for both devices was 0.81 eV (1.53 μm), and at this energy the value of b is close to 5. The quantities dn/dN and dg/dN are closely related. By simply differentiating the Kramers-Krönig relation with respect to carrier concentration N, we find

$$\frac{\mathrm{d}n(E)}{\mathrm{d}N} = -\frac{hc}{\pi}\int_0^\infty \frac{\mathrm{d}g(E')/\mathrm{d}N}{(E'^2 - E^2)}\mathrm{d}E' \tag{2.61}$$

where E is the photon energy and the other symbols have their usual meaning. Readers who are familiar with calculations of nonlinear refraction in passive semiconductor samples will recognize (2.61) from the corresponding equation with gain replaced by absorption.

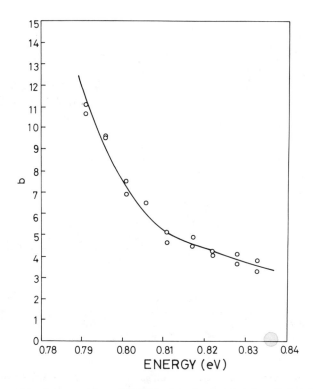

Figure 2.24(a) Linewidth enhancement factor b as a function of energy: a 1.5-μm InGaAsP LD. (From [35]. Reproduced by permission of IEE.)

The index change in the latter case comes about from the blocking of absorbing transitions by thermalized populations of electrons and holes that are created by absorption of an incident light beam. Thus, an increase of incident light intensity results in an increase of carrier concentrations and a consequent decrease of refractive index. In the case of the amplifier, on the other hand, the carrier concentrations are injected electrically into the active region of the device, and the incident light beam has the effect of depleting the carrier densities as a consequence of optical gain. Thus, for the amplifier, increasing optical input leads to an increase in the refractive index.

However, correctly speaking, in the case of nonlinear material susceptibility, the Kramers-Krönig relation does not hold [36]. Chow et al. explicitly demonstrated that in the nonlinear regime, the gain and refractive index in LDs saturate differently, with the index being a much weaker function of the laser flux [37]. Shore and Chan [38] presented the exact relationship between the real and imaginary parts of nonlinear susceptibility.

Arakawa et al. [39] and Burt [40] calculated the value of b in QW LDs based on the steplike density of state. It was predicted that the value of b at lasing frequency is appreciably reduced as compared to that in conventional DH lasers, since the differential

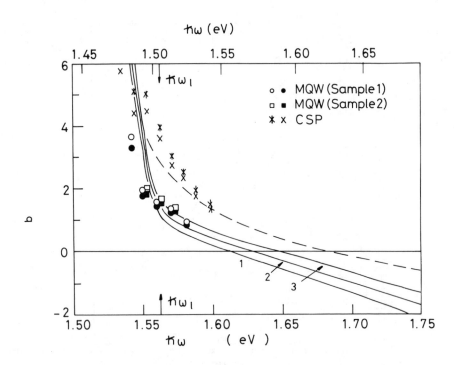

Figure 2.24(b) Linewidth enhancement factor *b* as a function of energy: GaAs/AlGaAs MQW LDs and a 0.8-μm CSP LD. (From [42]. Reproduced by permission of JJAP.)

gain dg/dN is enhanced in MQW LDs owing to the large density of state at the band edge. Westbrook and Adams [41] also showed that the optimum strategy for minimal linewidth enhancement is to employ narrow wells, low injected carrier densities, and photon energies much greater than the effective bandgap.

Ogasawara et al. [42] measured the dispersion of dn/dN, dg/dN, and b in MQW LDs. A comparison of the obtained results with those on a DH LD demonstrates that a reduction of b and an enhancement of dg/dN are realized in MQW LDs as predicted by the calculation. Experiments were performed on GaAs/AlGaAs MQW LDs with a self-aligned structure. The active layer consisted of four 4-nm-thick GaAs wells and four 5-nm-thick $Al_{0.3}Ga_{0.7}As$ barriers. Figure 2.24(b) shows spectra of the b factor determined experimentally for two MQW LDs as a function of photon energy $\hbar\omega$, which is shown in the lower end of the figure. The lasing photon energy $\hbar\omega_l = 1.562$ eV is denoted by the arrow. The asterisks in Figure 2.24(b) denote the spectra of b obtained experimentally for a 0.8-μm channeled-substrate-planar (CSP) stripe laser. The photon energy for the CSP laser is shown in the upper end of the figures, and the lasing energy of this sample $\hbar\omega_l = 1.506$ eV is also denoted by the arrow. The absolute values of both dn/dN and dg/dN for the MQW LDs are larger than those for the CSP LD. However, the enhancement in dg/dN is much more marked than that in dn/dN. This leads to the reduction of b in the MQW LDs. At the lasing energy $\hbar\omega_l$, b for the MQW LDs is approximately 1.6 and is about one-half of the value 3.8 for the CSP LD.

2.3.3 Carrier Effects: Intraband Transition

The third mechanism is due to the nonequilibrium carrier distribution in the bands, such as a spectral hole-burning or a carrier-heating effect. Relaxation is characterized with the intraband relaxation time on the order of 10^{-13} sec. This effect occurs even when the numbers of injected carriers in the active region are constant.

A spectral hole refers to a slight reduction in gain around the spectral region of the lasing wavelength, due to the finite intraband relaxation time of carriers. At the lasing wavelength λ_0, intense stimulation emission depletes carriers in the neighborhood of λ_0 faster than the rate at which carriers can "fill in" the "hole." The magnitude of gain compression produced by this effect depends critically on the intraband relaxation times of the carriers.

The dynamic carrier heating is explained as follows. The rapid carrier thermalization maintains electrons and holes in thermal equilibrium with each other at an elevated temperature T_e and T_h, respectively, where these temperatures are higher than that of the lattice T_L. This carrier heating is brought about by stimulated emission and free-carrier absorption. The heated carrier gas loses energy to the lattice by a combination of relaxation processes. In steady state, the heating is balanced by the cooling effects of relaxation and a hot carrier temperature is obtained. The higher carrier temperature results in a gain reduction since the active layer gain is a sensitive function of carrier temperature. This is thought to be one of the origins of the well-known gain compression effect observed in macroscopic device behavior. Several mechanisms are possible in bringing about energy loss (relaxation), such as electron-hole collisions and electron-lattice collisions.

The strong gain suppression was seen in a GaAs/AlGaAs [43] and InGaAsP/InP [44] LD amplifier by the injection of a femtosecond optical input. In GaAlAs LD amplifiers, 300-fs and 1.5-ps response mechanisms were observed. They were attributable to the e-h thermalization and the relaxation of the carrier distribution, respectively [43]. A 650-fs response mechanism, attributed to carrier heating, was observed in InGaAsP LD amplifiers [44]. The values of n_2 due to the intraband relaxation mechanisms have been reported as $n_2 = -2 \times 10^{-11}$ cm^2/W in InGaAsP [45,46] and $n_2 = -5 \times 10^{-12}$ cm^2/W in an AlGaAs optical amplifier [47]. These values are much greater than the nonresonant ultrafast response n_2 in AlGaAs ($n_2 = -3 \times 10^{-13}$ cm^2/W [48]) and are comparable to n_2 originating from the optical Stark effect ($n_2 = -2 \times 10^{-12}$ cm^2/W [49]) in a low-temperature GaAs/AlGaAs MQW.

Spectral hole burning has been suggested as a dominant source for the nonlinear gain, which strongly influences the dynamic properties of LDs [50]. By comparison of the measured K factors (ratio of the damping factor to the square of the resonance frequency) of DFB and Fabry-Perot lasers, it was found that the relaxation time associated with nonlinear gain for 1.3-μm InGaAsP LDs is about 0.1 ps [51]. This short time constant is consistent with spectral hole burning being the dominant process responsible for the nonlinear gain. Using a density-matrix formalism, gain saturation related to intraband relaxation of carriers was calculated to be about 1 cm^{-1} at 1–mW optical output power [52].

REFERENCES

[1] Casey, H. C., Jr., and M. B. Panish, *Heterostructure Lasers*, Orlando: Academic Press, 1978.

[2] Kressel, H., and J. K. Butler, *Semiconductor Lasers and Heterojunction LEDs*, Orlando: Academic Press, 1978.

[3] Thompson, G. H. B., *Physics of Semiconductor Laser Devices*, Chichester: John Wiley & Sons, 1980.

[4] Agrawal, G. P., and N. K. Dutta, *Semiconductor Lasers*, 2nd edition, New York: Van Nostrand Reinhold, 1993.

[5] Zory, P. S. Jr., ed., *Quantum Well Lasers*, San Diego: Academic Press, 1993.

[6] Bowers, J. E., and M. A. Pollack, "Semiconductor Lasers for Telecommunications," in *Optical Fiber Telecommunications II*, S. E. Miller, and I. P. Kaminow, eds., San Diego: Academic Press, 1988.

[7] For example, Acket, G. A., W. Nijiman, and H.'t Lam, "Electron Lifetime and Diffusion Constant in Germanium-Doped Gallium Arsenide," *J. Appl. Phys.*, Vol. 45, No. 7, July 1974, pp. 3033–3040.

[8] For example, Lamprecht, K. F., S. Juen, L. Palmetshofer, and R. A. Höpfel, "Ultrashort Carrier Lifetimes in H$^+$ Bombarded InP," *Appl. Phys. Lett.*, Vol. 59, No. 8, August 1991, pp. 926–928.

[9] Penty, R. V., H. K. Tsang, I. H. White, R. S. Grant, W. Sibbett, and J. E. A. Whiteaway, "Repression and Speed Improvement of Photogenerated Carrier Induced Refractive Nonlinearity in InGaAs/InGaAsP Quantum Well Waveguide," *Electron. Lett.*, Vol. 27, No. 16, August 1991, pp. 1447–1449.

[10] Panish, M. B., "Heterostructure Injection Lasers," *Proc. IEEE*, Vol. 64, No. 10, October 1976, pp. 1512–1542.

[11] Henry, C. H., "Measurement of Gain and Absorption Spectra in AlGaAs Buried Heterostructure Lasers," *J. Appl. Phys.*, Vol. 51, No. 6, June 1980, pp. 3042–3050.

[12] Paoli, T. L., "Near-Threshold Behavior of the Intrinsic Resonant Frequency in a Semiconductor Laser," *IEEE J. Quantum Electronics*, Vol. QE-15, No. 8, August 1979, pp. 807–812.

[13] Yariv, A., *Optical Electronics*, 4th edition, Philadelphia: Saunders College Publishing, 1991.

[14] Kogelnik, H., and C. V. Shank, "Coupled-Wave Theory of Distributed Feedback Lasers," *J. Appl. Phys.*, Vol. 43, No. 5, May 1972, pp. 2327–2335.

[15] Streifer, W., R. D. Burnham, and D. S. Scifres, "Effects of External Reflectors on Longitudinal Modes of Distributed Feedback Lasers," *IEEE J. Quantum Electronics*, Vol. QE-11, No. 4, April 1975, pp. 154–161.

[16] Sekartedjo, K., N. Eda, K. Furuya, Y. Suematsu, F. Koyama, and T. Tanbun-ek, "1.5 μm Phase-Shifted DFB Lasers for Single-Mode Operation," *Electron. Lett.*, Vol. 20, No. 2, January 1984, pp. 80–81.

[17] Buus, J., *Single Frequency Semiconductor Lasers*, Bellingham, Washington: SPIE Optical Engineering Press, 1991.

[18] Arakawa, Y., and A. Yariv, "Quantum Well Lasers--Gain, Spectra, Dynamics," *IEEE J. Quantum Electronics*, Vol. QE-22, No. 9, September 1986, pp. 1887–1889.

[19] Asada, M., A. Kameyama, and Y. Suematsu, "Gain and Intervalence Band Absorption in Quantum-Well Lasers," *IEEE J. Quantum Electronics*, Vol. QE-20, No. 7, July 1984, pp. 745–753.

[20] Yamanishi, M., and I. Suemune, "Comment on Polarization Dependent Momentum Matrix Elements in Quantum Well Lasers," *Japanese J. Appl. Phys.*, Vol. 23, No. 1, January 1984, pp. L35–L36.

[21] Kane, E. O., "Band Structure of Indium Antimonide," *J. Phys. Chem. Solids*, Vol. 1, 1957, pp. 249–261.

[22] Chemla, D. S., D. A. B. Miller, and S. Schmitt-Rink, "Nonlinear Optical Properties of Semiconductor Quantum Wells," in *Optical Nonlinearities and Instabilities in Semiconductors*, H. Haug, ed., Academic Press.

[23] Elliott, R. J., "Intensity of Optical Absorption by excitons," *Phys. Rev.*, Vol. 108, No. 6, December 1957, pp. 1384–1389.

[24] Shinada, M., and S. Sugano, "Interband Optical Transitions in Extremely Anisotropic Semiconductors. I. Bound and Unbound Exciton Absorption," *J. Phys. Soc. Japan*, Vol. 21, No. 10, October 1966, pp. 1936–1946.

[25] Okamoto, H., "Semiconductor Quantum-Well Structures for Optoelectronics--Recent Advances and Future Prospects," *Japanese J. Appl. Phys.*, Vol. 26, No. 3, March 1987, pp. 315–330.

[26] Miller, D. A. B., D. S. Chemla, and S. Schmitt-Rink, "Electric Field Dependence of Optical Properties of Semiconductor Quatum Wells: Physics and Applications," in *Optical Nonlinearities and Instabilities in Semiconductors*, H. Haug, ed., Academic Press.

[27] Miller, D. A. B., D. S. Chemla, and S. Schmitt-Rink, "Relation Between Electroabsorption in Bulk Semiconductors and in Quantum Wells: The Quantum-Confined Franz-Keldysh Effect," *Phys. Rev. B*, Vol. 33, No. 10, May 1986, pp. 6976–6982.

[28] Miller, D. A. B., D. S. Chemla, T. C. Damen, A. C. Gossard, W. Wiegmann, T. H. Wood, and C. A. Burrus, "Band-Edge Electroabsorption in Quantum Well Structures: The Quantum-Confined Stark Effect," *Phys. Rev. Lett.*, Vol. 53, No. 22, November 1984, pp. 2173–2176.

[29] Miller, D. A. B., D. S. Chemla, T. C. Damen, A. C. Gossard, W. Wiegmann, T. H. Wood, and C. A. Burrus, "Electric Field Dependence of Optical Absorption Near the Band Gap of Quantum-Well Structure," *Phys. Rev. B*, Vol. 32, No. 2, July 1985, pp. 1043–1060.

[30] Weiner, J. S., D. A. B. Miller, D. S. Chemla, T. C. Damen, C. A. Burrus, T. H. Wood, A. C. Gossard, and W. Wiegmann, "Strong Polarization-Sensitive Electroabsorption in GaAs/AlGaAs Quantum Well Waveguides," *Appl. Phys. Lett.*, Vol. 47, No. 11, December 1985, pp. 1148–1150.

[31] Miller, D. A. B., J. S. Weiner, and D. S. Chemla, "Electric-Field Dependence of Linear Optical Properties in Quantum Well Structures: Waveguide Electroabsorption and Sum Rules," *IEEE J. Quantum Electronics*, Vol. QE-22, No. 9, September 1986, pp. 1816–1830.

[32] Thijs, P. J. A., and T. van Dongen, "Strained-Layer InGaAs Multiple Quantum Well Lasers Emitting at 1.5 Micron Wavelength," *Extended Abstracts of the 22nd Int. Conf. on Solid State Devices and Materials*, Sendai, 1990, pp. 541–544.

[33] Magari, K., M. Okamoto, H. Yasaka, K. Sato, Y. Noguchi, and O. Mikami, "Polarization Insensitive Traveling Wave Type Amplifier Using Strained Multiple Quantum Well Structure," *IEEE Photonics Tech. Lett.*, Vol. 2, No. 8, August 1990, pp. 556–558.

[34] Osinski, M., and J. Buus, "Linewidth Broadening Factor in Semiconductor Lasers--An Overview," *IEEE J. Quantum Electronics*, Vol. QW-23, No. 1, January 1987, pp. 9–29.

[35] Westbrook, L. D., "Measurements of dg/dN and dn/dN and Their Dependence on Photon Energy in $\lambda = 1.5 \ \mu m$ InGaAsP Laser Diodes" *IEE Proc. Pt. J.*, Vol. 133, No. 2, April 1986, pp. 135–142.

[36] Hutchings, D. C., M. Sheik-Bahae, D. J. Hagan, E. W. Van Stryland, "Kramers-Krönig Relations in Nonlinear Optics," *Optical and Quantum Electronics*, Vol. 24, 1992, pp. 1–30.

[37] Chow, W. W., G. C. Dente, and D. Depatie, "Saturation Effects in the Carrier-Induced Refractive Index in a Semiconductor Gain Medium," *Opt. Lett.*, Vol. 12, No. 1, January 1987, pp. 25–27.

[38] Shore, K. A., and D. A. S. Chan, "Kramers-Kronig Relations for Nonlinear Optics," *Electron. Lett.*, Vol. 26, No. 15, July 1990, pp. 1206–1207.

[39] Arakawa, Y., K. Vahala, and A. Yariv, "Quantum Noise and Dynamics in Quantum Well and Quantum Wire Laser," *Appl. Phys. Lett.*, Vol. 45, No. 9, November 1984, pp. 950–952.

[40] Burt, M. G., "Linewidth Enhancement Factor for Quantum-Well Lasers," *Electron. Lett.*, Vol. 20, No. 1, January 1984, pp. 27–29.

[41] Westbrook, L. D., and M. J. Adams, "Explicit Approximations for the Linewidth-Enhancement Factor in Quantum-Well Lasers," *IEE Proc. J.*, Vol. 135, No. 3, June 1988, pp. 223–225.

[42] Ogasawara, N., R. Ito, and R. Morita, "Linewidth Enhancement Factor in GaAs/AlGaAs Multi-Quantum-Well Lasers," *Japanese J. Appl. Phys.*, Vol. 24, No. 7, July 1985, pp. L519–L521.

[43] Kesler, M. P., and E. P. Ippen, "Subpicosecond Gain Dynamics in GaAlAs Laser Diodes," *Appl. Phys. Lett.*, Vol. 51, No. 22, November 1987, pp. 1765–1767.

[44] Hall, K. L., J. Mark, E. P. Ippen, and G. Eisenstein, "Femtosecond Gain Dynamics in InGaAsP Optical Amplifiers," *Appl. Phys. Lett.*, Vol. 56, No. 18, April 1990, pp. 1740–1742.

[45] Grant, R. S., and W. Sibbett, "Observations of Ultrafast Nonlinear Refraction in an InGaAsP Optical Amplifier," *Appl. Phys. Lett.*, Vol. 58, No. 11, March 1991, pp. 1119–1121.

[46] Hall, K. L., A. M. Darwish, E. P. Ippen, U. Koren, and G. Raybon, "Femtosecond Index Nonlinearities in InGaAsP Optical Amplifiers," *Appl. Phys. Lett.*, Vol. 62, No. 12, March 1993, pp. 1320–1322.

[47] Hultgren, C. T., and E. P. Ippen, "Ultrafast Refractive Index Dynamics in AlGaAs Diode Laser Amplifiers," *Appl. Phys. Lett.*, Vol. 59, No. 6, August 1991, pp. 635–637.

[48] Anderson, K. K., M. J. LaGasse, H. A. Haus, and J. G. Fujimoto, "Femtosecond Studies of Nonlinear Optical Switching InGaAs Waveguides Using Time Domain Interferometry," *SPIE*, Vol. 1216, 1990, pp. 2–12.

[49] Hulin, D., A. Mysyrowicz, A. Antonetti, A. Migus, W. T. Masselink, H. Morkoc, H. M. Gibbs, and N. Peyghambarian, "Ultrafast All-Optical Gate With Subpicosecond ON and OFF Response Time," *Phys. Rev. Lett.*, Vol. 49, No. 13, September 1986, pp. 749–751.

[50] Asada, M., and Y. Suematsu, "Density-Matrix Theory of Semiconductor Lasers With Relaxation Broadening Model--Gain and Gain-Suppression in Semiconductor Lasers," *IEEE J. Quantum Electronics*, Vol. QE-21, No. 5, May 1985, pp. 434–442.

[51] Eom, J., C. B. Su, W. Rideout, R. B. Lauer, and J. S. LaCorse, "Determination of the Gain Nonlinearity Time Constant in 1.3 μm Semiconductor Lasers," *Appl. Phys. Lett.*, Vol. 58, No. 3, January 1991, pp. 234–236.

[52] Ogasawara, N., and R. Ito, "Longitudinal Mode Competition and Asymmetric Gain Saturation in Semiconductor Injection Lasers. II Theory," *Japanese J. Appl. Phys.*, Vol. 27, No. 4, April 1988, pp. 615–626.

Chapter 3
Absorptive Bistable Laser Diodes

3.1 OPTICAL BISTABILITY

Optical bistability, as the term implies, refers to the situation in which two stable optical output states are associated with a single optical input state (Figure 3.1(a,c)). Two general requirements must be satisfied for optical bistability to occur. The first is that there must exist an appropriate system parameter, such as the absorption coefficient or refractive index, which depends on optical input intensity. The second is the existence of a feedback mechanism. By setting the parameters to proper values, we can also obtain *differential-gain* characteristics (Figure 3.1(b,d)).

Various types of bistability appear in LDs, which depend on their configurations and operation conditions. Hysteresis occurs both in a counterclockwise sense (Figure 3.1(a)) and in a clockwise sense (Figure 3.1(c)). These types of hysteresis are called *S-shape bistability* in this book.

Pitchfork bifurcation is usually defined [1] as a form represented by the following differential equation, which depends on a single parameter μ:

$$\frac{dx}{dt} = \mu x - x^3 \tag{3.1}$$

The bifurcation diagram for this equation is depicted in Figure 3.2, in which the branches of equilibria are shown in (x,μ) space. Here, the only bifurcation point is $(x,\mu) = (0,0)$. It is easy to check that the unique fixed point $x = 0$ existing for $\mu \le 0$ is stable, that it becomes unstable for $\mu > 0$, and that the new bifurcating fixed points at $x = \pm \sqrt{\mu}$ are stable.

As will be described in Chapter 5, many types of bistability that have shapes similar to pitchfork bifurcation appear in two-mode LDs (Figure 3.2(b)). These are called *pitchfork bifurcation bistability* in this book.

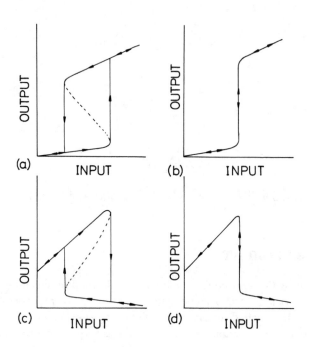

Figure 3.1 Typical output-versus-input characteristics of bistable optical device, called *S-shaped bistability*: (a) counterclockwise bistability; (b) differential gain; (c) clockwise bistability; (d) differential gain.

We will focus on the bistabilities in LDs and their applications. Details of the mathematical background can be found in a book by Guckenheimer and Holmes [1]. Other types of optical bistable devices, such as hybrid optical bistable devices and nonlinear etalons with unpumped semiconductors, can be found in books by Gibbs [2] and edited by Mandel et al. [3].

3.2 INHOMOGENEOUSLY CURRENT-INJECTED LASER DIODES

LDs that include saturable absorbers in their cavity show bistability in the optical-output-versus-current (L-I) curve and in the optical-output-versus-optical-input (L-P) curve. A saturable absorber is defined as a material whose absorption decreases with the increase in incident radiation intensity. In the OFF state of a BLD, there is only spontaneous emission from the gain region. Since the light level is low, light traveling in the waveguide between the reflecting ends will be strongly absorbed and there will be no laser action. In the ON state, the device operates as a laser. The population in the absorber is inverted by optical pumping from the gain region so that it is essentially transparent to the laser radiation. The quasi-Fermi level in the gain region decreases as it goes from OFF state

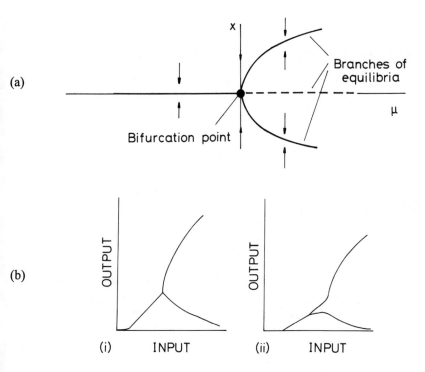

Figure 3.2 (a) Pitchfork bifurcation; (b) some forms of pitchfork bifurcation bistability.

to ON state, while, in the absorber, it increases. The two states of the device are stable, as will be shown later.

Figure 3.3 shows some examples of the schematic structure of absorptive BLDs [4]. In 1964, Lasher proposed a BLD in which the top electrical contact of the device was split to construct a gain region and a saturable absorption region, as shown in Figure 3.3(a) [5]. This was demonstrated experimentally with a Fabry-Perot laser cavity [6] and a DFB laser cavity (Figure 3.3(e)) [7,8]. The DFB BLDs have the advantages that their bistable properties can be controlled by detuning the Bragg wavelength from the wavelength at which the gain is the maximum [8], and they also allow single-frequency operation. Kawaguchi and Iwane used a distributed method [9] in which the device has alternate short sections of unbiased active waveguide and biased gain sections along the whole length of the device (Figure 3.3(b)). This device has a single contact, which provides improved reliability and allows for junction-side-down bonding and hence good thermal stability of the active region. A structure with a very short saturable absorber region with a small forward bias voltage applied to it has been used for faster switching (Figure 3.3(c)) [10]. Tarucha and Okamoto [11] have demonstrated voltage-controlled bistable operation using the nonlinear optical absorption of two-dimensional excitons in a GaAs-AlGaAs MQW LD (Figure 3.3(d)). They observed optical bistability in the light-output-

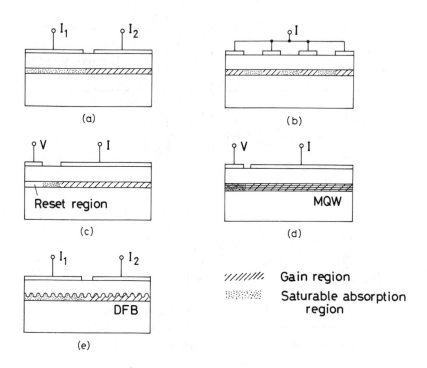

Figure 3.3 Absorptive BLDs: (a) two-section Fabry-Perot LD; (b) multisegment LD; (c) high-speed BLD with a short reset region; (d) two-section MQW LD; (e) two-section DFB LD. The output wavelength differs from the input wavelength in general. (From [4].)

versus-current curve of a tandem-electrode MQW LD. Using the quantum-confined Stark effect, this bistability can be controlled by a voltage applied to an electrode segment. Similar results were obtained in an InGaAs-InP MQW BLD [12]. Kucharska et al. observed bistability resulting from excitonic saturation in a tandem-type MQW LD without any control voltage into the absorbing region [13,14]. Yamada and Harris [15] and Frateschi et al. [16] demonstrated a strained InGaAs single-QW BLD with a saturable absorber section.

As bistable laser materials, GaAlAs systems (operating wavelength about 0.85 μm), InGaAs/GaAs systems (about 1.0 μm), and InGaAsP systems (about 1.3, 1.55 μm) as well as GaInAsP-GaAs systems (about 0.8 μm) [17] have been used.

The simplest model for absorptive bistability is the mean-field and single-mode one in the two-section LD with current biasing. The output characteristics can be analyzed by the following rate equations [18,19].

$$\frac{dn_{e1}}{dt} = P_1 - Bn_{e1}^2 - g_1(n_{e1})n_p v_g - \frac{n_{e1}}{\tau_{nr1}} \qquad (3.2)$$

$$\frac{dn_{e2}}{dt} = P_2 - Bn_{e2}^2 - g_2(n_{e2})n_p v_g - \frac{n_{e2}}{\tau_{nr2}} \tag{3.3}$$

and

$$\frac{dn_p}{dt} = n_p v_g [\gamma_1 g_1(n_{e1}) + \gamma_2 g_2(n_{e2})] - \frac{n_p}{\tau_p}$$

$$+ \beta_{sp} B(\gamma_1 n_{e1}^2 + \gamma_2 n_{e2}^2) \tag{3.4}$$

where $\gamma_1 + \gamma_2 = 1$. Subscripts 1 and 2 refer to parameters in regions I and II, respectively, while n_e and n_p are the numbers of injected carriers and photons. The pump rate per unit volume is $P = I/qV$, where I is the total input current and V is the volume. B is the recombination coefficient, v_g is the group velocity, β_{sp} is the spontaneous emission coefficient, γ_i is a ratio of the length of region i to the whole cavity length. We can obtain characteristics related to bistable operation, such as ON and OFF thresholds, hysteresis width, and switching dynamics between ON and OFF states using the above rate equations.

The following analysis shows how the pin junction with a low pump rate acts as a saturable absorber. We assume quasineutrality (electron density equals hole density) and an absorption constant α, which depends linearly on carrier density [20].

$$\alpha = -a(n - n_{tr}) \tag{3.5}$$

The proportionality factor a is the differential gain (or absorption) constant and n_{tr} is the carrier density corresponding to transparency. For current biasing, the carrier density is given by the static solution of the following rate equation, which is similar to (3.2) or (3.3).

$$\frac{dn}{dt} = P - \frac{n}{\tau_s} - a(n - n_{tr})\frac{I}{h\nu} \tag{3.6}$$

Here, I is the light intensity. Solving (3.5) and (3.6) for the absorption α as a function of the light intensity I, we obtain the following expression for $dn/dt = 0$.

$$\alpha = \frac{\alpha_0}{1 + I/I_s} \tag{3.7}$$

The absorption is α_0 at low intensities and becomes bleached above the saturation intensity I_s:

$$I_s = \frac{h\nu}{\tau_s a} \tag{3.8}$$

$$\alpha_0 = \frac{h\nu}{I_s}\left(\frac{n_{tr}}{\tau_s} - P\right) \tag{3.9}$$

Therefore, the pin junction acts as a saturable absorber if biased with a current source, and tandem-type LDs show bistability.

Steady-state rate equations are obtained by setting the time derivatives equal to zero in (3.2) to (3.4). The gain for the present laser model is written as

$$g(n_e) = an_e^2 + bn_e + c \tag{3.10}$$

As a numerical example we choose $a = 2.67 \times 10^{-35}$ cm^6 cm^{-1}, $b = 1.00 \times 10^{-16}$ cm^3 cm^{-1}, $c = -165$ cm^{-1}, $B = 1.33 \times 10^{-10}$ cm^3 s^{-1}, $\beta_{sp} = 10^{-5}$, $v_g = 6.7 \times 10^9$ cm s^{-1}, $\tau_{nr1} = \tau_{nr2} = \tau_{nr} = 10$ ns, $\tau_p = 3.73 \times 10^{-12}$ sec and $P_2 = 0$. L-I characteristics dependent on the gain model were also considered [18]. As a result, it can be concluded that the existence of bistability does not depend on the form of the gain dependence on n_e (i.e., linear dependence or superlinear dependence).

The computed variations in photon density as a function of the excitation rate are shown in Figure 3.4 [19]. The horizontal axis shows the pump rate, $\gamma_1 P_1$, implying the total injection current for the device. The figure clearly shows the bistability that exists

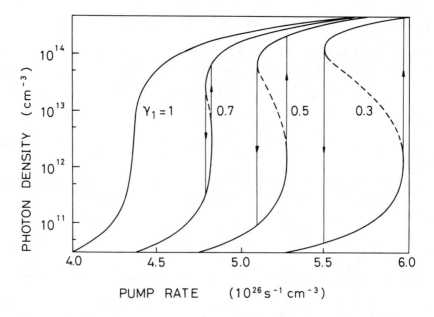

Figure 3.4 Calculated photon density versus excitation rate with γ_1 (ratio of the region 1 length to the cavity length) as the parameter. (Reprinted with permission from Chapman & Hall Ltd. From [19].)

n the output curves, except for at $\gamma_1 = 1$ (ordinary stripe laser excited uniformly). In practice, the negative-slope portion of the curve (broken line) corresponds to an unstable trajectory and bistability leads to a hysteresis loop. With decreasing γ_1, both the bistability current range and the threshold current increase.

More detailed analysis has been conducted by many researchers. The impact of electrical [21] and optical [22] coupling between adjacent sections has been analyzed using the single-mode, mean-field hypotheses. The spatial dependence of both the field and the gain was fully taken into account [23]. The traveling-wave multimode model [24,25] is useful for the simulation of the bistable characteristics, which depend on the positions of the saturable absorber in the laser cavity [26].

A detailed comparison between the results of single and multimode, mean-field and traveling-wave models has been reported in [25], taking into account the physical characteristics of the semiconductor, the electrical and optical couplings, and the effect of the reflection at the junction between pumped and unpumped sections. One example of the results is shown in Table 3.1 for a comparison of the mean-field and traveling-wave models [25]. Devices 400 μm long with 200-μm gain and 200-μm absorbing sections were chosen for the simulation. The table shows the calculated switch ON and OFF currents, from which we can observe that (1) for finely segmented structures (i.e., homogeneous gain/absorption regions 50 μm long), the mean-field approach is quite accurate; (2) with only two homogeneous regions, the difference between the mean-field and traveling-wave models becomes very pronounced for the switch ON current.

As one example of absorptive BLDs, the InGaAsP DFB BLD structure operating at 1.55 μm is shown in Figure 3.5(a) [7]. The p-type electrode was divided into three parts, and the resistance between the p-type electrodes was a few hundred ohms. Thus, the divided regions can be excited independently through the electrodes. If the current of

Table 3.1
Comparison Between Mean Field and Traveling-Wave Models

	Switch-on Current (mA)	Switch-off Current (mA)
Mean Field Model	49.8	21.2
Traveling-Wave Model		
	49.5	21.3
	26.5	20.8

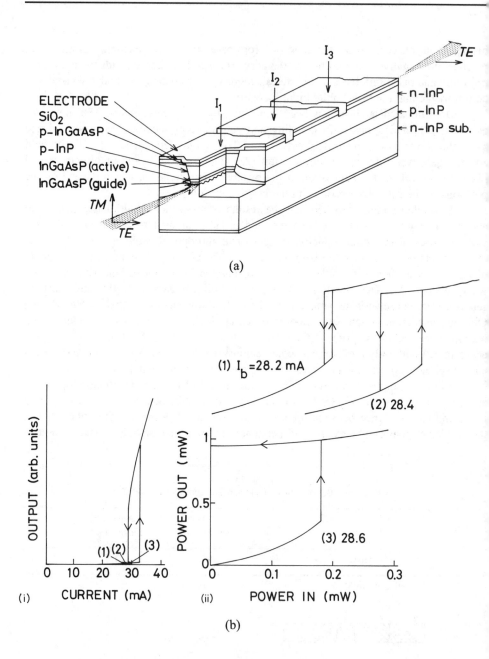

Figure 3.5 (a) Structure of absorptive bistable DFB LD. I_1 set at a low value; region I acts as a saturable absorber. (b) Output characteristics of absorptive bistable DFB LD: (i) light-output-versus-current curve; (ii) optical-output-versus-input curve. Bias current values are indicated by the closed circles (1) to (3) in (i). (©1988 IEEE. From [7].)

region I is set at zero or a low value, region I acts as a saturable absorber. It is then possible to obtain bistable characteristics in both the light-output-current and optical-output-input curves.

A typical measured light-output-versus-current curve is shown in Figure 3.5(b)(i). A noticeable hysteresis loop is seen in the curve. The current range over which bistability exists decreases with an increase in the current into the saturable absorber (not shown in the figure).

When the bias current is set at a value just below the turn-off threshold, the optical-output-input curve exhibits bistability resulting from an injection of optical power into the saturable absorber, as shown in Figure 3.5(b)(ii). The bias current values are indicated by the closed circles (1) to (3) in Figure 3.5(a)(i). An increase in the bias current markedly increases the bistability range of the input optical power. When the bias current is within the bistability range, the BLD emits coherent light when input optical power is injected. The laser stays in an ON state even when the input optical power is reset to zero. This means that the BLD acts as an optical memory device.

Generally, in absorptive BLDs, the output wavelength differs from the input wavelength. This characteristic suggests the use of an absorptive BLD as a wavelength converter as described in Section 8.1.

3.3 BISTABILITY AND SELF-PULSATION

3.3.1 Stability Analysis Under Constant Current Operation

Self-pulsation (i.e., repetitive strong intensity change in output power with a repetition rate of hundreds of megahertz to a few gigahertz) strongly connects with bistability.

The output stability depends on the distribution of carrier lifetime along the laser cavity [27]. When the carrier lifetime is broadly the same over the whole laser cavity, inhomogeneously excited LDs show bistability as described above. When the carrier lifetime of the unpumped or low-pumped region is shorter than the gain region, self-pulsation occurs even when excited with only a dc bias current.

The threshold pump rates for laser oscillation can be obtained by solving rate equations (3.2), (3.3), and (3.4) under the steady-state condition. The solid lines in Figure 3.6 indicate the calculated threshold pump rates [28]. In the case of $\tau_{nr2} > 2$ ns, the laser exhibits two threshold pump rates depending on the initial conditions. As shown in the inset, when the excitation level P_1 is increased from zero, the output abruptly jumps from the lower to the higher level at P_{th1}. However, as the excitation level is reduced, the output drops abruptly at the lower threshold level P_{th2}. This bistable characteristic in the L-P curve is attributed to the fact that the active layer of region II acts as a saturable absorber.

On the other hand, when $\tau_{nr2} < 2$ ns, the laser shows only one threshold pump rate. Furthermore, for $\tau_{nr2} < 0.5$ ns, the laser threshold gradually increases with decreasing τ_{nr2}, because the injected carriers for region I sink into the active laser in region II through the generation of spontaneous emission in region I and optical absorption in region II.

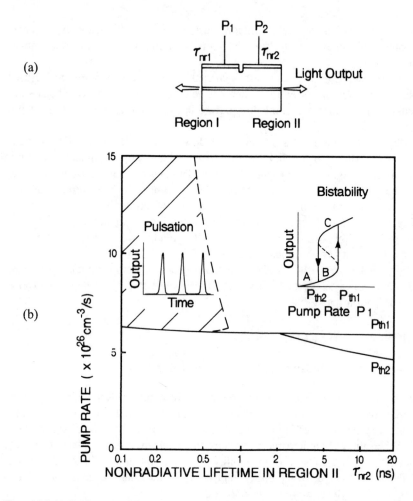

Figure 3.6 (a) Tandem-type LD structure; (b) threshold pump rate (full lines) and pulsation region (shaded region) versus nonradiative carrier lifetime in region II. (From [28].)

Optical output stability in a tandem-type LD is calculated from (3.2) to (3.4) using small-signal approximation [18].

It is assumed that the physical quantities can be written as the sum of a time-independent term and a small deviated term:

$$n_{e1} = \bar{n}_{e1} + \Delta n_{e1}, \; n_{e2} = \bar{n}_{e2} + \Delta n_{e2}, \; n_p = \bar{n}_p + \Delta n_p \tag{3.11}$$

where the electron and photon populations deviate slightly from their equilibrium values $\bar{n}_{e1}$, $\bar{n}_{e2}$, and $\bar{n}_p$ by Δn_{e1}, Δn_{e2}, and Δn_p, respectively.

Inserting (3.11) into (3.2) to (3.4) and neglecting second and high-order terms in Δn_{e1}, Δn_{e2}, and Δn_p, we obtain

$$\frac{d(\Delta n_{e1})}{dt} = \Delta n_{e1} - 2B\bar{n}_{e1} - \bar{n}_p v_g(2a\bar{n}_{e1} + b) - \frac{1}{\tau_{nr1}}$$

$$- \Delta n_p v_g(a\bar{n}_{e1}^2 + b\bar{n}_{e1} + c) \qquad (3.12)$$

$$\frac{d(\Delta n_{e2})}{dt} = \Delta n_{e2} - 2B\bar{n}_{e2} - n_p v_g(2a\bar{n}_{e2} + b) - \frac{1}{\tau_{nr2}}$$

$$- \Delta n_p v_g(a\bar{n}_{e2}^2 + bn_{e2} + c) \qquad (3.13)$$

and

$$\frac{d(\Delta n_p)}{dt} = \Delta n_{e1} \bar{n}_p v_g \gamma_1(2a\bar{n}_{e1} + b) + 2\beta_{sp} B \gamma_1 \bar{n}_{e1}$$

$$+ \Delta n_{e2} \bar{n}_p v_g \gamma_2(2a\bar{n}_{e2} + b) + 2\beta_{sp} B \gamma_2 \bar{n}_{e2}$$

$$+ \Delta n_p \left[v_g \gamma_1(a\bar{n}_{e1}^2 + b\bar{n}_{e1} + c) \right. \qquad (3.14)$$

$$\left. + v_g \gamma_2(a\bar{n}_{e2}^2 + b\bar{n}_{e2} + c) - \frac{1}{\tau_p} \right]$$

When (3.12) and (3.14) are inserted into (3.13), the result is a third-order differential equation:

$$a_e \Delta \dddot{n}_{e1} + b_e \Delta \ddot{n}_{e1} + c_e \Delta n_{e1} + d_e \Delta n_{e1} = 0 \qquad (3.15)$$

By setting

$$\Delta n_{e1} = A \exp(\lambda t) \qquad (3.16)$$

equation (3.14) can be written as

$$a_e \lambda^3 + b_e \lambda^2 + c_e \lambda + d_e = 0 \qquad (3.17)$$

The third-order equation can be solved by the well-known Cardano method. If the real part of λ is negative, then n_{e1} will be stable.

The calculated pulsation region is represented by the shadow region in Figure 3.6. For $\tau_{nr2} > 0.8$ ns, the optical output is stable under all conditions of excitation except for the negative-slope portion of the L-P curve in the bistable region. Under the laser threshold (region A in the inset of Figure 3.6), deviations decrease exponentially with time. For the negative-slope portion (region B), deviations increase; that is, the negative-slope portion of the curve corresponds to an unstable trajectory. Region C shows that the deviations decrease exponentially with damped oscillation.

Around $\tau_{nr2} = 0.8$ ns, pulsation occurs only in the limited range slightly above the laser threshold. The pulsation region rapidly increases with decreasing τ_{nr2}. For $\tau_{nr2} < 0.5$ ns, pulsation occurs in the wide range from just above the laser threshold to a very high excitation level.

The carrier lifetime can be shortened by impurity doping [29], proton bombardment [30], or reverse bias [31]. For example, the carrier lifetime in an InGaAs/InGaAsP MQW diode is reduced to approximately 50 ps under reverse bias, as described in Section 2.3 [31].

Nonlinearity in the gain-versus-carrier-density relation was taken into consideration for the stability analysis [32]. The results indicate that bistable operation, self-sustained pulsation, and stable operation can be obtained by switchable combinations of three key parameters, namely τ_a/τ_b, g_a/g_b, and β. Here, τ_a and τ_b are the carrier lifetime in regions a and b, respectively. g_a and g_b are the differential gain defined by $g_{a(b)} = dG_{a(b)}(n_{a(b)})/dn_{a(b)}$. Here, $G_{a(b)}(n_{a(b)})$ is the gain value for a carrier density of $n_{a(b)}$. β is relative unsaturated absorption. Figure 3.9 depicts bistable region, unstable region, and stable region on the τ_a/τ_b-versus-g_a/g_b plane. The bistable region is defined by

$$\frac{\tau_a}{\tau_b}\frac{g_a}{g_b} < \frac{-\beta}{1-\beta} \tag{3.18}$$

and hysteresis phenomena occur in this region. Here, $\beta = hG_{ob}(n_b)/\Gamma$, where h is the volume fraction of the absorbing region in the cavity, Γ is the cavity loss, and $G_{ob}(n_b)$ are the unsaturated gain values related to carrier density in region b. Normalized hysteresis width ΔX, which is defined as the difference between the pumping level at the threshold and at the point of optical output drops, is depicted in Figure 3.7 by the contour lines. It can be seen that contour lines for ΔX are along the border line of the bistable region and that hysteresis width becomes wider with a decrease in the product of τ_a/τ_b and g_a/g_b.

On the other hand, an unstable region, in which self-sustained pulsations are expected to take place within some pumping regions, is subdivided into two regions, namely unstable region I, where the minimum photon density for the self-sustained pulsation is zero, and unstable region II, where the minimum photon density is larger than zero. Unstable region I, which occupies the major portion of the unstable region is defined by

$$\frac{\tau_a}{\tau_b}\frac{g_a}{g_b} > \frac{-\beta}{1-\beta} \quad \text{and} \quad \frac{\tau_a}{\tau_b} > \frac{1}{1-h}\frac{1-\beta}{-\beta}\frac{g_a}{g_b} + 1 \tag{3.19}$$

The normalized maximum photon density N_{max}/N_{sa} within the unstable region is depicted in Figure 3.7 by the contour lines. It is revealed that contour lines for N_{max}/N_{sa} are almost

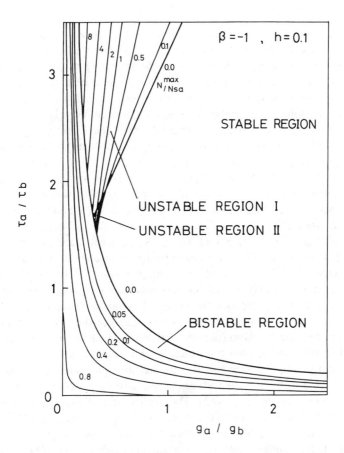

Figure 3.7 Bistable region, unstable region, and stable region in the τ_a/τ_b-versus-g_a/g_b plane, with $\beta = -1$ and $h = 0.1$. The unstable region is subdivided into two regions, namely unstable region I and unstable region II. Normalized hysteresis width ΔX in the bistable region and normalized maximum photon density N_{max}/N_{sa} in the unstable region, respectively, are indicated by contour lines. (From [32]. Reproduced by permission of AIP.)

straight lines, and asymptotic to the border line between unstable region and bistable region, and that maximum photon density in the unstable region becomes larger with decreasing g_a/g_b.

Taking the direct optical coupling between two sections into account, a theory on the stability of two-section LDs has also been established [33].

3.3.2 Effects of Load Resistance

When the pin junction is biased with a voltage source, the emission characteristics are strongly influenced by the external drive circuits [34]. Negative differential resistance is obtained in the current-voltage characteristic (I_2–V_2) of the absorbing section shown in Figure 3.8(a). This negative differential resistance is optoelectronic in origin, and the mechanism causing it can be understood with the following simple model as illustrated in Figure 3.8(b). For simplicity, assume the BLD consists of two parts, a gain section pumped with a constant current I_1 plus an absorber section that acts like a pin photodiode within the optical cavity. The current I_2 through the photodiode consists of the following two terms: the normal diode current that depends exponentially on the applied voltage V_2 and a negative photocurrent I_{ph}, which is roughly proportional to the photon density in the active region as written in the following equations: $I_2 = I_s\{\exp(V_2/V_T) - 1\} - I_{ph}$ and $I_{ph} = \eta \times P$. Here, V_T is the thermal voltage. These photons are generated under the gain contact and guided via the BH waveguide to the absorbing region under the second contact. This photodiode characteristic is drawn in Figure 3.8(b) for eight different normalized photon densities, $P = 0$ to $P = 7$. We now explain how the measured I_2–V_2 characteristic is produced. The gain section is biased with a fixed current I_1 and the absorbing section is biased with a voltage source V_2, thus introducing an unsaturable amount of absorption. For zero voltage, $V_2 = 0$, the absorption of the photodiode is strong, thus suppressing stimulated emission, and only a small photocurrent due to spontaneous emission in the gain section is generated. Increasing the voltage V_2 reduces absorption and therefore the cavity losses. As soon as the total cavity losses becomes smaller than the gain, stimulated emission is initiated and the photon density increases, generating, in turn, a larger negative photocurrent in the absorbing section. Increasing V_2 further reduces the losses, which increases the stimulated emission and also the negative photocurrent I_{ph}, thus producing the negative slope. Finally, at large voltages V_2, the absorbing section becomes transparent and the photocurrent decreases to zero. The positive exponential term representing the normal diode behavior now starts to dominate and I_2 increases very quickly. The specific curve as shown in Figure 3.8(b) is obtained for a fixed gain current I_1. We shall now show that an LD that contains such an absorbing pin diode can either be bistable or self-pulsating, depending on the biasing circuit. The I_2–V_2 characteristic is shown in Figure 3.8(c) again along with the characterization of the source driving this section, the load line. This load line shows the voltage available to the absorber section as a function of the current through it. The state of the system satisfying all static circuit equations is given by the intersection of the load line with the characteristic of the device. The state P_1 in Figure 3.8(c) is obtained by intersecting the characteristic of the absorber with a load line corresponding to a voltage source of $V = -0.85$V and a series resistance of $R_L = 20$ kΩ. The gain section is driven with a normalized current of $I_1/I_{th} = 1.29$, with I_{th} being the threshold current of the homogeneously pumped laser. In the state P_1, the laser is switched off. Increasing the pump current I_1 causes the intersection point to move along the load line from P_1 to P_2 and at $I_1/I_{th} = 1.56$ to jump to P_3, since this is the only stable intersection of the load line with the characteristic of the absorbing section. In this state, the laser is switched on. A decrease of I_1 now causes the state to move back to P_4 and then jump back to P_5, and the laser switches off again. A laser with a segmented contact whose absorbing section is biased with such a large load resistance will display a hysteresis

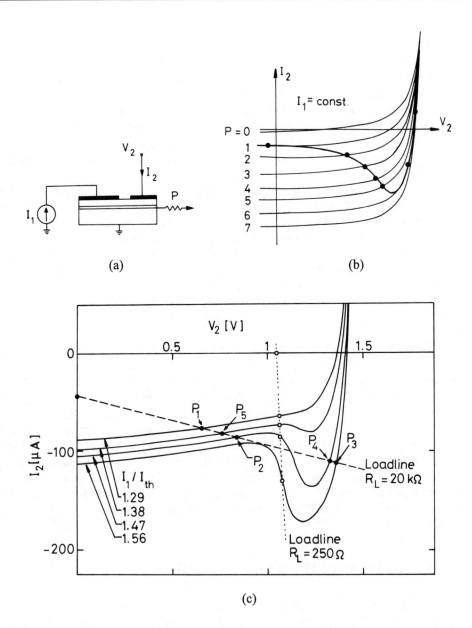

(a)

(b)

(c)

Figure 3.8 (a,b) Simplified photodiode model to explain the current-voltage characteristic I_2-V_2 of the absorber diode for a fixed current I_1 through the gain section. (c) Current-voltage (I_2-V_2) characteristic of the absorbing diode for different normalized currents I_1/I_{th} through the gain section. I_{th} is the threshold current for homogeneous pumping (both anode contacts are connected). Also shown are two load lines corresponding to a biasing source of −0.85V and R_L = 20 kΩ and 1.05V and R_L = 250Ω, respectively. (©1982 IEEE. From [34].)

in the light-versus-current characteristic, and it is bistable. This is also obvious from the fact that the load line has three intersections, one unstable state and two stable ones, P_2 (switched off) and P_4 (switched on) with the I_2–V_2 characteristic for $I_1/I_{th} = 1.47$.

The load line corresponding to a voltage source of $V = 1.05$V and a series resistance of $R_L = 250\Omega$ is also shown in Figure 3.8(c). This load line always has only one intersection with the characteristic, and consequently the laser will not display bistability. With such a small load resistance on the absorbing section, the laser can be biased in the region of negative differential resistance of the absorber. This leads to electrical microwave oscillation and concomitant light-intensity pulsations.

The effective load resistance for the absorber was shown to be an important factor to be considered if self-pulsations were to be observed. This load resistance is dependent on the parasitic resistance between the sections. Farrell et al. have defined a self-pulsation operating regime within the V-I characteristics of the absorber section of a two-section LD [35]. An active load is used that both controls the absorber voltage and reduces dependence on the parasitic resistance between the laser sections. The self-pulsating regime exists only when gain section currents produce a decrease in the value of the absorber current at the threshold rather than an increase as observed by Harder et al. [34]. This decrease in the absorber current at the threshold is shown to be consistent with an S-shaped V-I characteristic.

They have also shown that it is not necessary to bias the absorber in a region of negative resistance to observe self-pulsation. Above the threshold over a range of absorber voltages, the self-pulsation is maintained even when the absorber is operating at a point that has a positive differential resistance. Thus, unlike the observations of [34], negative resistance can be interpreted only as an indication of the existence of a larger region of self-pulsation within the absorber V-I characteristic. The absorber voltage range over which self-pulsation occurs increases with the gain section current. This absorber voltage range and the gain section current determine the limits of a complete self-pulsation operating regime. This regime is shown in Figure 3.9 as a crosshatched area overlaid on the normalized absorber V-I characteristics for gain section currents between 65 mA and 90 mA. In this figure, the voltage on the horizontal axis is normalized to the absorber voltage at the threshold.

3.4 BISTABILITY IN OPTICAL INPUT-OUTPUT

Inhomogeneously excited LDs with bias current also show bistability in optical-output-versus-optical-input (P_O-P_I) characteristics [36]. The optical input can drive the device through many operating modes as follows: (1) incoherent optical pumping of the absorbing region, (2) incoherent optical pumping of the gain region, and (3) coherent optical pumping of the absorbing and gain region. For cases (1) and (2), the device shows the same characteristics as that of injection current switching by substituting current pumping for optical pumping. For case (3), the P_O-P_I characteristics can be calculated by the following

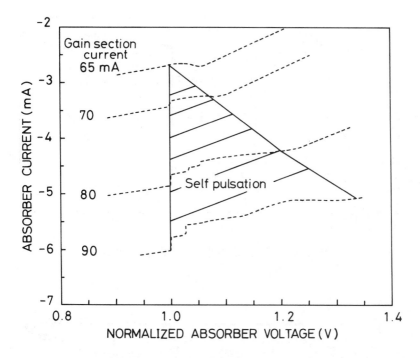

Figure 3.9 Self-pulsation operating regime overlaid on normalized absorber V-I characteristics for gain section currents above 65 mA. Self-pulsation operating regime shown as crosshatched area. (From [35]. Reproduced by permission of IEE.)

equation in which a term representing the optical input power P_{in} is added to (3.4), which expresses the case in which the wavelength of the optical input is equal to that of the bistable laser output [19].

$$\frac{dn_p}{dt} = n_p v_g [\gamma_1 g_1(n_{e1}) + \gamma_2 g_2(n_{e2})] - \frac{n_p}{\tau_p}$$
$$+ \beta_{sp} B(\gamma_1 n_{e1}^2 + \gamma_2 n_{e2}^2) + P_{in} \tag{3.20}$$

The computed variation in output photon density is shown in Figure 3.9 as a function of the excitation rate. Figure 3.10(a) clearly shows bistability for all γ_1 except $\gamma_1 = 1$ (ordinary stripe laser excited uniformly). The bias currents are set at $0.999 P_{th}$, where P_{th} is the turn-off threshold current. With decreasing γ_1, the bistability input optical power range increases. Figure 3.10(b) shows the P_0-P_I characteristics for various bias currents with $\gamma_1 = 0.5$. The P_0-P_I characteristics are very sensitive to the bias current, and a large bistable region is obtained with a bias current closer to the laser threshold.

If there is a frequency mismatch, then the rate equation is modified as follows [37].

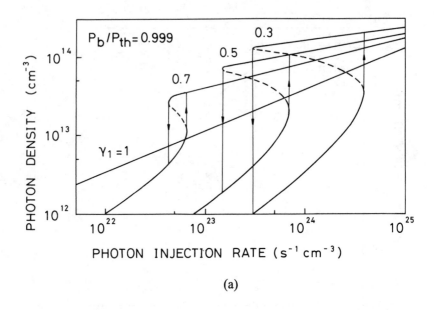

(a)

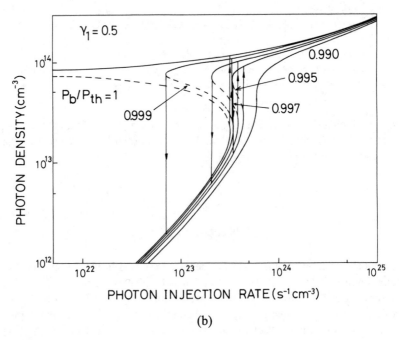

(b)

Figure 3.10 (a) Calculated photon density versus photon injection rate with γ_1 as the parameter and bias pumping (P_b) set at $0.999P_{th}$; (b) calculated photon density versus photon injection rate with P_b as the parameter and $\gamma_1 = 0.5$. (Reprinted with permission from Chapman & Hall Ltd. From [19].)

$$\frac{dn_p}{dt} = -\Delta\Gamma n_p + \beta_{sp}B[\gamma_1 n_{e1}^2 + \gamma_2 n_{e2}^2] + \frac{\Delta\Gamma(t_0/T)^2 C}{(\Delta\Gamma/2)^2 + \delta\nu_0^2} \tag{3.21}$$

where $-\Delta\Gamma = v_g[\gamma_1 g_1(n_{e1}) + \gamma_2 g_2(n_{e2})] - 1/\tau_p$ and $C = \int d\nu s_\nu^* s_\nu$ is proportional to the total intensity of the injected radiation. t_0 is the amplitude transmission coefficient and T is the round-trip time. $\delta\nu_0 = \nu_0 - \nu_R$, where ν_0 is the peak angular frequency of injecting radiation and ν_R is the resonant angular frequency of the BLD. The bistable optical input region decreases with increasing Δf. For more authenticity, however, the wavelength shift during the optical switching of BLDs based on the carrier-dependent refractive index must be taken into account. This was analyzed theoretically by Adams [38].

3.5 BISTABLE SWITCHING

3.5.1 Bistable Switching by Current Pulses

Among the variety of possible functions of an optical bistable device, one of the most fundamental is optical switching between the lower (OFF) and upper (ON) stable states.

To obtain the time response for a BLD, the rate equations (3.2) to (3.4) have been solved numerically by the Runge-Kutta-Gill method [18]. The parameters used are the same as those described in Section 3.2. Figure 3.11 shows the transient response to a step-current pulse of a BLD with $\gamma_1 = 0.5$.

A theoretical calculation of the photon response to an ON current pulse is shown in Figure 3.11(a). The injection current changes with a rectangular pulse from I_b to $1.05I_b$, $1.10I_b$, and $1.17I_b$, where I_b is at the middle value in the current range for bistability, as shown in the inset. The initial laser condition ($t = 0$) is an OFF condition (the lower branch in the photon-density/excitation rate curve). In Figure 3.11(b), injected-carrier-density-versus-time responses are shown for the gain and absorbing regions. When a fast-rise-time current pulse is applied to a BLD, the light output power is delayed a few subnanoseconds and is then characterized by damped oscillation. The short delay time and frequency of the damped oscillation depend on the height of the current in the same manner as conventional stripe-geometry lasers. However, unlike conventional LDs, the time responses for BLDs show a component that gradually increases with time, and the optical output approaches the final output exponentially with damped oscillation. This distinctive phenomenon is attributable to carrier generation by the laser field in the noninjected region.

The calculated responses to an OFF pulse are shown in Figure 3.11(c,d). The initial condition is an ON condition. The injection current changes from I_b to $0.95I_b$, $0.90I_b$, and $0.83I_b$ at $t = 0$. If the OFF current pulse is sufficiently large, the light output decreases without delay when a fast-rise-time current pulse is applied to a device. Conversely, when the pulse height is low, the light output gradually decreases with oscillation damping. The time response for carrier densities corresponding to Figure 3.11(c) are shown in Figure 3.11(d).

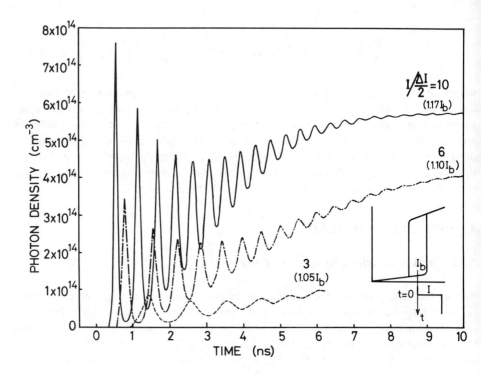

Figure 3.11(a) Calculated transient response to step current pulse of device with $\gamma_1 = 0.5$. Time evolution of photon with ON pulse. Injection current changes from I_b to $1.05I_b$, $1.10I_b$, and $1.17I_b$. (From [18].)

The influence of the height and width of trigger current pulses for switching from an OFF to an ON and from an ON to an OFF state have also been examined. Photon and carrier responses to $5 \times \Delta I$ height (from I_b to $1.17\ I_b$) trigger current ON pulses, with 1- and 2-ns widths, are shown in Figure 3.12(a,b),, respectively. Pumping-rate time dependence is also shown in Figure 3.12(a). With a 1-ns current pulse, switching cannot be carried out. This is due to a lack of sufficient carrier accumulation in the noninjection region. Conversely, switching from OFF to ON can be achieved with a 2-ns pulse. Responses to the OFF pulse are shown in Figure 3.12(c,d). When $I_t = 5 \times \Delta I$ (from I_b to $0.83I_b$), an 8-ns-wide current is necessary for switching.

It can be seen from the calculations that the ON trigger pulsewidth needed for the switching is about one-third or one-quarter of the carrier lifetime in the noninjection region (τ_{nr}). The OFF trigger pulsewidth needed is slightly less than τ_{nr}. Therefore, if a crystal with a short carrier lifetime (a few nanoseconds or less) were used, then this device could be operated at a rate of several hundred megabits per second or more. The switching speed also depends on the current pulse height as expected. The device can be operated at a higher speed when larger current pulses are used.

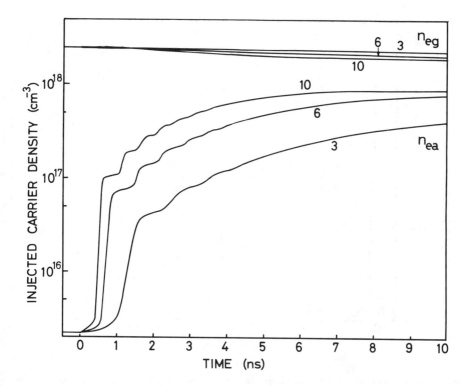

Figure 3.11(b) Calculated transient response to step current pulse of device with $\gamma_1 = 0.5$. Time evolution of carriers with ON pulse. Injection current changes from I_b to $1.05I_b$, $1.10I_b$, and $1.17I_b$. (From [18].)

Switching between the two states of bistable operation was investigated by applying an electrical pulse to the device, as shown in Figure 3.13. The structure of the laser is shown in Figure 3.13(a). The lasers were prepared using an LPE growth technique for an InP/InGaAsP/InP DH structure. The current channel was made by a planar zinc diffusion method based on designed stripe geometry. The stripe width was about 10 μm, the injection region length was 20 μm, and the noninjection region length was about 10 μm. The cavity length was about 200 μm. The excited regions are shaded in the figure. A micrograph of the stripe pattern fabricated is also shown in the inset of Figure 3.13(a). The bias current was set at 148 mA, which is the middle value of the current range for bistability, as shown in Figure 3.13(b). The switching pulses were formed by the sum of the positive pulse and 200-ns delayed negative pulses (rise time about 3.5 ns). Light from the device was focused on a fast-response avalanche photodiode (APD) (response time about 0.3 ns) using a microscope objective lens. The output from the APD amplified was observed by a sampling oscilloscope. The response time for this detection system was less than 0.3 ns.

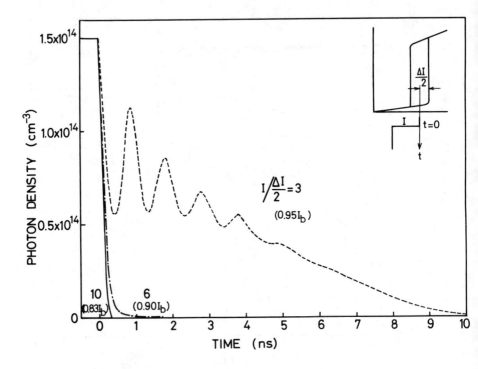

Figure 3.11(c) Calculated transient response to step current pulse of device with $\gamma_1 = 0.5$. Time evolution of photon with OFF pulse. Injection current changes from I_b to $0.95I_b$, $0.90I_b$, and $0.83I_b$. (From [18].)

Figure 3.13(c) shows the light output (upper trace) and trigger current (lower trace). The amplitude for both the ON and OFF current pulses was about 2.5 mA. The light-output trace shows typical characteristics of the bistable multivibrator, which can be induced to make an abrupt transition from one state to the other by means of external trigger excitation. When the ON current pulse is applied, the device emits coherent light (ON state). It stays ON until the OFF current pulse is applied. The smallest necessary amplitude for the trigger pulse depends mainly on the stability of the device temperature. If the temperature is sufficiently stabilized, the operation conditions can be chosen where ΔI is small.

3.5.2 Bistable Switching by Optical Pulses

The switching time and energy needed for bistable switching strongly depend on how close the bias current is to the laser threshold in an absorptive BLD. Therefore, these

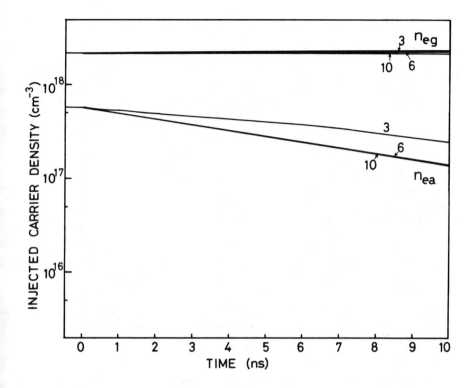

Figure 3.11(d) Calculated transient response to step current pulse of device with $\gamma_1 = 0.5$. Time evolution of carriers with OFF pulse. Injection current changes from I_b to $0.95I_b$, $0.90I_b$, and $0.83I_b$. (From [18].)

may depend on the stability of the system in which BLDs are used, such as the temperature and the driving currents, as well as on their intrinsic performance. Li and Wang have reported results in which the switch-on was achieved with a very small optical input (200 nW) [39].

The wavelength characteristic obtained experimentally for optical set pulse power in BLDs is shown in Figure 3.14 [40]. The optical set pulse power is lowest when the input signal wavelength is in between the lasing wavelengths for BLDs with optical input (ON in the figure) and the gain maximum wavelength for BLDs without optical input (OFF in the figure). This wide wavelength region available for optical switching (about 10 nm) is the great advantage of absorptive bistability over the dispersive bistability described in Chapter 4. Perkins et al. [24] also analyzed the effect of optical injection into the active and passive segments as a function of the wavelength of the injected light, and found that the device accepts a wide range of wavelengths of the input flux to initiate switching. If we measure the wavelength dependence more precisely, we can see that

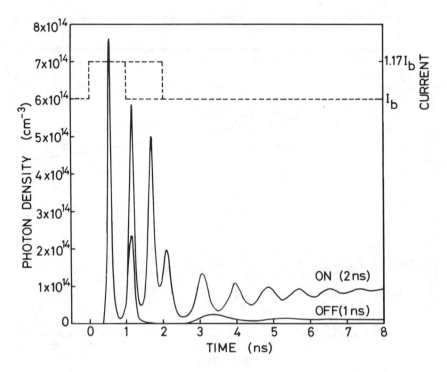

Figure 3.12(a) Calculated transient response to trigger current pulse for device with $\gamma_1 = 0.5$. Time evolution of photon with $5 \times \Delta I$ height ON pulse of 1 and 2 ns. (From [18].)

threshold set light intensity strongly depends on its wavelength and has minima around the resonant wavelength of the electroluminescent (EL) state [41]. This will be described later in detail.

Usually, the BLDs are set by optical pulses and reset by electrical pulses. This method has the advantage that the speed of the OFF switching can be made fast by a sweep-out of the accumulated carriers in the LD active region.

Öhlander et al. have measured minimum switching energy as a function of current bias level for different input pulse wavelengths. This lowest bistable switching energy and the fastest rise time they recorded were 23 fJ and less than 100 ps, respectively [42]. Blixt and Öhlander have obtained bistable switching with a switch-on time of 19 ps, a turn-on delay time of 20 ps, and an input coupled optical energy of 30 fJ [43]. Switch-off was obtained by an electrical reset pulse, and the switch-off time was 94 ps, which was probably limited by the duration of the electrical reset pulse. However, a repetition rate of 500 MHz was used, and the maximum repetition rate has not, as yet, been clarified in this experiment. The repetition frequency is limited mainly by the recovery time of the carrier density after a reset and set operation. This recovery time may be shortened

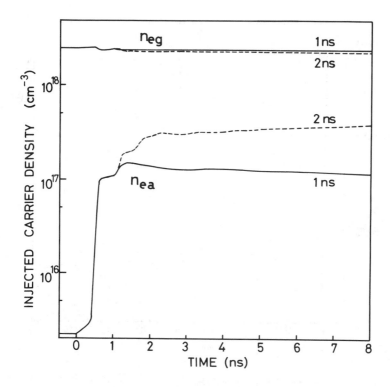

Figure 3.12(b) Calculated transient response to trigger current pulse for device with $\gamma_1 = 0.5$. Time evolution of carriers with $5 \times \Delta I$ height ON pulse of 1 and 2 ns. (From [18].)

by reducing the carrier density changes. They have also reported 0.7-fJ minimum input coupled switch-on energy and 60-ps switch-on time for a bistable three-section laser [44].

Flip-flop operation at 2.5-Gbps repetition rate has been achieved in a BLD with a semi-insulating layer, which realized low carrier density in the reset region (16 μm long) and low parasitic capacitance (3 pF over 300 μm) [10]. The repetition rate has been improved to 5 Gbps using a BLD with a short (10 μm) saturable absorption region reset by 0.2V electrical pulses [45].

In order to make a flexible device and to open up the possibility of using a BLD as an all-optical logic device, it seems highly likely that it will be necessary to control them solely with optical pulses. Therefore, many methods have been proposed, shown schematically in Figure 3.15. The detailed transient dynamics are not drawn in the figure. Three-level optical input can be used for such a purpose (Figure 3.15(a)) [46].

Many attempts have been made to achieve all-optical switching using only "positive" optical pulses. In this case, the device does not require holding optical power. Optical reset of a BLD using the beat between injected light and the lasing light of the BLD has been achieved (Figure 3.15(b)) [47]. This idea was first demonstrated by the

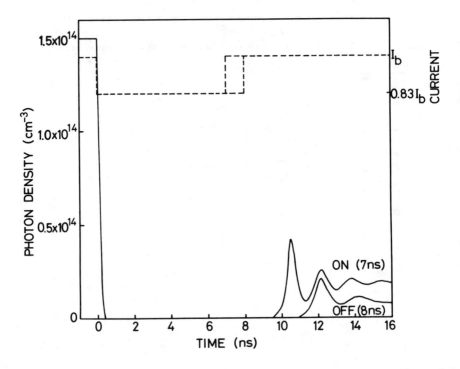

Figure 3.12(c) Calculated transient response to trigger current pulse for device with $\gamma_1 = 0.5$. Time evolution of photon with $5 \times \Delta I$ height OFF pulse of 7 and 8 ns. (From [18].)

same author in a dispersive BLD, which will be described in Chapter 4, using two light inputs of different frequencies [48]. The mechanism is considered as follows [48]. When a light beam with a slightly different frequency from the BLD lasing frequency is injected into the BLD, the total light intensity oscillates at the beat frequency between the BLD lasing light and the externally injected light. Responding to this oscillation, the medium loss in the saturable absorber oscillates, causing the lasing threshold value to oscillate also. With appropriate input power, during the oscillation a time exists when the lasing threshold value of the BLD exceeds the bias current injected into the BLD. At this time the BLD ceases to lase and does not lase again, even after the injected light is removed. Thus, the BLD can be optically switched from ON to OFF. Therefore, the input light frequency has to be set within a few gigahertz of the lasing frequency. The threshold analysis was formulated for a dispersive BLD at the steady-state condition [48]. The dynamic behavior is not yet clearly understood.

Gain quenching by injecting light with a wavelength that is longer than the bistable lasing wavelength has been used for the reset (Figure 3.15(c)) [49]. Optical set and reset operation with different wavelengths at a repetition rate of 100 MHz with transition times of less than 120 ps has been reported [50]. Both set and reset were also achieved with

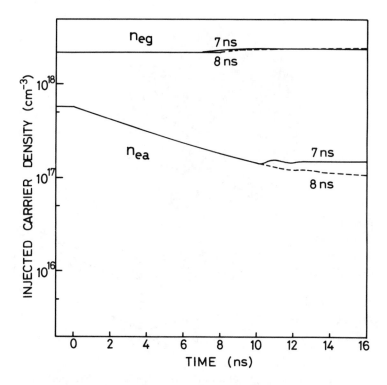

Figure 3.12(d) Calculated transient response to trigger current pulse for device with $\gamma_l = 0.5$. Time evolution of carriers with $5 \times \Delta I$ height OFF pulse of 7 and 8 ns. (From [18].)

the same wavelength (Figure 3.15(d)) [51]. Here, to optically set and reset the BLD using single-wavelength light signals, the wavelength dependence of the threshold set and reset light intensity is discussed in detail. The sensitivity for set operation is high around the lasing wavelength, and the sensitivity for reset is high around the wavelength slightly longer than the lasing wavelength. In addition, the BLD has a Fabry-Perot cavity. So when light is collinearly injected into the BLD, the wavelength dependence of the threshold set or reset light intensity will be modulated by the resonant characteristics of the Fabry-Perot cavity as shown in Figure 3.16(a) [52]. The solid line shows the threshold set light intensity and the dashed line shows threshold reset light intensity. They are different because the carrier densities of the gain and saturable absorption regions change between the EL and the lasing state, and, as a result, the refractive index of the regions changes. The threshold set light intensity has minima near the resonant wavelengths in the EL state, and the threshold reset light intensity has minima near those in the lasing state. In short, they have minima near the resonant wavelengths in the initial state before light injection.

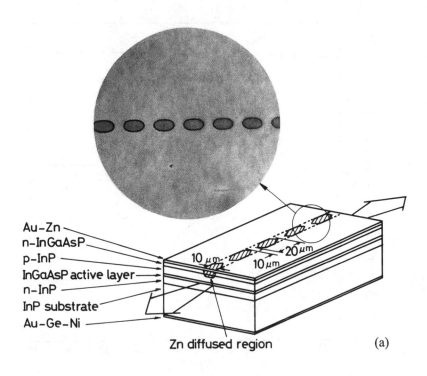

Figure 3.13(a) Structure of InGaAsP/InP DH LD with a periodic excitation-stripe geometry.

A light at point A is used for setting and a light at point B for resetting. Thus, by changing the wavelength of injection light, optical pumping and gain quenching were achieved. However, if a light at point C acts as the set light and a light at point D acts as the reset light, single-wavelength light will pump the saturable absorption region and quench the gain by changing the intensity of the light. As shown in Figure 3.16(b), the gain spectrum in the saturable absorption region will probably be changed by injection light with a wavelength at points C and D. The lower dashed line shows the gain spectrum in the EL state. In this state, the region has a loss coefficient even at a wavelength slightly longer than the lasing wavelength. So when light at the wavelength is injected into the BLD, the light pumps the saturable absorption region and sets the LD. The gain spectrum then changes as shown by the upper solid line. In this state, because the saturable absorption region is pumped by the lasing light, it has a gain coefficient at a wavelength longer than the lasing wavelength. However, when the injection light intensity is further increased, the light will quench the gain of the saturable absorption region as shown by the middle line. Thus the light stops the lasing. Though when the light is then decreased slowly, the BLD starts to lase again, when the light is decreased faster than the recovery time of the net gain, the BLD will enter the EL state. Therefore, it is possible to set and reset the BLD with a single-wavelength light having a wavelength slightly longer than the lasing

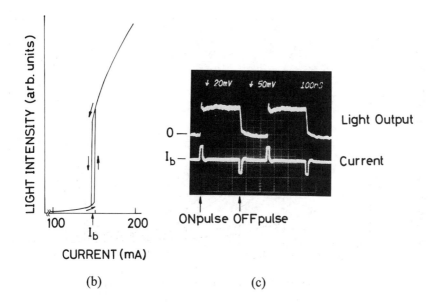

(b)   (c)

Figure 3.13(b) Typical variations of the laser output power observed for CW operation at 208K; (c) sampling oscilloscope trace of light output and trigger current for optical bistable multivibrator; amplitude for both ON current pulse and OFF current pulse was about 2.5 mA. (Reprinted with permission from Chapman & Hall Ltd. From [19].)

wavelength. However, greater optical energy (higher peak power or longer duration) is needed for the reset operation than for the set operation.

Okada et al. analyzed the optical set-reset operation of inhomogeneously excited LDs using the rate equations, taking into account the carrier-induced refractive index change and the wavelength-dependent optical gain [53]. As the detuning of the optical input (λ_{in}) from LD oscillation wavelength (λ_1) is increased, the power required for the set increases and that for the reset decreases with the condition that $\lambda_1 < \lambda_{in}$.

BLDs can also be switched by using their transient response. Undershoot switching (i.e., downward switching from the upper to the lower stable state by undershooting during the rapidly diminishing part of the optical input) is possible, even when the optical input is not decreased to the critical level necessary for quasistationary downward switching (Figure 3.15(e)) [54].

3.6 VOLTAGE-CONTROLLED ABSORPTIVE BISTABILITY IN MQW LASERS

One of the most significant properties of semiconductor QW structures is that optical absorption caused by an exciton is clearly observed at room temperature. The exciton absorption peak shifts to the lower energy side by the presence of an applied electric

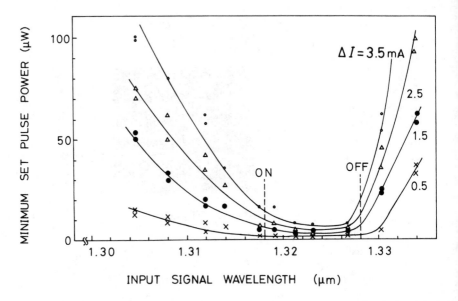

Figure 3.14 Wavelength characteristic for BLD set optical pulse power. The ON wavelength shows the lasing wavelength with input signal. The OFF wavelength shows the amplified spontaneous emission wavelength without input signal. (©1986 IEEE. From [40].)

field, and this phenomenon is QCSE (see 2.2.3). Bistability that originates from nonlinear optical absorption associated with the room temperature exciton is observed in MQW LDs with tandem electrode configuration (Figure 3.3(d)). Using QCSE, the hysteresis loop in the light-output-versus-injection-current curve can be controlled by applying a bias voltage to one segment of electrodes V_B.

A GaAs-AlGaAs MQW LD, which consisted of 16 periods of undoped GaAs (8 nm) $-Al_{0.25}Ga_{0.75}As$ (5 nm), was used to demonstrate bistability [11]. The device has a 20-μm-wide high-mesa stripe geometry. The waveguide was electrically separated into two sections (A and B). The segmented electrode MQW LD is operated by current injection into segment A (the current is denoted by I_A) while voltage is applied to segment B (the voltage is denoted by V_B).

At room temperature, a hysteresis loop was observed when V_B was lying between −0.1V and −0.4V. The width of the hysteresis loop was changed with V_B. In order to understand the characteristics and the relevant mechanism more consistently, this device was operated under dc conditions at 77K. This is because the CW operation at room temperature was difficult for this device, mainly due to a heat dissipation problem. Figure 3.17(a) shows the L-I_A curves for three different values of V_B. At $V_B = +1.3V$ (the built-in voltage at room temperature), no hysteresis appeared. At $V_B = +0.7$ and +0.2V, clear hysteresis loops were observed. The hysteresis loop was observed at a range of V_B from +1.2V to 0V.

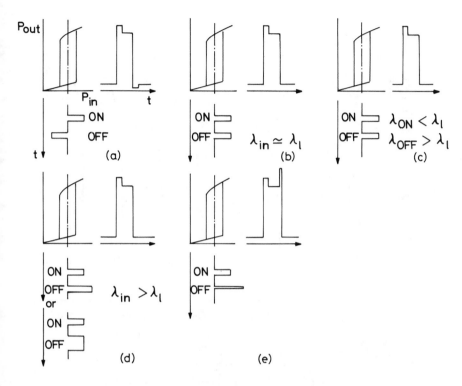

Figure 3.15 Methods of all-optical set-reset operation of absorptive BLDs (detailed transient responses are not shown): (a) three-level optical input; (b) optical reset by the beat between injected light and the lasing light; (c) gain quenching by injecting light with longer wavelength than lasing wavelength; (d) set-reset by the inputs with same wavelength; (e) undershoot switching. (From [4].)

In order to confirm the nonlinear absorption responsible for the bistability, nonlinear optical absorption in section B was investigated. For this purpose, the two sections were isolated not only electrically but also optically by making a deep, 2.6-μm-wide groove penetrating into the lower clad layer. Then the structure were given a laser section A and an external detector section B, schematically shown in the inset of Figure 3.17(b). Light emitted from the laser section was introduced into the detector section. The introduced light intensity P_i can be varied with the injection current I_A independently of the bias voltage V_B. Figure 3.17(b) shows the V_B dependence of photocurrent responsivity, which is equivalent to the V_B dependence of the optical absorption spectrum measured at a fixed wavelength (λ_{LD} of section A). The larger V_B side (left-hand side) on the horizontal axis corresponds to the lower energy side of the optical absorption spectrum, since the absorption spectrum shifts to the lower energy side when V_B is decreased towards the negative direction. λ_{LD} was constant within ± 1 nm for different values of P_i. Two peaks observed around $V_B = +0.2$ and -0.5V are due to the $n = 1$ heavy-hole and light-hole excitons,

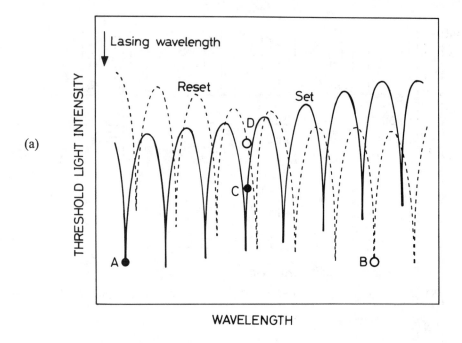

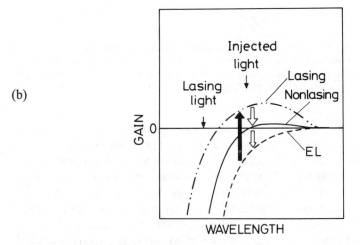

Figure 3.16 (a) Wavelength dependence of threshold set (solid line) and reset (dashed line) light in intensity. Points A and B show conditions of set and reset using light beams of two different wavelengths. Points C and D show those using light beams of identical frequency but two different intensities. (b) Mechanism of optical set and reset using single-wavelength light signals. As the injection light intensity increases, the gain spectrum changes from the lower line (the EL state) to the upper line (the lasing state) and to the middle line. When the injection light is decreased quickly, the gain spectrum goes to the lower line (EL state) again. (From [52].)

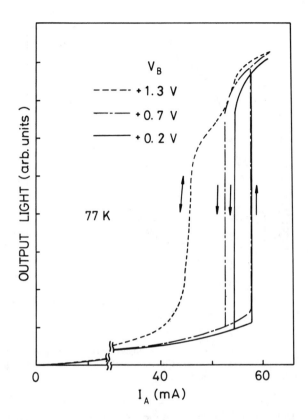

igure 3.17(a) *L-I_A* curves of GaAs/AlGaAs MQW LD with two segments of electrode at 77K under various values of V_B.

spectively. As clearly observed, both of the peaks decrease as P_i increases. The reduction larger in the larger V_B side (lower energy side) of the peaks. The V_B range where the exciton absorption decreases as P_i increases was from +0.8V to 0V. This range agrees asonably well with the range where the bistability was observed as shown in Figure .17(a). This is the third piece of evidence of the nonlinear exciton absorption responsible or the voltage-controlled optical bistability. The nonlinearity of the exciton absorption ay be much more enhanced in the BLD, because the light intensity is larger by a factor f 10 or more in the tandem electrode structure than in the structure with a deep groove, hown in the inset of Figure 3.17(b).

Voltage-controlled optical bistability was also demonstrated in $In_{0.53}Ga_{0.47}As$/InP IQW LDs [12]. Optical bistability was obtained in a wide range of control voltages from 0.7V to −0.6V. Switching operation is achieved by injecting a set light pulse and applying verse-bias reset voltage to the saturable absorption region. Turn-on time is 300 ps with put light of 6 mW, and turn-off time of 260 ps is obtained by reducing the stray apacitance in the saturable absorption region.

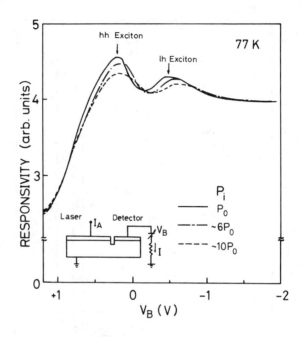

Figure 3.17(b) Photocurrent responsivity as a function of applied bias voltage V_B in a structure shown in the inset; P_i is optical power introduced into detector section B from laser section A. (From [11] Reproduced by permission of AIP.)

    The turn-off time strongly depends on the switching reverse voltage in MQW LDs. Figure 3.18 represents turn-off time versus control voltage V_c. The circles show the experimental data. The solid line shows the calculation results obtained by solving coupled rate equations. The ratios of the differential gain coefficient in the gain region g_g to the saturable absorption region g_a (g_g/g_a) of 0.7 and 0.5 were used in the calculation. As shown by the solid circles, turn-off time decreases with increasing reverse-bias voltage and is saturated about 500 ps with a control voltage V_c of less than −3.5V. On the other hand, the calculation yields a turn-off time of less than 100 ps. This fact indicates that the stray capacitance of the device limits the turn-off time because it is as high as 5 pF. In order to reduce the stray capacitance, the bonding pad of the saturable absorption region was formed from polyimide. The resultant turn-off time is shown by open circles in Figure 3.18. Because of the small dielectric constant of polyimide, the turn-off time here is reduced to about half of the previous one. It is 260 ps at the control voltage of −1.6V.

    The optical triggering speed of a BLD with a saturable absorber is usually limited to several hundred megabits per second by long fall time owing to long carrier lifetime in the absorber section, as described before. The carrier density in this section should be smaller than that needed for transparency. A solution was proposed, consisting of bombard-

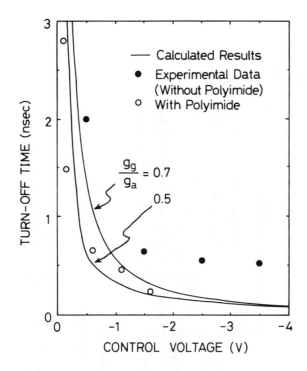

Figure 3.18 Turn-off time of InGaAs/InP MQW BLD as a function of control voltage V_c. (From [12]. Reproduced by permission of JJAP.)

ng the saturable absorber region with a proton beam [55,56]. This bombardment creates defects and increases the carrier nonradiative-recombination rate in this region. The BLD used is a two-section MQW Fabry-Perot LD [55]. The MQW BLD has two 200-μm-long active sections, biased independently and separated by a 20-μm-long saturable absorber section. The latter has been bombarded by a proton beam, which gives a very short carrier lifetime and provides an electrical isolation as high as 162 kΩ between the two sections. When the MQW BLD is biased under the downward threshold of the bistable regime, it can be switched on by an injected optical signal. Under the optimal operation condition, the rise and fall times are about 200 ps. A 2.5-Gbps operation with a penalty of 0.5 dB at a bit error rate (BER) of 10^{-9} was demonstrated by using this optically triggered BLD.

As described in the previous section, Fabry-Perot BLDs have strong sensitivity depending on input light wavelength, which corresponds to each cavity mode. To avoid being affected by the cavity mode, a BLD with a subwaveguide through which input light is injected into the saturable absorption region of such BLD is shown in Figure 3.19(a) [57]. It consists of the main waveguide MQW laser and a subwaveguide optical amplifier with 3- and 6-μm-wide ridge structures, respectively, which are arranged in an orthogonally

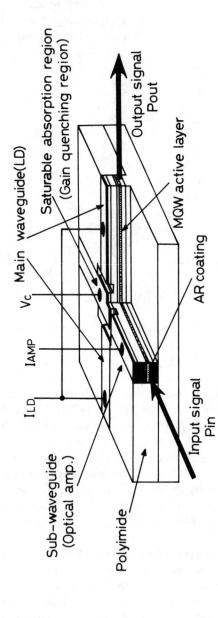

Figure 3.19(a) Schematic view of side-light-injection-type MQW BLD with gain quenching region and saturable absorption region.

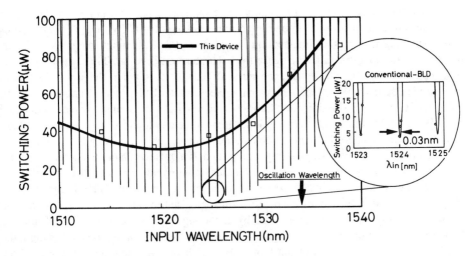

Figure 3.19(b) Switching optical power dependence on input wavelength. Heavy line is for this device and thin line is for conventional BLD. (©1993 IEEE. From [57].)

crossed configuration. The subwaveguide conducts and amplifies light to the absorption region in the main laser cavity. Electrodes are formed in three segments on the laser waveguide, optical amplifier waveguide, and saturable absorption region. The length of the saturable absorption region was designed to be 6 μm to provide sufficient nonlinear response of the InGaAs/InP MQW layer with good input light coupling between the subwaveguide and saturable absorption region. AR coating was formed on the incident edge surface of the subwaveguide. The AR coating prevents the subwaveguide from having a cavity structure.

The subwaveguide is excited by a 20-mA current injection to amplify the input light. Differential-gain and bistable characteristics are observed for the applied voltage at saturable absorber conditions of 0.67V and 0.63V, respectively. The switching power dependence on input light wavelength for the differential-gain operation is shown in Figure 3.19(b). Minimum switching power of the device was 30 μW at a wavelength of around 1.52 μm when input light was coupled from a single-mode fiber to the subwavegu-ide. The switching power dependence on input light wavelength is very small, and a wavelength region yielding a sensitivity of less than 3 dB was 28 nm wide. By contrast, conventional Fabry-Perot BLDs have very discrete sensitivity, with bandwidth of about 0.03 nm corresponding to each cavity mode, though the peak sensitivity is several micro-watts. No leakage of input light into output light was observed in the output spectrum; the input/output isolation ratio was over 30 dB.

Set-on and set-off operations by input light of the same wavelength have been observed in a side-light-injection MQW BLD [58]. The main BLD was located perpendicu-lar to two waveguides for amplifying input light. Two intersections, the gain quenching region and the saturable absorption region, were spatially separated. Input light of 40 μW results in saturable absorption in one intersection biased at +0.65V, and 570 μW causes gain quenching in the other intersection biased at +0.93V. As input intensity increases, the turn-on and turn-off times decrease [59]. The turn-on time is 200 ps when the input light peak intensity is 1 mW and the turn-off time is 2 ns when the input light peak intensity is 200 mW. Wavelength conversion (see Chapter 8) has been demonstrated [60] using a 1.56-μm DFB side-injection light-controlled BLD with an input wavelength range of 1.496 to 1.597 μm. Insensitivity to input light wavelength and a large separation ratio of output from input (more than 30 dB) are advantages for a wavelength converter.

Time-division optical signal processing such as a pulsewidth conversion and a pulse delay control can be obtained by the photonic memory switch, which monolithically integrates a voltage-controlled BLD and an optical gate on an InP substrate (Figure 3.20(a)) [61]. The BLD section is composed of a gain region and a saturable absorption region. As schematically shown in Figure 3.20(b), if the signal light pulse is injected into the BLD section when the applied voltage to the saturable absorption region (Vc) is high, the BLD section changes its state from luminescent to lasing and continues lasing until the next reset electrical pulse is applied to the saturable absorption region. Then the memory

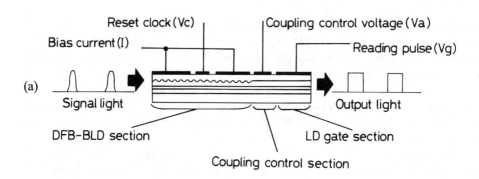

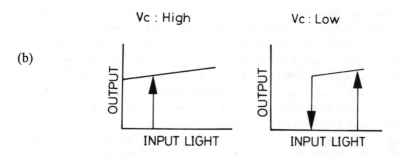

Figure 3.20 Photonic memory switch: (a) schematic structure; (b) output-versus-input-light curves.

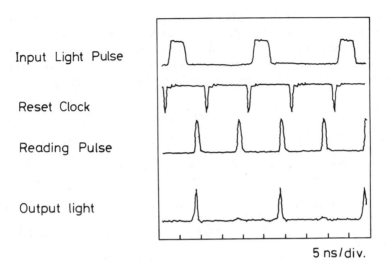

Input Light Pulse

Reset Clock

Reading Pulse

Output light

5 ns/div.

Figure 3.20(c) Optical-pulsewidth conversion using the photonic memory switch: 4-ns input light pulses were converted to 1-ns light pulses. (©1992 IEEE. From [61].)

state can be read out by opening the LD gate at an arbitrary delay and width in a clock cycle. If Vc is low, the photonic memory switch does not oscillate by light injection and no light pulses are emitted through the LD gate section. The coupling control voltage Va is applied to the coupling control section to optically separate the BLD section from the LD gate section. The length of the BLD section, the saturable absorption region positioned beside the $\lambda/4$ shift of the grating, the coupling control section, and the LD gate are 400, 10, 40, and 145 μm, respectively. The grown layers of the photonic memory switch consist of an n-InP cladding layer, an InGaAsP guide layer, an InGaAs/InAlAs MQW active layer, an InGaAsP guide layer, a p-InP cladding layer, and a p^+ – InGaAsP contact layer. The grating was formed only in the BLD section between the InGaAsP guide layer and the p-InP cladding layer by chemical etching. The segmented electrodes were formed by removing the InGaAsP contact layer. The separation resistances among those electrodes were about 1 kΩ. An AR coating layer (<0.1% at 1.5 μm) was deposited on both facets of the photonic memory switch.

Optical-pulsewidth conversion using the photonic memory switch is obtained as shown in Figure 3.20(c). The upper curve represents the input light pulses (the peak power was 120 μW). The second curve signifies the 100-MHz reset clock voltage to the DFB BLD section (the pulsewidth was 500 ps, the high level was +0.20V and the low level was −0.50V). The third curve represents the reading pulse applied to the LD gate section (the high level was 2.15V and the low level was −0.15V). The fourth curve indicates the output light from the photonic memory switch. When no input light signals are injected into the memory switch, the BLD section does not oscillate and no output light is emitted through the LD gate; when input signals are injected into the switch, the BLD section becomes lasing and output light is emitted from the memory switch when the LD gate is opened by applying reading signals. Input light pulses of 500 ps and 4 ns were successfully converted to 4- and 1-ns light pulses at an arbitrary delay in a clock cycle.

REFERENCES

[1] Guckenheimer, J., and P. Holmes, *Nonlinear Oscillations, Dynamical Systems, and Bifurcations of Vector Fields*, New York, Berlin, Heidelberg, Tokyo: Springer-Verlag, 1983.

[2] Gibbs, H. M., *Optical Bistability: Controlling Light With Light*, New York: Academic, 1985.

[3] Mandel, P., S. D. Smith, and B. S. Wherrett, *From Optical Bistability Towards Optical Computing*, Amsterdam: North-Holland, 1987.

[4] Kawaguchi, H., "Bistable Laser Diodes and Their Applications for Photonic Switching," *Int. J. Optoelectronics*, Vol. 7, No. 3, 1992, pp. 301–348.

[5] Lasher, G. J., "Analysis of a Proposed Bistable Injection Laser," *Solid-State Electronics*, Vol. 7, 1964, pp. 707–716.

[6] Harder, Ch., K. Y. Lau, and A. Yariv, "Bistability and Pulsations in CW Semiconductor Lasers With a Controlled Amount of Saturable Absorption," *Appl. Phys. Lett.*, Vol. 39, No. 5, September 1981, pp. 382–384.

[7] Kawaguchi, H., K. Magari, H. Yasaka, M. Fukuda, and K. Oe, "Tunable Optical-Wavelength Conversion Using an Optically Triggerable Multielectrode Distributed Feedback Laser Diode," *IEEE J. Quantum Electronics*, Vol. QE-24, No. 11, November 1988, pp. 2153–2159.

[8] Shoji, H., Y. Arakawa, and Y. Fujii, "Theoretical Analysis of Bistable Distributed Feedback Lasers With Detuning Effect," *J. Lightwave Technology*, Vol. LT-8, No. 10, October 1990, pp. 1630–1637.

[9] Kawaguchi, H., and G. Iwane, "Bistable Operation in Semiconductor Lasers With Inhomogeneous Excitation," *Electron. Lett.*, Vol. 17, No. 4, February 1981, pp. 167–168.

[10] Odagawa, T., T. Machida, T. Sanada, K. Nakai, K. Wakao, and S. Yamakoshi, "High Repetition Rate Operation of Bistable Laser Diodes," *IEE Proc. J.*, Vol. 138, No. 2, April 1991, pp. 75–78.

[11] Tarucha, S., and H. Okamoto, "Voltage-Controlled Optical Bistability Associated With Two-Dimensional Exciton in GaAs-AlGaAs Multiple Quantum Well Lasers," *Appl. Phys. Lett.*, Vol. 49, No. 10, September 1986, pp. 543–545.

[12] Uenohara, H., H. Iwamura, and M. Naganuma, "Switching Characteristics of InGaAs/InP Multiquantum Well Voltage-Controlled Bistable Laser Diodes," *Japanese J. Appl. Phys.*, Vol. 29, No. 12, December 1990, pp. L2442–L2444.

[13] Kucharska, A. I., P. Blood, E. D. Fletcher, and P. J. Hulyer, "Bistability in Inhomogeneously Pumped Quantum Well Laser Diodes," *IEEE Proc. Pt. J*, Vol. 135, No. 1, February 1988, pp. 31–33.

[14] Middlemast, I., J. Sarma, K. A. Shore, A. I. Kucharska, E. D. Fletcher, and P. Blood, "Absorptive Bistability in Inhomogeneously Pumped Quantum Well Laser Diodes," *IEE Proc. J.*, Vol. 138, No. 5, October 1991, pp. 301–308.

[15] Yamada, N., and J. S. Harris, Jr., "Strained InGaAs Single Quantum Well Lasers With Saturable Absorbers Fabricated by Quantum Well Mixing," *Appl. Phys. Lett.*, Vol. 60, No. 20, May 1992, pp. 2463–2465.

[16] Frateschi, N. C., H. Zhao, J. Elliot, S. Siala, M. Govindarajan, R. N. Nottenburg, and P. D. Dapkus, "Three-Terminal Bistable Low-Threshold Strained InGaAs/GaAs Laser Grown on Structured Substrates for Digital Modulation," *IEEE Photonics Tech. Lett.*, Vol. 5, No. 3, March 1993, pp. 275–278.

[17] Ishikawa, J., T. Ito, N. S. Takahashi, and S. Kurita, "Bistable Operation of 0.8 μm GaInAsP/GaAs Lasers," *Electron. Lett.*, Vol. 24, No. 16, August 1988, pp. 1014–1016.

[18] Kawaguchi, H., "Optical Bistable-Switching Operation in Semiconductor Lasers With Inhomogeneous Excitation," *IEE Proc. Pt. I*, Vol. 129, No. 4, August 1982, pp. 141–148.

[19] Kawaguchi, H., "Absorptive and Dispersive Bistability in Semiconductor Injection Lasers," *Optical and Quantum Electronics*, Vol. 19, 1987, pp. S1–S36.

[20] Harder, Ch., and A. Yariv, "Bistability in Semiconductor Laser Diodes," in *Optical Nonlinearities and Instabilities in Semiconductors*, H. Haug, ed., Academic Press.

[21] Öhlander, U., and O. Sahlén, "Influence of Carrier Leakage on Bistability in an Inhomogeneously Pumped Semiconductor Laser," *IEEE J. Quantum Electronics*, Vol. QE-23, No. 5, May 1987, pp. 487–498.

[22] Liu, H. F., T. Kamiya, and B. X. Du, "Temperature Dependence of Bistable InGaAsP/InP Lasers," *IEEE J. Quantum Electronics*, Vol. QE-22, No. 9, September 1986, pp. 1579–1586.

[23] Thedrez, B. J., and C. H. Lee, "Effect of the Spatial Gain and Intensity Variations on a Two-Section Fabry-Perot Semiconductor Laser: An Analytical Study," *IEEE J. Quantum Electronics*, Vol. QE-29, No. 3, March 1993, pp. 864–876.

[24] Perkins, M. C., R. F. Ormondroyd, and T. E. Rozzi, "Analysis of Absorptive Bistable Characteristics of Multisegment Lasers," *IEE Proc. Pt. J*, Vol. 133, No. 4, August 1986, pp. 283–292.

[25] Paradisi, A., and I. Montrosset, "Numerical Modeling of Bistable Laser Diodes With Saturable Absorbers," *IEEE J. Quantum Electronics*, Vol. 27, No. 3, March 1991, pp. 817–823.

[26] Öhlander, U., and O. Sahlén, "Bistable Operation of InGaAsP Lasers Using Different Absorber Positions," *Appl. Phys. Lett.*, Vol. 54, No. 13, March 1989, pp. 1198–1200.

[27] Kawaguchi, H., "Optical Bistability and Chaos in a Semiconductor Laser With a Saturable Absorber," *Appl. Phys. Lett.*, Vol. 45, No. 12 , December 1984, pp. 1264–1266.

[28] Kawaguchi, H., "Optical Nonlinearities in Semiconductor Lasers and Their Applications for Functional Devices," *Int. J. Nonlinear Optical Physics*, Vol. 1, No. 1, 1992, pp. 203–221.

[29] For example, Acket, G. A., W. Nijiman, and H.'t Lam, "Electron Lifetime and Diffusion Constant in Germanium-Doped Gallium Arsenide," *J. Appl. Phys.*, Vol. 45, No. 7, July 1974, pp. 3033–3040.

[30] For example, Lamprecht, K. F., S. Juen, L. Palmetshofer, and R. A. Höpfel, "Ultrashort Carrier Lifetimes in H+ Bombarded InP," *Appl. Phys. Lett.*, Vol. 59, No. 8, August 1991, pp. 926–928.

[31] Penty, R. V., H. K. Tsang, I. H. White, R. S. Grant, W. Sibbett, and J. E. A. Whiteaway, "Repression and Speed Improvement of Photogenerated Carrier Induced Refractive Nonlinearity in InGaAs/InGaAsP Quantum Well Waveguide," *Electron. Lett.*, Vol. 27, No. 16, August 1991, pp. 1447–1449.

[32] Ueno, M., and R. Lang, "Conditions for Self-Sustained Pulsation and Bistability in Semiconductor Lasers," *J. Appl. Phys.*, Vol. 58, No. 4, August 1985, pp. 1689–1692.

[33] Baoxun, D., "Stability Theory of Double Section Lasers," *IEEE J. Quantum Electronics*, Vol. 25, No. 5, May 1989, pp. 847–849.

[34] Harder, Ch., K. Y. Lau, and A. Yariv, "Bistability and Pulsations in Semiconductor Lasers With Inhomogeneous Current Injection," *IEEE J. Quantum Electronics*, Vol. 18, No. 9, September 1982, pp. 1351–1361.

[35] Farrell, G., P. Phelan, and J. Hegarty, "Self-Pulsation Operating Region for Absorber of Two Section Laser Diode," *Electron. Lett.*, Vol. 27, No. 16, August 1991, pp. 1403–1404.

[36] Kawaguchi, H., "Bistable Operation of Semiconductor Lasers by Optical Injection," *Electron. Lett.*, Vol. 17, No. 20, October 1981, pp. 741–742.

[37] Lang, R., and K. Kobayashi, "Suppression of the Relaxation Oscillation in the Modulated Output of Semiconductor Lasers," *IEEE J. Quantum Electronics*, Vol. QE-12, No. 3, March 1976, pp. 194–199.

[38] Adams, M. J., "Theory of Two-Section Laser Amplifiers," *Optical and Quantum Electronics*, Vol. 21, 1989, pp. S15–S31.

[39] Li, J., and Q. Wang, "A Common-Cavity Two-Section InGaAsP/InP Bistable Laser With a Low Optical Switching Power," *Optics Communications*, Vol. 83, No. 1,2, May 1991, pp. 71–75.

[40] Suzuki, S., T. Terakado, K. Komatsu, K. Nagashima, A. Suzuki, and M. Kondo, "An Experiment on High-Speed Optical Time-Division Switching," *J. Lightwave Technology*, Vol. LT-4, No. 7, July 1986, pp. 894–899.

[41] Kondo, K., M. Kuno, S. Yamakoshi, and K. Wakao, "A Tunable Wavelength-Conversion Laser," *IEEE J. Quantum Electronics*, Vol. QE-28, No. 5, May 1992, pp. 1343–1348.

[42] Öhlander, U., P. Blixt, and O. Sahlén, "Subnanosecond Switching of Bistable Tandem Lasers by Subpico-joule Optical Triggering," *Appl. Phys. Lett.*, Vol. 53, No. 14, October 1988, pp. 1227–1229.

[43] Blixt, P., and U. Öhlander, "19 ps Switching of a Bistable Laser Diode With 30 fJ Optical Pulses," *IEEE Photonics Tech. Lett.*, Vol. 2, No. 3, March 1990, pp. 175–177.

[44] Blixt, P., and U. Öhlander, "Femtojoule Bistable Optical Switching of Inhomogeneously Pumped Laser Diode at 500 MHz Using Mode-Locked Tunable Diode Laser," *Electron. Lett.*, Vol. 25, No. 11, May 1989, pp. 699–700.

[45] Machida, T., T. Odagawa, K. Tanaka, H. Nobuhara, K. Wakao, and N. Okazaki, "High Speed Operation of Bistable Laser Diode by Voltage-Reset Method," *IECE Japan*, 1990, C-132 (in Japanese).

[46] Kawaguchi, H., "Optical Input and Output Characteristics for Bistable Semiconductor Lasers," *Appl. Phys. Lett.*, Vol. 41, No. 8, October 1982, pp. 702–704.

[47] Inoue, K., and K. Oe, "Optically Triggered Off-Switching in a Bistable Laser Diode Using a Two-Electrode DFB-LD," *Electron. Lett.*, Vol. 24, No. 9, April 1988, pp. 512–513.

[48] Inoue, K., "All-Optical Flip-Flop Operation in an Optical Bistable Device Using Two Lights of Different Frequencies," *Opt. Lett.*, Vol. 12, No. 11, November 1987, pp. 918–920.

[49] Odagawa, T., T. Sanada, and S. Yamakoshi, "All Optical Flip-Flop Operation of Bistable Laser Diode," *Extended Abstracts of the 20th Int. Conf. on Solid State Devices and Materials*, August 1988, Tokyo, Japan, pp. 331–334.

[50] Blixt, P., and U. Ohlander, "All-Optical Set-Reset Operation of a Bistable Laser Diode at a Repetition Rate of 100 MHz," *Tech. Dig. Photonic Switching*, Kobe, Tokyo: IEICE, 14C-4, pp. 249–251.

[51] Odagawa, T., and S. Yamakoshi, "Optical Set-Reset Operations of Bistable Laser Diode With Single-Wavelength Light," *Electron. Lett.*, Vol. 25, No. 21, October 1989, pp. 1428–1429.

[52] Odagawa, T., T. Sanada, and S. Yamakoshi, "Bistable Laser Diode for Optical Signal Processing," *Fujitsu Sci. Tech. J.*, Vol. 26, No. 2, June 1990, pp. 138–148.

[53] Okada, M., H. Kikuchi, K. Takizawa, and H. Fujikake, "The Effects of a Detuned Optical Input on Bistable Laser Diodes With Inhomogeneous Current Injection," *IEEE J. Quantum Electronics*, Vol. QE-29, No. 1, January 1993, pp. 109–120.

[54] Okada, M., K. Takizawa, H. Kikuchi, and H. Fujikake, "Undershooting and Set-Reset Operation in Bistable Laser Diodes With Inhomogeneous Excitation," *IEEE J. Quantum Electronics*, Vol. QE-26, No. 5, May 1990, pp. 850–857.

[55] Landais, P., G.-H. Duan, C. Chabran, and J. Jacquet, "2.5 Gbit/s Low-Penalty Operation of an Optically Triggered Bistable Laser Incorporating a Proton Bombarded Absorber," *Electron. Lett.*, Vol. 29, No. 15, July 1993, pp. 1363–1364.

[56] Landais, P., G.-H. Duan, E. Gaumont-Goarin, P. Garabédian, and J. Jacquet, "Transition Time and Turn-On Jitter of Optically Triggered Bistable Lasers Incorporating a Proton Bombarded Absorber," *Appl. Phys. Lett.*, Vol. 63, No. 19, November 1993, pp. 2615–2617.

[57] Nonaka, K., H. Tsuda, H. Uenohara, H. Iwamura, and T. Kurokawa, "Optical Nonlinear Characteristics of a Side-Injection Light-Controlled Laser Diode With a Multiple-Quantum-Well Saturable Absorption Region," *IEEE Photonics Tech. Lett.*, Vol. 5, No. 2, February 1993, pp. 139–141.

[58] Uenohara, H., Y. Kawamura, H. Iwamura, K. Nonaka, H. Tsuda, and T. Kurokawa, "Side-Light-Injection MQW Bistable Laser Using Saturable Absorption and Gain Quenching," *Electron. Lett.*, Vol. 28, No. 21, October 1992, pp. 1973–1975.

[59] Uenohara, H., Y. Kawamura, H. Iwamura, K. Nonaka, H. Tsuda, and T. Kurokawa, "Set and Reset Operation Dependence on Input Light Intensity of a Side-Light-Injection MQW Bistable Laser," *Electron. Lett.*, Vol. 29, No. 18, September 1993, pp. 1609–1611.

[60] Tsuda, H., K. Nonaka, K. Hirabayashi, H. Uenohara, H. Iwamura, and T. Kurokawa, "Wide Range Wavelength Conversion Experiments Using a Side-Injection Light-Controlled Bistable Laser Diode," *Appl. Phys. Lett.*, Vol. 63, No. 23, December 1993, pp. 3116–3118.

[61] Tsuda, H., T. Kurokawa, H. Uenohara, and H. Iwamura, "Photonic Memory Switch; Monolithic Integration of a Voltage-Controlled Bistable Laser Diode and an Optical Gate," *IEEE Photonics Tech. Lett.*, Vol. 4, No. 7, July 1992, pp. 760–762.

Chapter 4
Dispersive Bistable Laser Diodes

4.1 BISTABILITY IN LASER DIODE AMPLIFIERS

4.1.1 General Consideration of Bistability in Nonlinear Cavity

Figure 4.1(a) shows the schematic of a typical nonlinear, plane-parallel Fabry-Perot interferometer. The cavity medium is assumed to have a nonlinear refractive index, which is proportional to the cavity, and hence the transmitted, intensity. This nonlinear index may be intrinsic to the material or may be artificially induced by feeding back the output intensity as a voltage across the cavity medium.

The transmission of a Fabry-Perot resonator τ can be written

$$\tau = \frac{P_t}{P_i} = \frac{(1 - R)^2}{(1 - R)^2 + 4R \sin^2(\delta)} \tag{4.1}$$

where P_i is the incident single-frequency light power, P_t is the transmitted light power, R is the reflectivity of the resonator mirrors (we have assumed both mirrors have equal reflectivities), and δ is the single-pass phase shift (see (2.20)). If we insert the material having a nonlinear refractive index, the refractive index can be written $n = n_0 + n_2|E|^2$, where E is the electric field in the cavity. For the case of the device shown in Figure 4.1(a), the phase shift can be written

$$\delta = \frac{2\pi n_0 l}{\lambda} + \frac{2\pi n_2 l}{\lambda} |E|^2$$

$$= \delta_0 + \gamma P_t \tag{4.2}$$

where $\gamma = 2\pi n_2 l/(\lambda T)$, $n_0 l$ is the optical length of the resonator with no light in the cavity, λ is the wavelength of the incident light, and $T = 1 - R$. The transmission of this device

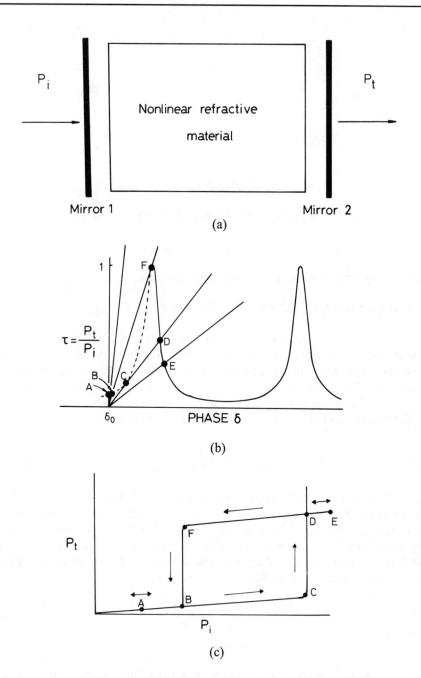

Figure 4.1 (a) Nonlinear refractive Fabry-Perot interferometer; (b) graphical solution of (4.1) and (4.3); (c) corresponding output-versus-input characteristics. (©1978 IEEE. From [1].)

for a given input power can be found from the simultaneous solution of (4.1) and (4.2). These solutions can be displayed graphically in a simple fashion, as shown in Figure 4.1(b) [1]. The heavy solid line is a plot of (4.1) and is the well-known Fabry-Perot response curve. Equation (4.2) can be written

$$\tau = \frac{P_t}{P_i} = \frac{\delta - \delta_0}{\gamma P_i} \tag{4.3}$$

This shows a linear dependence of transmission on the phase shift for values larger than δ_0. The slope of this line is $1/(\gamma P_i)$ and thus depends inversely on the incident light power. Several plots of (4.3) corresponding to different incident light powers are shown in Figure 4.1(b). It can be seen that for large incident power, there are several points of intersection of the two curves and thus several potential operating points for the device.

The "power-out-versus-power-in" characteristic of a device tuned as shown in Figure 4.1(b) can be deduced from the intersection of the two plots. At a low incident power, the linear plot has a large slope and the point of intersection is near the origin (A). As the incident power is increased, the point of intersection moves to point B and then C. As the power is further increased, the operating point jumps discontinuously to D and then travels down the far side of the Fabry-Perot characteristic (point E). If the incident power is now reduced, the operating point moves up the Fabry-Perot characteristic from E to D and then on up to F before dropping discontinuously down to B and then back to A as the incident power is reduced to zero. This behavior is illustrated in Figure 4.1(c), which shows the predicted power-out-versus-power-in characteristic. In general, the switching time will be governed by the response time of the optical nonlinearity and the resonator response time (= $nl/c(1 - R)$, where c is the velocity of light).

Garmire discussed the effect of saturation of the nonlinear refractive index on the performance of a nonlinear lossy etalon [2]. She calculated the requirement on the ratio of the saturated value of the optically induced index change δn_s to the linear loss per unit length α, and showed that operation of a bistable device requires $\delta n_s/\alpha > \lambda/\pi$, where λ is the free-space wavelength.

4.1.2 Dispersive Bistability in Laser Diode Amplifiers

A resonant-type LD amplifier, consisting of a normal LD biased below the laser oscillation threshold, can act as a nonlinear cavity and shows bistability in the optical input-output characteristics. This is because the active layer refractive index changes due to gain saturation by light injection (see Section 2.3.2 for the relation between refractive index change and gain saturation). This type of bistability was first discussed and demonstrated by Otsuka et al. with a Fabry-Perot cavity-type LD amplifier [3,4]. The structure is shown in Figure 4.2(a) [5]. Values of around 30 μW for switching in GaAs laser amplifiers [6] and as low as 1 μW for InGaAsP devices at 1.3 μm [7] and 1.55 μm [8] have been

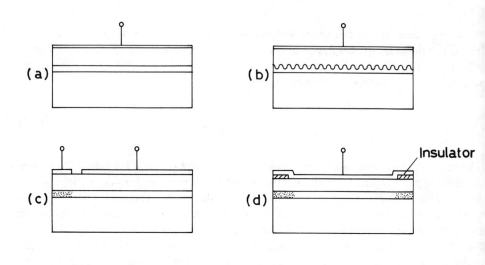

◼◼◼◼ Saturable absorption region

Figure 4.2 Dispersive BLDs: (a) Fabry-Perot amplifier; (b) DFB amplifier; (c) two-section Fabry-Perot amplifier with saturable absorber; (d) three-region Fabry-Perot amplifier with saturable absorber. (From [5].)

reported. Sharfin and Dagenais have also demonstrated the very low switching power operation of an optical switch requiring less than 1 fJ (<7,000 photons) at 1.3 μm [9]. Recovery times are predicted to be on the order of carrier lifetimes (i.e., a few nanoseconds at room temperature) [10]. It is predicted that incorporating MQWs into the active region of the amplifier might reduce the optical and electrical power requirements and enhance the high-frequency performance [11]. Bistability has been observed in InGaAs strained QW amplifiers [12] and InGaAs/InGaAsP MQW amplifiers [13].

Bistability is also observed in a DFB LD amplifier (Figure 4.2(b)). This was first demonstrated by Kawaguchi et al. in a 1.5-μm DFB LD [14]. Adams and Wyatt have predicted theoretically that, as a consequence of the spectral asymmetry of the bistable DFB LD amplifier, the hysteresis loops on either side of the stop band are expected to exhibit somewhat different shapes [15]. This difference in the nonlinearity in the two passbands was found to be enhanced by the mechanism of asymmetric facet reflection [16].

The two-section laser amplifier, which includes a saturable absorber in its cavity, also shows bistability and allows the freedom to optimize the bistable characteristics (Figure 4.2(c)) [17,18]. There are two separated modes of operation in a two-section laser: one for which the output wavelength is the same as that of the input, and the other for which the output wavelength is different from that of the input. In the former mode, the device may be properly described as an amplifier, whereas in the second the device is acting as a laser with some degree of optical pumping, as described in Chapter 3. In the amplifier mode, Marshall et al. have observed a nonlinear transfer function with a maximum gain of 26 dB, a minimum input power of 1 μW, and a maximum pulse repetition frequency of 700 MHz [17]. Barnsley et al. reported nonlinear amplification with more than a 15-nm wavelength range in a three-region InGaAsP laser amplifier shown in Figure 4.2(d) [19].

The analysis is carried out by using the relationships derived by Adams [10] to describe laser amplifier behavior. For the region of interest for bistable operation, the input light intensity I_{in} is much larger than spontaneous intensity. Therefore, it is permissible to neglect spontaneous emission in the equations. The appropriate mean optical intensity I_{av} obtained by averaging the axial intensity distributions within the Fabry-Perot cavity is used.

The transient evolution of the electron concentration n within the cavity is described by the rate equation [10,20–22]

$$\frac{dn}{dt} = \frac{j}{ed} - \frac{n}{\tau} - \frac{\Gamma g_m}{E} I_{av} \tag{4.4}$$

where j is the current density, e is the electronic change, d is the active layer thickness, τ is the electron lifetime, Γ is the confinement factor, E is the photon energy, and g_m is the material gain.

In order to find the mean optical intensity I_{av}, a straightforward longitudinal averaging procedure is carried out on the forward- and backward-traveling waves. This is illustrated in Figure 4.3 where the solid lines correspond to the traveling-wave intensities and the total intensity as functions of position along the cavity, and the broken line denotes I_{av}. For this case of uncoated facets with reflectivities 30% and 40%, the total intensity varies only slowly along the cavity length. A self-consistent position-dependent calculation shows that the corresponding variation of electron concentration n along the length for this case is even smaller, in fact less than 0.2%. Hence, the approximation of a position-independent rate equation (4.4) is justified.

The material gain is assumed to be linearly dependent on the carrier concentration and follows the expression

$$g_m = a(n - n_0) \tag{4.5}$$

where a is the gain coefficient and n_0 is the carrier concentration at transparency. The phase change ϕ in a single pass through the cavity (length L) is given by

$$\phi = \phi_0 + \frac{2\pi L}{\lambda}(n - n_1)\frac{dN}{dn} \tag{4.6}$$

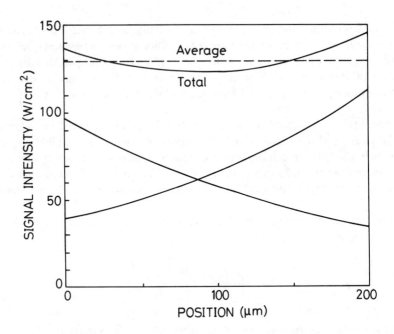

Figure 4.3 Signal intensity at resonance as a function of position along the cavity for forward- and backward-traveling waves. $R_1 = 0.4$, $R_2 = 0.3$, $L = 200$ μm, $\Gamma = 0.5$, $\alpha = 5 \times 10^{-16}$ cm^2, $n_0 = 10^{18}$ cm^{-3}, $a =$ 50 cm^{-1}, $j/j_{th} = 0.98$. The input intensity is taken as 1 W/cm^2 and the output intensity is 78.5 W/cm^2. (From [20]. Reproduced by permission of IEE.)

where ϕ_0 is the phase detuning, λ is the wavelength, N is the refractive index, and n_1 is the carrier concentration in the absence of an input signal, given, from (4.4), by $n_1 = j\tau/ed$. Straightforward elimination of n in favor of ϕ from (4.6) produces

$$\tau\frac{d\phi}{dt} = (\phi_0 - \phi)\left(1 + \frac{I_{av}}{I_s}\right) - \frac{I_{av}}{I_s}\frac{2\pi L}{\lambda}(n_1 - n_0)\frac{dN}{dn} \tag{4.7}$$

where I_s is defined by

$$I_s = \frac{E}{\Gamma a\tau} \tag{4.8}$$

Equation (4.7) can be cast in a more convenient form by using the definitions of b and g_0 in the equations

$$b = -\frac{4\pi}{\lambda a}\frac{dN}{dn} \tag{4.9}$$

and

$$g_0 = a(n_1 - n_0) \tag{4.10}$$

whence the relationship

$$\tau\frac{d\phi}{dt} = (\phi_0 - \phi)\left(1 + \frac{I_{av}}{I_s}\right) + \frac{g_0 L b}{2}\frac{I_{av}}{I_s} \tag{4.11}$$

holds true. In this result, g_0 is the unsaturated material gain per unit length and b is the ratio of real to imaginary index changes. The corresponding result for material gain g_m from (4.5) becomes

$$g_m = g_0 + \frac{2}{bL}(\phi_0 - \phi) \tag{4.12}$$

The connection with the net gain per unit length g is given by the equation

$$g = \Gamma g_m - \alpha \tag{4.13}$$

where α is the effective loss coefficient. Substitution of (4.12) into (4.13) yields the following relationship between the net gain g and the unsaturated gain g_0:

$$gL = \Gamma g_0 L + \Gamma(\phi_0 - \phi)\frac{2}{b} - \alpha L \tag{4.14}$$

In many cases of interest, the adiabatic elimination of the optical fields is a sufficiently good approximation. The mean field I_{av} relates to the output and input intensities, I_{trans} and I_{in}, respectively, by

$$\frac{I_{av}}{I_s} = \frac{I_{trans}}{I_s}\frac{(1 + R_2\,e^{gL})}{(1 - R_2)}\frac{(e^{gL} - 1)}{(e^{gL}\,gL)} \tag{4.15}$$

and

$$\frac{I_{trans}}{I_s} = \frac{I_{in}}{I_s}\frac{(1 - R_1)(1 - R_2)e^{gL}}{\left(1 - \sqrt{R_1 R_2}\,e^{gL}\right)^2 + 4\,\sqrt{R_1 R_2}\,e^{gL}\,\sin^2\phi} \tag{4.16}$$

The corresponding result for reflected intensity I_{ref} is

$$\frac{I_{ref}}{I_s} = \frac{I_{in}}{I_s} \frac{\left(\sqrt{R_1} - \sqrt{R_2} \ e^{gL}\right)^2 + 4\sqrt{R_1R_2} \ e^{gL} \sin^2\phi}{\left(1 - \sqrt{R_1R_2} \ e^{gL}\right)^2 + 4\sqrt{R_1R_2} \ e^{gL} \sin^2\phi} \tag{4.17}$$

In (4.15) to (4.17), R_1 and R_2 are the reflectivities of the input and output cavity mirrors, respectively. The unsaturated gain g_0 is normalized using the lasing threshold value g_{th}, based on the lasing threshold condition relationship

$$g_{th}L = -\frac{1}{2\Gamma} \ln (R_1R_2) + \frac{\alpha L}{\Gamma} \tag{4.18}$$

The characteristics of dispersive BLDs strongly depend on their bias current and the detuning from the resonant frequency ϕ_0. Figure 4.4(a) shows plots of the normalized transmitted output intensity as a function of the normalized input intensity for a bistable amplifier [23]. The parameters used in the calculations are as follows: $\Gamma = 0.5$, $\alpha = 25$ cm^{-1}, $b = 3$, and $g = 0.95g_{th}$. The input power for switching increases as the detuning from the resonance frequency is increased.

In the reflected light, a rich variety of behavior may be observed [24]. Figure 4.4(b) shows the three main forms of hysteresis loop exhibited by a Fabry-Perot amplifier in reflection. For values of the gain close to the threshold, the loop is the same as that for transmission, and the hysteresis occurs in a counterclockwise sense (Figure 4.4(b)(i)). In contrast, for lower values of the gain, the loop is described in a clockwise sense and is rather similar to that for reflection from a passive Fabry-Perot etalon (Figure 4.4(b)(ii)). For intermediate values of gain, the loop is traversed in the clockwise direction for increasing input, and in the counterclockwise direction for decreasing input (Figure 4.4(b)(iii)). However, there is no experimental report for such characteristics of the reflected light.

Optical bistability in an LD amplifier can be achieved by the layout shown schematically in Figure 4.5(a) [22]. The single-frequency optical beam was generated by a 1.3-μm InP/InGaAsP DFB LD with doubly buried heterostructure on a p-type InP substrate. This laser emitted more than 15 mW of optical power at a single frequency under 200-mA dc biasing. The laser beam was intensity-modulated without optical-frequency shifting by using an acousto-optic (AO) modulator, and injected into a 1.3-μm BH InP/InGaAsP LD amplifier with a long Fabry-Perot cavity through a Faraday isolator. The cavity is 1.35 mm long with an uncoated 32% reflectivity facet at each end. The polarization of the DFB LD and the Fabry-Perot LD amplifier was matched. The temperature stability of the two devices was controlled to within ±0.05°C. The optical input power, divided by a beam splitter, and the optical output power from the LD amplifier were detected with a germanium pin photodiode, and the input-output curve was monitored using an oscilloscope.

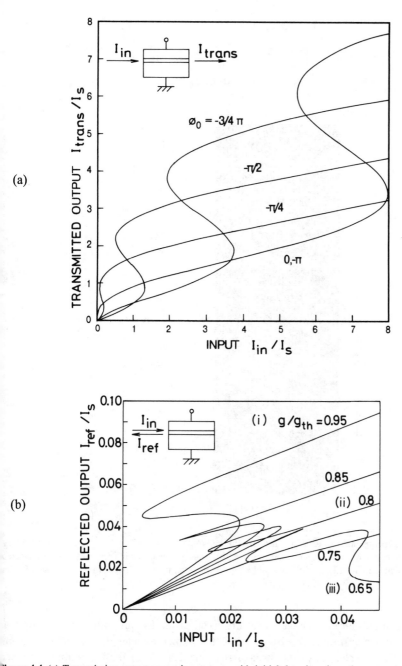

Figure 4.4 (a) Transmission-output-versus-input curve with initial detuning ϕ_0 as the parameter for a Fabry-Perot amplifier. $b = 3$, $g_0 = 0.95 g_{th}$. (Reprinted with permission from Chapman & Hall Ltd. From [23].) (b) Reflected-output-versus-input curve with g/g_{th} as the parameter. $\phi_0 = -0.3\pi$. (Reprinted with permission from Chapman & Hall Ltd. From [24].)

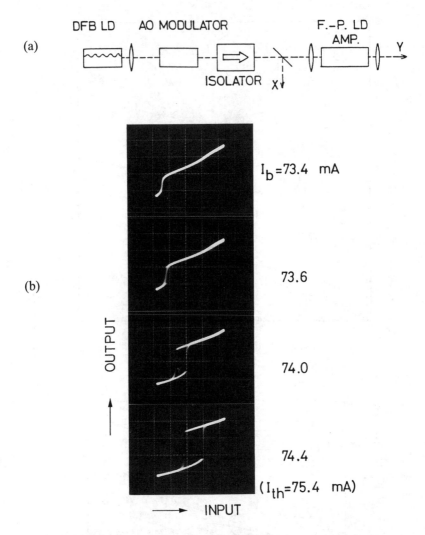

Figure 4.5 (a) Experimental layout. A 1.3-μm BH InP/InGaAsP LD amplifier with a 1.35-mm-long Fabry Perot cavity was used as a nonlinear device. (b) Observed input-output characteristics with different bias currents for LD amplifier. Detuning increases with increasing bias current at the rate of about 3.6 GHz/mA. (From [22].)

Typical input-output characteristics observed in the LD amplifier for various bias currents are shown in Figure 4.5(b). The injected signal wavelength was about 5.5 nm shorter than the wavelength of the strongest resonant mode of the LD amplifier biased at the laser threshold. At $I_b = 73.4$ mA, the wavelength of the injected light was slightly shorter than that of the nearest LD amplifier's resonant mode without optical injection

The resonant mode of the LD amplifier shifted toward the shorter wavelength side at about 3.6 GHz mA^{-1} as an increasing current reduced the carrier-dependent refractive index. The wavelength difference therefore increases with increasing current. Consequently, it can be predicted from Figure 4.4(b) that the amount of hysteresis increases with increasing current. This was very clearly confirmed by the experimental results shown in Figure 4.5(b).

Accurate measurements of bistable input-output characteristics provide a value of the nonlinear refractive index coefficient n_2 and a ratio of real to imaginary refractive index changes b. Calvani and Caponi measured the value of n_2 for a bistable Fabry-Perot GaAlAs amplifier based on a linear fit of the single-pass Fabry-Perot phase derived from the inverse transmittivity data, in the saturation zone of the hysteresis cycle, and obtained the result $n_2 = 5.9 \times 10^{-10}$ m^2/W (corresponding to $\chi^{(3)} = 1.2 \times 10^{-3}$ [esu]) [25]. Pan et al. obtained a value of 3.1 for b and a value for the change of the refractive index versus carrier density of -1.8×10^{-20} cm^3 in a GaAs LD [26].

In order to clarify the dynamic properties of LD amplifiers, the transient response curve to step-function inputs was numerically analyzed [10,27]. Ogasawara and Ito showed the existence of output spiking due to self-tuning at the onset of light injection [27].

Pan and Dagenais [28] considered the dynamic frequency chirp associated with the turn-on spike at the onset of light injection. Such a frequency chirp exists because in dispersive bistability the Fabry-Perot resonance (or the DBR/DFB transmission peak) sweeps through the injected wavelength as a result of a light-induced refractive index change. Figure 4.6(a) shows the calculated output power in response to a step input of 20 and 5 μW. Turn-on spike and ringing, as well as critical slowing down, is clearly seen. In the case of 20-μW input, the full width at half maximum (FWHM) of the spike is about 130 ps. Figure 4.6(b) shows the calculated frequency chirp that accompanies the turn-on spike. Across most of the pulse, the instantaneous frequency is downshifted from the incident frequency by as much as 4 GHz. The chirp is almost linear over the central portion of the spike. The instantaneous frequency then increases quickly; however, little energy is contained in this part of the spike. The frequency deviation goes to zero as the output power approaches the steady-state level. The magnitude of the frequency chirp is smaller when the input power is 5 μW, which can easily be understood in terms of critical slowing down.

4.1.3 Flip-Flop Operation Using Two Elements

The clockwise bistability will open up new possibilities, such as an all-optical flip-flop, by the combination of two bistable etalons based on thermal nonlinearity [29,30]. All-optical flip-flops constructed using dispersive BLDs have advantages over passive bistable etalons, such as high-speed operation and optical gain.

Figure 4.7(a) shows the schematic view of the optical flip-flop constructed by coupling two optical triode switches using a dispersive BLD. Hold$_1$ ($H^{(1)}$) and a set beam

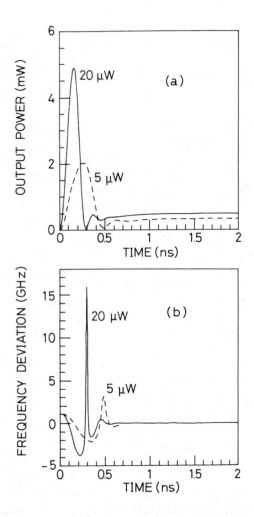

Figure 4.6 (a) Calculated output response of bistable LD amplifier to a step-function input of 20 and 5 μW (b) calculated instantaneous frequency deviation. (From [28]. Reproduced by permission of OSA.

($P_{in}^{(1)}$) are incident on BLD_1, and $hold_2$ ($H^{(2)}$) and a reset beam ($P_{in}^{(2)}$) are incident on LD. The reflected beams ($I_{ref}^{(1)}$ and $I_{ref}^{(2)}$) are then incident on each other with both BLDs working as an inverter gate in reflection mode. When LD_1 is in the high-reflection state and LD in the low-reflection state, and when the trigger pulse is incident on BLD_1, BLD_1 changes to the low-reflection state. This causes BLD_2 to change to the high-reflection state, and so on, and the circuit continues operating as a flip-flop. Two complementary binary outputs suited to signal processing are obtained from the transmitted hold beams, $I_{out}^{(1)}$ and $I_{out}^{(2)}$.

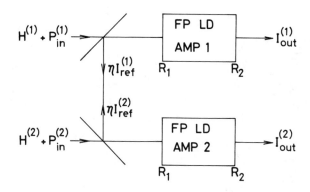

Figure 4.7(a) Optical flip-flop by coupling between two dispersive BLDs. (From [5].)

We can compute the flip-flop operation by using (4.4) through (4.18) with the following relation for the input beam intensity $I_{in}^{(i)}(t)$ to the BLD_i,

$$I_{in}^{(i)}(t) = H^{(i)} + P_{in}^{(i)}(t) + \eta I_{ref}^{(j)}(t - t_r) \qquad (4.19)$$

$$(i,j) = (1,2) \text{ and } (2,1)$$

where $P_{in}^{(i)}(t)$ is the trigger pulse, $I_{ref}^{(j)}(t - t_r)$ is the reflected hold beam from the BLD_j, t_r is the feedback delay time, and η is the coupling constant.

Figure 4.7(b) shows the calculated variation of the normalized transmitted output (i) and the reflected-output-versus-input intensities (ii) for a BLD without coupling to another BLD [5]. The parameters used in the calculation are as follows: $g_0 = 0.85g_{th}$, $R_1 = 0.05$, $R_2 = 0.3$, $\Gamma = 0.5$, $b = 5$, $\alpha L = 0.5$, $\phi_0 = -0.5\pi$, where R_1 and R_2 are the reflectivities for input side and output side, respectively. For both the transmitted and reflected outputs, we can obtain optical gain (i.e., I_{trans}, $I_{ref} > I_{in}$) at the input condition we observe flip-flop operation.

An example of the calculated flip-flop operation is shown in Figure 4.7(c), where trigger pulsewidths and heights are 5τ (τ is the material response time and is on the order of nanoseconds for dispersive BLDs) and 0.25, respectively, and $H^{(i)} = 0.07$, $\eta = 0.41$, and $\tau_r = 0$ [5]. Other parameters are the same as those for Figure 4.7(b). The holding input $H^{(i)}$ and the effective bias input ($I_{in0}^{(i)} = H^{(i)} + \eta I_{ref}^{(j)}$) are shown in Figure 4.7(b)(ii). The output power is greater than the trigger pulse power, as shown in the figure.

4.4 Bistability With Two Optical Inputs

Clockwise bistable characteristics can also be obtained by using two optical inputs with different frequencies [31]. Here, we consider two cases. The first is that the initial detuning

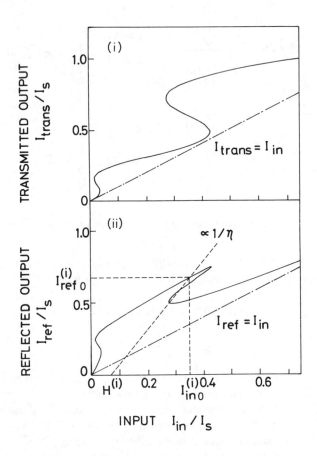

Figure 4.7(b) Calculated variation of (i) normalized transmitted output and (ii) reflected output. Optical gain was obtained for both the outputs. (From [5].)

of a signal beam 2 is set to π; that is, the signal beam wavelength coincides with the resonance wavelength of the LD amplifier when $P_{in1} = 0$ as shown in Figure 4.8(a)(i). The second is that wavelength detuning of the signal beam from the resonance wavelength is greater than that of the control beam (Figure 4.8(a)(ii)). When P_{in1} is increased, the carrier density decreases, causing the refractive index of the active layer to increase. This results in the lowering of the cavity resonance frequency; that is, the peaks in Figure 4.8(a)(i) move to the left-hand side. Therefore, bistable characteristics are observed in the P_{in1}–P_{out1} curve. In the P_{in1}–P_{out2} curve, reverse hysteresis is obtained as shown in Figure 4.8(b)(i). On the other hand, when $|\pi - \delta_2| > |\pi - \delta_1|$, counterclockwise hysteresis is observed in both the P_{in1}–P_{out1} and the P_{in1}–P_{out2} curves, as shown in Figure 4.8(b)(ii).

The experimental results are shown in Figure 4.8(c). In Figure 4.8(c)(i), the signal beam detuning is estimated to be π rad from a comparison between the calculated and

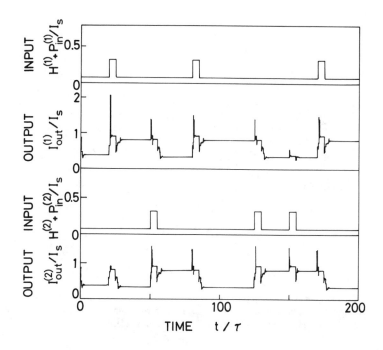

igure 4.7(c) Calculated time diagram showing flip-flop operation with two dispersive BLDs. (From [5].)

xperimental results of the P_{in1}–P_{out2} curve. By increasing the LD2 current, the wavelength f LD2 is shifted toward the longer wavelength side. Then, in Figure 4.8(c)(ii), a 2.2-rad ignal beam detuning state is observed. Therefore, counterclockwise hysteresis is observed a both the P_{in1}–P_{out1} and the P_{in1}–P_{out2} curves as shown in Figure 4.8(c)(ii). The experimental esults agree well with the theoretical results shown previously.

By changing the initial detunings, δ_1 and δ_2, and the powers of the two input beams, $_{in1}$ and P_{in2}, we can obtain many kinds of optical-input-versus-output characteristics [32].)kada et al. have studied bistability in a Fabry-Perot LD amplifier, which is biased slightly bove the laser oscillation threshold, with two optical inputs of different frequencies [33]. 'hey showed that such a branch becomes remarkable as P_{in2} is increased and can be used or the set and reset of the optical output by the application of two optical pulse trains f different detuning.

.1.5 Multistability and Cascade Connection

mong the various types of BLDs studied previously, only Fabry-Perot-type LD amplifiers ppear to have the potential of well-controlled multiple bistability or multistability. In the D amplifier, there are many constructive interference transmission peaks, and if the

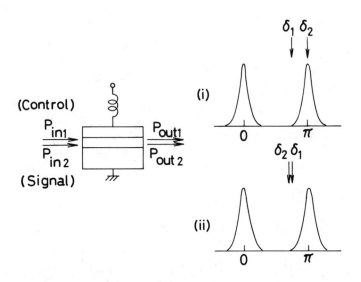

Figure 4.8(a) Bistable Fabry-Perot amplifier having two optical inputs with different wavelengths: (i) sign beam wavelength coincides with the resonance wavelength ($\delta_2 = \pi$ rad); (ii) signal beam wave length is longer than control beam wavelength. (From [5].)

cavity Q stays constant, the device shows multistability for strong light input intensitie However, the cavity Q decreases with increasing injected optical intensity through gai saturation. It is known that saturation intensity depends on the cavity length and cavit reflectivity, as well as on the percentage of mode confinement to the active region [34 Therefore, the parameters of an LD amplifier device have to be optimized to achiev multistability or multiple bistability. Optical multiple bistability was demonstrated usin a long Fabry-Perot cavity LD amplifier [22].

Here, multiple bistability is defined as two or more bistable regions existing in inpu versus-output curves, and multistability is defined as three or more stable output state existing for one input power level. Optical multistability has been demonstrated in passive GaAlAs waveguide [35] as a result of the thermal effect. Multistability has als been observed in hybrid electro-optic devices [1]. Such devices may find use in digitizir incoming light pulses. In addition, the top of their optical input-output curves is very fla that is, the devices act as efficient optical limiters. Another application would be f multilevel optical logic. Optical multistable devices are compatible and synergistic wit multiple-valued logic. The potential benefits of multiple-valued logic are increased spee and reliability, higher information storage density, decreased size, reduced cost and pow requirements, and the fact that only a few optical beams are required for signal transmissio

In this section, multistable operation in Fabry-Perot cavity LD amplifiers is studie theoretically and experimentally [22]. The effects of various device parameters, such cavity length and cavity reflectivity, on the input-output characteristics are studied throug

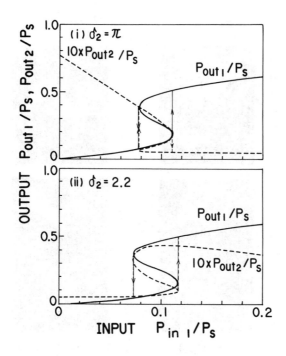

Figure 4.8(b) Calculated normalized output power, P_{out1}/P_s and P_{out2}/P_s, versus input power P_{in}/P_s for Fabry-Perot LD amplifier with two holding beams: (i) $\delta_2 = \pi$ rad; (ii) $\delta_2 = 2.2$ rad. (From [5].)

numerical analysis. The experimental observation of multiple bistability in an LD amplifier with a long Fabry-Perot cavity is also described.

Optical input-output characteristics are calculated under steady-state conditions using the same parameters as those described in the previous section. The initial detuning ϕ_0 is set at $-\pi/4$. The ratio of real to imaginary index change b has been assessed by a number of authors, but the published values are scattered from 2 to 7 for 1.3- to 1.5-μm InGaAsP lasers [36], as described in Section 2.1. These values seem to depend on the laser structure parameters. Therefore, b is set at values from 3 to 7 in the calculations.

To explain the terms, Figure 4.9(a) shows schematically the optical input-output curve, which has both bistability and multistability. N is the branch number--that is, $N = 1$ represents the lowest branch of bistability/multistability and $N = 2$ represents the second branch. To show the size of the bistable region, we have introduced a new indication, m. For the lowest branch, m is defined as $m_1 = I_{inu}/I_{ind}$, where I_{inu} and I_{ind} are the optical upward switching input from the lower output to the higher output and the optical downward switching input from the higher output to the lower output, respectively, as shown in the figure. For $N \geq 2$, m_N is defined by

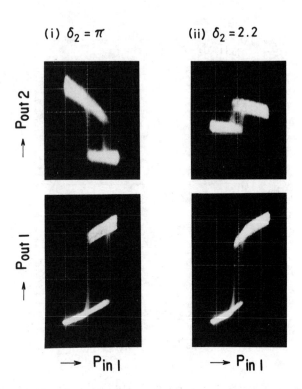

(i) $\delta_2 = \pi$ (ii) $\delta_2 = 2.2$

Figure 4.8(c) Observed P_{out1} versus P_{in1} and P_{out2} versus P_{in1} curves: (i) $\delta_2 = \pi$ rad; (ii) $\delta_2 = 2.2$ rad. (From [31].)

$$m_N = \frac{I_{inu}(N) - I_{inu}(N - 1)}{I_{ind}(N) - I_{inu}(N - 1)} \qquad (4.20)$$

Therefore, $m_N(N \geq 2)$ becomes negative for multistability.

The input-output characteristics have been calculated up to the third bistable branch. Figure 4.9(b) shows the calculated m_1 values as contour lines versus cavity reflectivities ($R_1 = R_2 = R$) and cavity lengths (L) when b is 7. From the definition of m_1, the larger m_1 is, the larger the bistable region. Clear bistability can be seen in the shorter-cavity-length LD amplifiers with cavity reflectivities of around 50%. This can be explained by the two main factors that influence bistable characteristics. The first is the cavity length L. In the calculation, the initial phase detuning ϕ_0 was kept constant at $-\pi/4$. Therefore, the frequency difference between the LD amplifier resonant mode and the injected light increases with decreasing cavity length. The second is the cavity reflectivity R. The lasing threshold current decreases with increasing R; consequently, the optical input level for gain saturation is lowered.

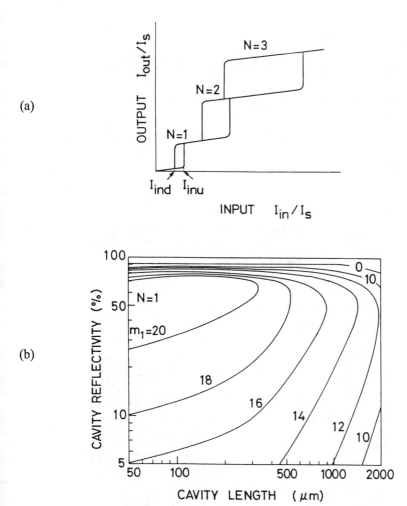

Figure 4.9(a,b) Calculated results of cavity length versus cavity reflectivity in regions where bistability or multistability are seen in the optical input-output characteristics, where $b = 7$, $g_0 = 0.99g_{th}$, $\phi_0 = -\pi/4$: (a) Schematic optical input-output curve; branch numbers N, I_{ind}, and I_{inu} are defined in the figure and m_N values are defined by (4.20) in the text; (b) calculated m_1 values.

The calculated m_2 values are shown in Figure 4.9(c). The LD amplifier having 30% to 50% cavity reflectivity and 500- to 1,000-μm cavity length shows remarkable bistability on the second branch. In this case, the two factors R and L, as well as the m_1 values play important roles in determining bistable characteristics. However, because of having a small ϕ for reaching the second resonance frequency, the optimum cavity length for

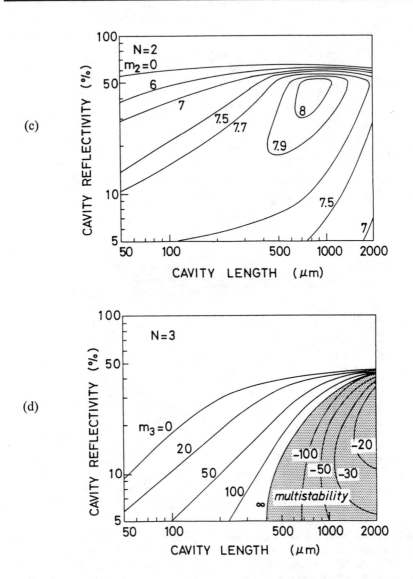

Figure 4.9(c,d) (c) Calculated m_2 values; (d) calculated m_3 values; multistability is seen in the shaded region. (From [22].)

achieving significant bistability shifts toward the longer side compared to that with $N = 1$.

Figure 4.9(d) shows calculated m_3 values. m_3 is negative in the shaded region in the figure. This means that an LD amplifier having a long cavity shows multistability on the third branch.

One example of the calculated input-output characteristics is shown in Figure 4.10. In the calculation, we set $b = 7$, $R_1 = R_2 = 0.05$, and $L = 1$ mm. For $I_{in}/I_s = 0.025$, indicated by the broken line, there are three stable output states. With $b < 7$, multistability was not found for any value of R or L. For example, the input-output characteristics of the LD amplifier with $b = 3$, $R_1 = R_2 = 0.3$, $g_0 = 0.95g_{th}$, and $L = 300$ μm shows multiple bistability, but multistability is not seen. This result seems to be similar to the experimental results described below.

Optical multiple bistability has been obtained for the first time by the combination of a long Fabry-Perot cavity LD amplifier and a high-power single-mode DFB LD using an experimental layout shown in Figure 4.5(a). Typical optical multiple bistability, as observed in the input-output characteristics, is shown in Figure 4.11(a). The LD amplifier was dc-biased at $0.95I_{th}$, where I_{th} is the threshold current. The injected signal wavelength was 5.5 nm shorter than that of the LD amplifier biased at the laser threshold. Optical input peak power was about 2 mW just in front of the input lens of the LD amplifier. By taking the coupling efficiency into account, the internal injected power can be estimated to be within a few hundred microwatts. The longitudinal-mode spacing is about 0.092 nm (15.8 GHz) for the 1.35-mm-long cavity. This long cavity enables multiple bistability to be achieved.

The multiple bistability can be seen more clearly in Figure 4.11(b), where the optical output from the LD amplifier is contrasted with the triangular optical input. The upper

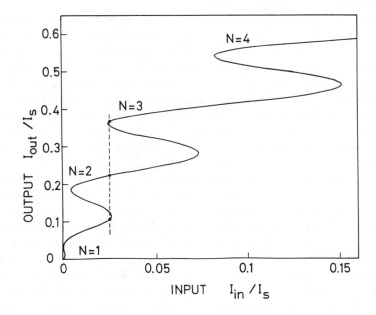

Figure 4.10 Calculated input-output characteristics. $b = 7$, $g_0 = 0.99g_{th}$, $\phi_0 = -\pi/4$, $R_1 = R_2 = 0.05$, $L = 1$ mm. Multistability is seen in the $N = 3$ branch. (From [22].)

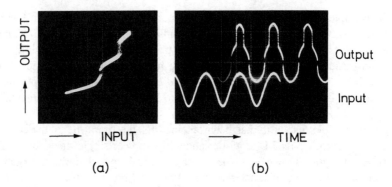

Figure 4.11 Experimentally observed multiple bistability in the input-output characteristics: (a) input-output curve; (b) (upper trace) optical output and (lower trace) optical input versus time; the switch-on and switch-off transitions occur at different instantaneous input powers and indicate that hysteresis is present. (From [22].)

trace is the optical output and the lower trace is the optical input versus time. The output waveform has two jump regions for each optical input increasing region and decreasing region. The switch-on and switch-off transitions occur at different instantaneous input powers, indicating the presence of hysteresis.

As described previously, we have achieved multiple bistability experimentally only up to the second branch in the LD amplifier. From the calculated results shown in Figure 4.9, multistability could be more easily obtained with reduced cavity reflectivity. Multistable characteristics also depend strongly on the ratio of real to imaginary index change b. The injected signal wavelength was 5.5 nm shorter than that of the LD amplifier biased at the laser threshold in the experiment. Conversely, the longer the wavelength is, the larger α is near the gain-maximum wavelength. Therefore, an optical input with longer wavelength than that used in the experiment is favorable for achieving multistability. The enhancement of the value of b by modification of the LD amplifier structure, as described in Section 2.1, provides another possibility for achieving multistability in LD amplifiers.

A cascade connection of bistable optical devices will become an important function for advanced optical signal processing. Cascadability in bistable LD amplifiers has been demonstrated [37]. The optical input to the first BLD amplifier is modulated by a mechanical chopper at frequency Ω_1. The output from the first BLD amplifier is then modulated at $\Omega_2 = 2\Omega_1$ using a second mechanical chopper. The light is thereafter used as the input to the second bistable device. When the input intensity to the second device is higher than a certain threshold, the output switches. The switching power in the data beam can be as small as 25 nW. Gains larger than 1,000 were demonstrated, indicating the large fan-out capability of the device. Severe wavelength matching would be unnecessary if absorptive BLDs are used.

4.2 BISTABILITY IN INJECTION-LOCKING LASER DIODE

For the injection-locking properties of an LD, Lang [38] predicted the existence of an asymmetry between locked longitudinal-mode intensity and wavelength detuning (i.e., the difference between the original longitudinal-mode wavelength and the injecting light wavelength). His prediction was based on the injected-carrier-density-dependent refractive index in the active region of the LD. This asymmetric characteristic was first observed experimentally by Kobayashi et al. [39] for 1.27-μm planar BH LDs. The same characteristic was also observed for GaAlAs lasers by Goldberg et al. [40].

Bistable characteristics in the locked-output-versus-detuning curve at one edge of the injection lock-in region have also been predicted for the bias current near the laser threshold by Otsuka and Kawaguchi [41] from the analysis based on generalized van der Pol equations, including the injected-carrier-density-dependent refractive index. However, the reason why bistable characteristics have not been experimentally observed in the semiconductor laser injection lock-in curve until very recently is thought to be the existence of many longitudinal modes rather than one mode in LDs near the laser threshold when there is no optical injection. Kawaguchi et al. [14] demonstrated, for the first time, bistability in the locking curve using single longitudinal-mode lasers.

In this section, the bistable output characteristics are delineated using accurate calculations of the equations described by Lang [38]. The conceptual model of an LD with external light injection is illustrated in the inset in Figure 4.12(a). The injection-locking properties of semiconductor lasers can be expressed by

$$\frac{d}{dt}E_0(t) = \frac{1}{2}[G(n) - \Gamma]E_0(t) - [\omega(n,\Omega) - \nu]E_0(t) + \kappa E_{\text{ext}} \tag{4.21}$$

$$\frac{d}{dt}N_u(t) = [G(n) - \Gamma]N_u(t) + \beta\gamma_s n \tag{4.22}$$

and

$$\frac{d}{dt}n = -\gamma_s n - G(n)[|E_0(t)|^2 + N_u(t)] + P \tag{4.23}$$

where the electric field in the longitudinal mode of the LD is expressed as $E_0 \exp(-i\nu t)$ [38]. In these equations, Γ is the cavity loss, κ is the proportionality constant, and $G(n)$ and $\omega(n,\Omega)$ are the modal gain and resonant frequency, respectively, where Ω is the oscillation frequency of the light-injected mode. N_u is the unlocked photon density, β is the spontaneous emission factor, γ_s is the inverse carrier lifetime, n is the carrier density, and P is the carrier injection rate per unit volume.

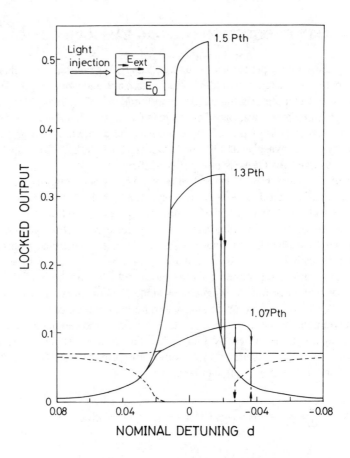

Figure 4.12(a) Computed locked output versus nominal detuning for various pumping levels.

$G(n)$ and $\omega(n,\Omega)$ are approximated with Taylor's series up to the first order for n and Ω, which can be expressed as

$$G(n) = G(n_{th}) + g\delta n = \Gamma + g\delta n \qquad (4.24)$$

$$\omega(n,\Omega) = \omega_{th} + h\delta n - (\eta_{eff}/\eta - 1)(\Omega - \omega_{th}) \qquad (4.25)$$

where $\delta n = n - n_{th}$, $g = \partial G/\partial n$, and $h = \partial\omega/\partial n$ [36]. η_{eff} is the effective index defined as $\eta_{eff} = \eta + \Omega(\partial\eta/\partial\Omega)$. Parameter R, defined as $R = -2(\partial\omega/\partial n)/(\partial G/\partial n)$, is introduced to express the carrier-density-dependent refractive index.

Equations (4.21) to (4.24) are the same as equations 1 to 4 in [38]. From the accurate calculations of these equations, bistable characteristics are clearly seen in the locking curve. An example of the computed locked output versus nominal detuning $d = (\nu - \omega_{th})$I

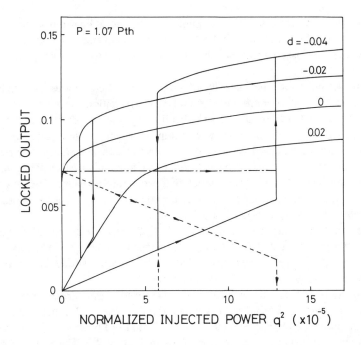

Figure 4.12(b) Computed locked output versus normalized light injection level for various nominal detunings with $P = 1.07P_{th}$. Broken lines and dash-dotted lines indicate unlocked output and total output, respectively. (Reprinted with permission from Chapman & Hall Ltd. From [23].)

is depicted by solid lines in Figure 4.12(a) for the various normalized pumping levels. For $1.07P_{th}$, the unlocked output and total output are shown by broken lines and dash-dotted lines, respectively. In these examples, the adopted parameters are $q^2 = 5 \times 10^{-5}$, $\eta_{eff}/\eta = 1.25$, $\beta = 10^{-5}$, $m = 1$, and $R = -2$, where the normalized injected power q^2 is defined as $(\kappa E_{ext})^2/\Gamma P_{th}$ and $m = gP_{th}/\gamma_s\Delta$. From the figure, the locked output clearly indicates the bistable characteristic at the negative side edge of the injection lock-in region under the low-pump-rate condition.

The computed locked output versus normalized light injection level q^2 is shown in Figure 4.12(b) by solid lines for the various nominal detunings d. The pumping level was set at $1.07P_{th}$, while the other parameters were the same as those in Figure 4.12(a). For $d = -0.04$, unlocked output and total output are shown with broken lines and dash-dotted lines, respectively. For small q^2, curves with a certain d coincide with those with $-d$. Bistability can be clearly seen in the locked-output-versus-input characteristics having negative nominal detuning when the input light wavelength remains constant.

Theoretically, bistable characteristics can be seen only under the low-pump-rate condition. Conventional LDs having Fabry-Perot resonators tend to operate in a multilongitudinal mode at a bias current near the laser threshold, even when operated in the CW mode. Moreover, conventional LDs exhibit a multilongitudinal-mode operation when intensity fluctuates. DFB LDs are suitable for the experiment because a single-longitudinal-mode operation avoids complexities that are not taken into account in the theoretical calculations.

Figure 4.13(a) plots the intensity of output from LD2 versus the injecting light wavelength under CW operation for several LD2 excitation currents above the laser threshold. While LD1 was set at a desired oscillating wavelength by changing the heat-sink temperature, the LD2 temperature was kept constant. In the figure, the individual points were measured after the LD1 temperature reached constant values. The wavelengths presented as the horizontal axis were calibrated using a scanning Fabry-Perot interferometer. In the figure, 10 GHz corresponds to a 0.9°C temperature change. The power injected into LD2 can therefore be considered to be approximately constant within the 40-GHz frequency shift of LD1 shown in Figure 4.13(a).

Under the low-pump-rate condition ($1.01I_{th}$), the bistable characteristics are clearly seen in the figure. This bistability, however, disappears when the LD2 drive current is increased ($1.11I_{th}$). The center frequency of the locking curves shifted at the rate of about 1 GHz/mA with change in the LD2 drive current.

When the LD2 bias current was set just below the laser threshold without optical injection, the system exhibited bistable characteristics in the LD2-output-versus-nominal-frequency-detuning curve. The characteristics indicated in Figure 4.13(b) are the same as those of the optical bistability in LD amplifiers in Section 4.1. The physical images are

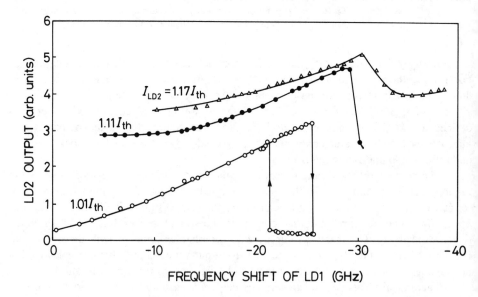

Figure 4.13(a) Measured optical output intensity from DFB-type slave LD versus injecting light wavelength: $I > I_{th}$. (©1985 IEEE. From [14].)

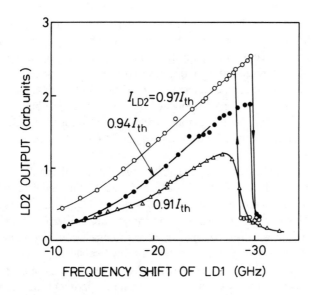

Figure 4.13(b) Measured optical output intensity from DFB-type slave LD versus injecting light wavelength: $I < I_{th}$. (©1985 IEEE. From [14].)

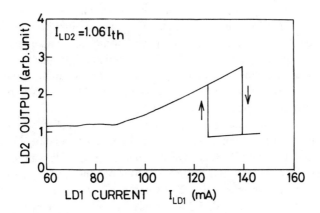

Figure 4.13(c) Measured optical output characteristics versus LD1 driving current. LD2 driving current was set at $1.06 I_{th}$. (©1985 IEEE. From [14].)

quite different for injection locking and resonant-type amplification with very weak light injections. However, in the case of strong light injection, which causes notable carrier saturation, the difference between injection locking and resonant-type amplification becomes unclear. The reason is that the laser threshold with strong light injection is difficult to define. From the experimental results shown in Figure 4.13(a), it is clear that the bistability in the output-versus-frequency-detuning curve is obtained at both above and below the no-light injection threshold current. It is also found that the more the LD2 bias current approaches the threshold from both above and below, the clearer the bistable characteristics become.

In the experimental results shown in Figure 4.13(a), the individual points were measured after the LD1 heat-sink temperatures had reached constant values. This technique represents the fundamental measurement methods for clarifying the injection-locking characteristics. This is not a practical method for obtaining bistable characteristics, however, because it requires a considerably long time to reach constant temperatures.

One example of locked-output characteristics is shown versus the driving current of LD1 in Figure 4.13(c). The driving current of LD2 was set at $1.06I_{th}$. Both LD1 and LD2 were maintained at constant heat-sink temperatures in the experiment. Measurement using a Fabry-Perot scanning interferometer shows that the LD1 wavelength was lengthened by the bias current at the rate of 0.45 GHz/mA. The bistable characteristic was obtained in the 125- to 140-mA range. Considering the shape of the curve and the lock-in bandwidth, it can be concluded that these characteristics were achieved through the LD1 wavelength shift, which occurred by heating the active region. This locked-output-versus-LD1-driving-current characteristic was very sensitive to the initial detuning (i.e., the temperature of LD1).

Bistability in the injected laser output versus detuning, which varies with change in LD1 temperature, has been discussed up to this point. Another mechanism, the injected carrier density dependence on the refractive index of the LD1 active region, will also be useful for varying the detuning. This mechanism exhibits a high-speed response of several gigahertz, the same as that of semiconductor laser direct modulation. Bistable characteristics can also be obtained by keeping the wavelength of LD1 a constant and changing the power of LD1 using an optical modulator.

As described above, since there is no discontinuity at the threshold of an LD, the optical bistability should be continuous from a point below to a point above the threshold [42]. The squares in Figure 4.14(a) give the measured bistable loop width versus the relative bias level of the slave laser from below to above the threshold, while the injected optical power is kept at approximately −23 dBm. Two identical DFB LDs with an emission wavelength of 1.554 μm were used. Below the threshold, the optical bistable loop increases its width with the increase of the bias level, while above the threshold, the loop width decreases with the bias level. The maximum optical bistable loop width is obtained, in this case, at about 1.03 times the free-running laser threshold, and this value is generally dependent on the amount of injected optical power.

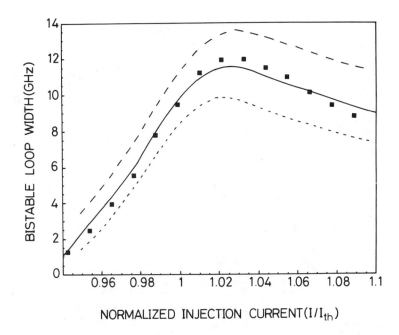

Figure 4.14(a) Measured bistable loop width versus the normalized injection current with the optical injection level at approximately −23 dBm (squares), together with the calculated results with an injected optical power P_{in} of −25 dBm (short-dashed curve), −23 dBm (solid curve), and −21 dBm (long-dashed curve). I_{th} is the free-running laser threshold.

In order to simulate the optical bistable operation, a unified treatment was presented, which allows one to consider the laser biased from below to above the threshold [42]. We directly solve the rate equations (4.21) through (4.23) numerically in the time domain using the fourth-order Runge-Kutta method. In order to make the result stable enough in the static properties' analysis, the calculated data in the time domain were averaged within 10 ns after 30 ns from turn-on. With a definite input optical power and by sweeping the input signal frequency detuning up and down, we can obtain optical bistability for the output optical power.

The calculated bistable loop width versus the normalized injection current of the slave laser, from below to above the threshold, is given in Figure 4.14(a) for three different values of injection optical power. The calculation shows that the best optical bistable results are obtained when the slave laser is pumped at a level a little above its free-running threshold. The calculation also revealed that the spontaneous emission coefficient β_{sp} also plays an important role in determining the optical bistable properties near the threshold, and this is shown in Figure 4.14(b).

Hui et al. [43] have also investigated theoretically and experimentally the dynamic properties of dispersive optical bistability in an LD biased from below to above the

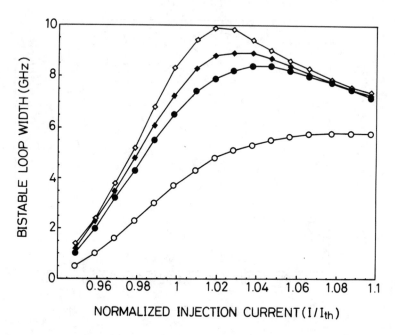

Figure 4.14(b) Calculated bistable loop width versus the normalized injection current with the slave laser's spontaneous emission coefficient β_{sp} as the parameter: $\beta_{sp} = 3.5 \times 10^{-5}$ (open diamonds), $\beta_{sp} = 1.75 \times 10^{-4}$ (filled diamonds), $\beta_{sp} = 3.5 \times 10^{-4}$ (filled circles), $\beta_{sp} = 1.75 \times 10^{-3}$ (open circles). I_{th} is the free-running laser threshold. (From [42]. Reproduced by permission of OSA.)

threshold. The optical bistability switch-off time is found to decrease continuously from below to above the threshold. A fast switch-off in less than 100 ps has been observed when the laser operates in the injection-locked condition.

Otsuka and Kawaguchi [41] predicted that detuned-laser systems that have anomalous dispersion effects at the lasing wavelength have period-doubling bifurcation for injected light signals. Figure 4.15(a) shows the injection-locking curve for which the following dimensionless parameters are introduced. $w = (P - P_{th})/P_{th}$ (P_{th}: threshold pump rate): normalized excess pump rate; $S_0 = E^2/\tau_p P_{th}$: normalized photon density; $S_i = (\omega_0 E_i/ 2Q)^2 \tau_p/P_{th}$: normalized injected photon density; $\Delta\omega = (n_e/n)[\omega_i - \omega_0(N_{th})]\tau_p$: normalized frequency detuning. Calculations were carried out assuming $w = 0.125$, $S_i = 10^{-4}$, and $R = -2$. The asymmetric nature of the tuning curve is apparent from this figure, and detuning characteristics were found to be dividable into the following regions: (I) a bistable region with hysteresis, (II) a stable lock-in region without bistability and instability, (III) a dynamically unstable region having pulsation solutions, and (IV) a self-modulating region outside the lock-in range.

Figure 4.15(b) shows the steady-state stability diagram for injection locking as a function of the normalized excess pump rate w, assuming $R = -2$ and $S_i = 10^{-4}$. For a low pump

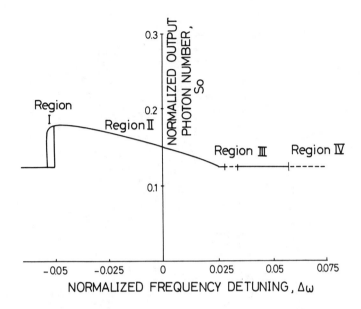

Figure 4.15(a) Injection-locking curve: $w = 0.125$, $S_i = 10^{-4}$, $R = -2$.

rate $w \leq 0.13$, bistable region I exists. With an increase in the pump rate, the stable lock-in region II becomes narrower and the bistable region I disappears where the dynamic unstable region III appears instead. For $w \geq 0.4$, subharmonic bifurcations begin to take place in region III. For $w < 0.4$, however, the system shows only self-sustained pulsations, and bifurcations were found to be absent. The asymmetrical tuning curve, as well as the hysteresis properties, results from the nonlinear change in the resonant frequency coming from the dependence of the refractive index on population-inversion density.

Sacher et al. [44] confirmed the route to chaos by numerical simulations. They demonstrated the importance of the linewidth enhancement factor and nonlinear gain for the nonlinear dynamical behaviors. Lee et al. [45] also confirmed the period-doubling route to chaos with the variation of the injection level and frequency detuning.

4.3 WAVELENGTH BISTABILITY DUE TO NONLINEAR REFRACTIVE INDEX

The external feedback can, when it is sufficiently strong, cause LDs to show bistability (Figure 4.16(a)). The origin of this bistability is considered to be the strong dependence of the refractive index of the laser active region on the carrier density [46]. Glas and Müller reported the two types of hysteresis for short and long delay times in a GaAlAs LD coupled to an external cavity [47]. In the case of a short optical delay, the low-power

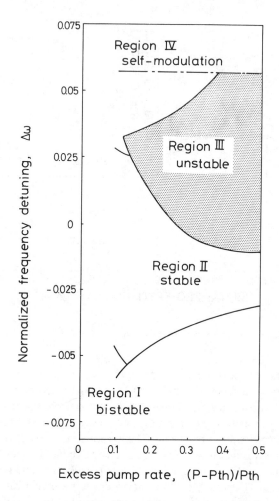

Figure 4.15(b) Stability diagram of an injection locking in detuned lasers: $S_i = 10^{-4}$, $R = -2$. (From [41].)

hysteresis branch corresponds to the noise level below the lasing threshold, whereas for a long delay, regular or irregular pulsations of considerable mean power occur. On the upper level, CW single-mode emission is dominant for both of the hysteresis types.

When an external cavity or an LD itself has a strong frequency dependence, the output shows clear bistable characteristics, even if a frequency-selective detection system is not used. A GaAlAs LD coupled to an external grating [47] and a DFB LD coupled to an external flat mirror [30] have been studied. Figure 4.17 shows typical light-output-versus-current curves of a DFB LD under external optical feedback. The device was an asymmetric-structure, 1.5-μm InGaAsP-InP DFB LD with a 500-μm cavity length. The coupling coefficient κ of the grating was estimated from the stop bandwidth to be 20

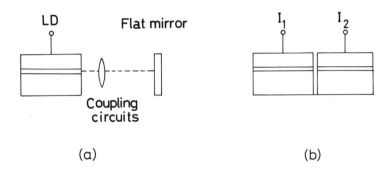

Figure 4.16 Structures of wavelength BLDs due to nonlinear refractive index: (a) Fabry-Perot LD with external cavity; (b) cleaved-coupled-cavity (C^3) LD. (From [5].)

cm^{-1}. A periodic undulation appeared on the light-output-versus-current curves when the feedback amount into the laser was weak ($\eta \sim 1\%$), as shown in Figure 4.17(a) [48]. On the contrary, a clear hysteresis appeared when $\eta \sim 3\%$, as shown in Figure 4.17(b) [5].

Spectral bistability in the output from a 1.5-μm buried crescent cleaved-coupled-cavity (C^3) laser has been demonstrated with a rise time of about 1 ns (Figure 4.16(b)) [49]. The C^3 laser has two optically coupled but electrically isolated sections formed by recleaving an LD chip approximately in the middle. It has been shown by analyzing the coupled rate equations for the C^3 laser and taking into account the carrier density dependence of the refractive index that such bistable regions exist and do not depend on saturable absorption. Large optical-power differences (about 1 mW) have been reported in 1.3-μm InGaAs-InP devices [50]. Bistability has also been demonstrated using GaAs-AlGaAs C^3 lasers. The mechanism responsible for the bistability is essentially dispersive [51].

4.4 POWER AND WAVELENGTH BISTABILITY IN DBR LASER DIODES [52]

A specially designed DBR LD exhibits both power and wavelength bistability and can be operated as an all-optical flip-flop. The mechanism governing the operation of the laser results from an interplay between the wavelength of a lasing mode and the wavelength-dependent reflectivity function of the Bragg section. This interplay, coupled with conventional laser nonlinearities, gain saturation, and thermally and carrier-induced frequency-shifting effects, results in bistable characteristics.

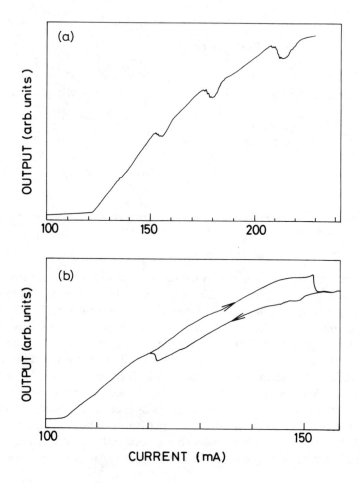

Figure 4.17 Light-output-versus-current curves of a DFB LD with external feedback: (a) weak feedback; (b) strong feedback. (From [5].)

The operation of the device relies on two properties of the specially designed two-section 1.5-μm DBR structure depicted in Figure 4.18(a): (1) a 250-μm-long Bragg reflector consisting of a low-loss shallow grating that yields a reflection function with a relatively narrow bandwidth of approximately 225-GHz FWHM, and (2) a short (130 μm) strained-layer QW gain medium that defines a large modal spacing of approximately 300 GHz. The optical coupling between the two sections is essentially lossless and reflectionless. This laser operates such that the Bragg reflector overlaps spectrally one cavity mode at most (Figure 4.18(b)); thus, it can be turned on and off by controlling the phase condition of the cavity. The latter can be accomplished because the cavity can be

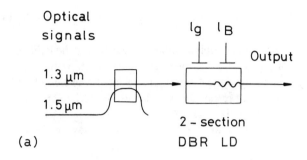

(a)

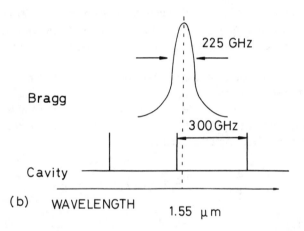

(b)

Figure 4.18 Bistable two-section DBR LD: (a) schematic diagram of the LD; (b) conceptual description of mode spacing (showing the mode-DBR overlap on the blue side of the Bragg reflection function). (From [52]. Reproduced by permission of OSA.)

adjusted so that no frequency, where there is sufficient gain, satisfies the $N \times 2\pi$ phase condition necessary for lasing. The laser can be turned into and out of this condition by an electrical drive or by optical signal injection. The key point in obtaining bistable operation of the laser is the adjustment of the spectral overlap such that the single cavity mode is placed on the short-wavelength side (the blue side) of the Bragg reflection function at low injection levels. This particular type of mode-DBR overlap, together with conventional gain nonlinearities and carrier effects, causes a self-adjusting mechanism of the phase condition that leads to bistable characteristics.

Optically controlled turn-on and turn-off with 1-pJ pulses at 1.5 μm and 2-pJ pulses at 1.3 μm, respectively, was demonstrated, as well as all-optical flip-flop operation [52].

4.5 POLARIZATION BISTABILITY USING NONLINEAR CAVITY

When transverse magnetic (TM) optical input is injected into the Fabry-Perot laser oscillating with TE polarization (i.e., $I_b > I_{th}$), the laser oscillation stops through the gain quenching effect caused by the strong optical input. If the input wavelength is set to be slightly longer than that of the cavity resonance wavelength for TM polarization, one can observe dispersive bistability in both TE and TM polarization output [53–55].

The relationship between the effective refractive index for TM mode $\tilde{n}_{TM}$ and carrier density n is written as

$$\tilde{n}_{TM} = \tilde{n}_{TM0} + \Gamma_{TM} \alpha \lambda_{TM} \frac{1}{4\pi} \frac{dg_{TM}}{dn}(n - n_{TE}) \qquad (4.26)$$

where $\tilde{n}_{TM0}$ is $\tilde{n}_{TM}$ at $n = n_{TE}$, α is the linewidth enhancement factor, λ_{TM} is the wavelength of the TM mode in a vacuum, g_{TM} is the TM gain coefficient, and n_{TE} is the carrier density when the TE gain coefficient g_{TE} is equal to zero. dg_{TM}/dn is approximately constant $(= a_{TM})$.

Since TE oscillation is generated, the rate equations for carrier density n and TE photon density S_{TE} can be used. For simplification of the calculation, we assume that TE oscillation operates in a single longitudinal mode and TM input consists of a single wavelength. The term of TM photon density S_{TM} should be included in the equations because TM photons also interact with carriers. The rate equations are

$$\frac{dn}{dt} = I - \frac{n}{\tau_s} - v_{TE} g_{TE} S_{TE} - v_{TM} g_{TM} S_{TM} \qquad (4.27)$$

and

$$\frac{dS_{TE}}{dt} = v_{TE}(\Gamma_{TE} g_{TE} - \alpha_0)S_{TE} + \Gamma_{TE}\beta_{sp} \frac{n}{\tau_s} \qquad (4.28)$$

where t is time, I is the pumping rate, and τ_s is the carrier lifetime. v_{TE} and v_{TM} are the group velocities of TE and TM modes, respectively, and are assumed to be constant. α_0 is defined as $\alpha_s + (1/L) \ln(1/R_{TE})$. β_{sp} is the spontaneous emission factor.

The relationship between the photon density of TE output in front of the laser facet $|E_{out}(TE)|^2$ and the TE photon density in the cavity S_{TE} is represented as

$$|E_{out}(TE)|^2 = \frac{G_{TE}L \exp(G_{TE}L)(1 - R_{TE})}{\{\exp(G_{TE}L) - 1\}\{1 + R_{TE} \exp(G_{TE}L)\}} S_{TE} \qquad (4.29)$$

where G_{TE} is the TE gain of the laser waveguide and is represented as $\Gamma_{TE} g_{TE} - \alpha_s$.

TM amplification can be represented by using the equation for the transmittance of the TM wave through the laser cavity.

$$T = \frac{|E_{out}(TM)|^2}{|E_{in}(TM)|^2}$$

$$= \frac{(1 - R_{TM})^2 \exp(G_{TM}L)}{\{1 - R_{TM}\exp(G_{TM}L)\}^2 + 4R_{TM}\exp(G_{TM}L)\sin^2\phi} \tag{4.30}$$

where T is the transmittance, $|E_{out}(TM)|^2$ is the photon density of the TM output, $|E_{in}(TM)|^2$ is the photon density of the TM input, and G_{TM} is the TM gain of the laser waveguide and is represented as $\Gamma_{TM}\, g_{TM} - \alpha_s$. ϕ means the phase detuning of TM wave from the cavity resonance. It is represented as

$$\phi = \frac{2\pi\tilde{n}_{TM}L}{\lambda_{TM}} \tag{4.31}$$

$\tilde{n}_{TM}$ varies with carrier density n as (4.26).

The transmittance T can be expressed by the other equation as

$$T = \frac{G_{TM}L(1 - R_{TM})}{\{1 - \exp(G_{TM}L)\}\{1 + R_{TM}\exp(G_{TM}L)\}} \times \frac{S_{TM}}{|E_{in}(TM)|^2} \tag{4.32}$$

Equations (4.30) and (4.32) should be satisfied simultaneously, because (4.30) is obtained from the boundary condition at the laser facets alone, while (4.32) relates input photon density $|E_{in}(TM)|^2$ to the inner photon density S_{TM}. This equation can be obtained from (4.30) and (4.33).

To calculate the light-output-versus-input characteristics, we should use the equations relating $|E_{out}(TM)|^2$ to S_{TM} and relating S_{TM} to $|E_{in}(TM)|^2$; that is,

$$|E_{out}(TM)|^2 = \frac{G_{TM}L\exp(G_{TM}L)(1 - R_{TM})}{\{\exp(G_{TM}L) - 1\}\{1 + R_{TM}\exp(G_{TM}L)\}} S_{TM} \tag{4.33}$$

and

$$S_{TM} = \frac{\{\exp(G_{TM}L) - 1\}\{1 + R_{TM}\exp(G_{TM}L)\}(1 - R_{TM})\,|E_{in}(TM)|^2}{G_{TM}L[\{1 - R_{TM}\exp(G_{TM}L)\}^2 + 4R_{TM}\exp(G_{TM}L)\sin^2\phi]} \tag{4.34}$$

respectively. Equation (4.34) is obtained by combining (4.30) and (4.33). We assume that g_{TM} and g_{TE} linearly vary with the carrier density; that is, $g_{TM} = a_{TM}(n - n_{TM})$ and $g_{TE} = a_{TE}(n - n_{TE})$.

The calculated results with TM injection under the bias current of $1.05 \times I_{th}$ are shown below. The wavelength of TE oscillation is determined to be 1.3052 μm, which is the resonant wavelength at the peak of the gain profile under the threshold. The wavelength of the TM mode is varied around 1.310 μm. Numerical analysis is based on the Runge-Kutta-Gill method.

Dynamic hysteresis loops are shown in Figure 4.19. The initial detuning is $0.3 \times \Delta\lambda$, where $\Delta\lambda$ is the longitudinal-mode spacing. TM input corresponding to the horizontal axis is sinusoidally modulated at the frequency of 5 MHz. The vertical axes of Figure

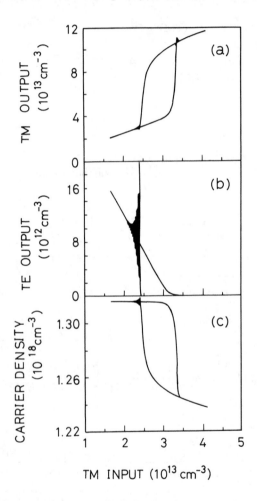

Figure 4.19 Calculated dynamic hysteresis loops. TM input is sinusoidally modulated at the frequency of 5 MHz: (a) relationship between TM output and TM input; (b) relationship between TE output and TM input; (c) relationship between carrier density and TM input. (©1991 IEEE. From [54].)

4.19(a–c) show TM output, TE output, and the carrier density, respectively. In the lower state of Figure 4.19(a), TM output increases in proportion to TM input so that the gain used for TM amplification increases. On the other hand, the total gain is constant because the carrier density is kept constant as shown in the upper state of Figure 4.19(c). Therefore, the gain that can be used for TE oscillation is reduced and TE output decreases linearly as shown in the upper state of Figure 4.19(b).

When TM input is larger than 3×10^{13} cm^{-3}, the carrier density in the higher state decreases because the carrier consumption rate for TM amplification becomes larger than the pumping rate. In this condition, TE oscillation stops as shown in Figure 4.19(b), since the carrier density is less than the threshold. By decreasing the carrier density, the detuning is reduced. This reduction produces the positive feedback between the increase of TM photon density in the cavity and the decrease of the carrier density. Therefore, the carrier density rapidly decreases to the lower state as shown in Figure 4.19(c). Simultaneously, TM output increases to the higher state. During transit, TM output shows the spike in the higher state because the detuning $\Delta \phi$ becomes zero momentarily. After transit, the detuning is very small and the TM photon density in the cavity is high so that TM output is kept in the higher state. In this condition, TE output and the carrier density remain in the lower states.

When TM input decreases, the carrier density cannot be kept low and the detuning increases because the carrier consumption rate decreases. This increase produces the positive feedback between the decrease of TM photon density in the cavity and the increase of the carrier density. Therefore, the carrier density rapidly increases to the higher state, the threshold density, as shown in Figure 4.19(c). Therefore, TE oscillation occurs again after the relaxation oscillation, as shown in Figure 4.19(b). These hysteresis loops are wider than those of calculated static characteristics because switch-up and switch-down take a few nanoseconds. The static characteristics are shown in [56] in detail.

REFERENCES

[1] Smith, P. W., E. H. Turner, and P. J. Maloney, "Electrooptic Nonlinear Fabry-Perot Devices," *IEEE J. Quantum Electronics*, Vol. QE-14, No. 3, March 1978, pp. 207–216.

[2] Garmire, E., "Criteria for Optical Bistability in a Lossy Saturating Fabry-Perot," *IEEE J. Quantum Electronics*, Vol. QE-25, No. 3, March 1989, pp. 289–295.

[3] Otsuka, K., and H. Iwamura, "Analysis of a Multistable Semiconductor Light Amplifier," *IEEE J. Quantum Electronics*, Vol. QE-19, No. 7, July 1983, pp. 1184–1186.

[4] Otsuka, K., and S. Kobayashi, "Optical Bistability and Nonlinear Resonance in a Resonant-Type Semiconductor Laser Amplifier," *Electron. Lett.*, Vol. 19, No. 7 March 1983, pp. 262–263.

[5] Kawaguchi, H., "Bistable Laser Diodes and Their Applications for Photonic Switching," *Int. J. Optoelectronics*, Vol. 7, No. 3, 1992, pp. 301–348.

[6] Nakai, T., N. Ogasawara, and R. Ito, "Optical Bistability in a Semiconductor Laser Amplifier," *Japanese J. Appl. Phys.*, Vol. 22, No. 5, May 1983, pp. L310–L312.

[7] Sharfin, W. F., and M. Dagenais, "Room-Temperature Optical Bistability in InGaAsP/InP Amplifiers and Implications for Passive Devices," *Appl. Phys. Lett.*, Vol. 46, No. 9, May 1985, pp. 819–821.

[8] Adams, M. J., H. J. Westlake, M. J. O'Mahony, and I. D. Henning, "A Comparison of Active and Passive

Optical Bistability in Semiconductors," *IEEE J. Quantum Electronics*, Vol. QE-21, No. 9, September 1985, pp. 1498–1504.

[9] Sharfin, W. F., and M. Dagenais, "Femtojoule Optical Switching in Nonlinear Semiconductor Laser Amplifiers," *Appl. Phys. Lett.*, Vol. 48, No. 5, February 1986, pp. 321–322.

[10] Adams, M. J., "Time Dependent Analysis of Active and Passive Optical Bistability in Semiconductor," *IEE Proc. Pt. J*, Vol. 132, No. 6, December 1985, pp. 343–348.

[11] Adams, M. J., and L. D., Westbrook, "Optical Bistability and Nonlinear Switching in Quantum Well Amplifiers," in *Optical Switching in Low-Dimensional Systems*, H. Haug and L. Banyai, eds., NATO ASI Series B: Physics, Vol. 194, Plenum, 1989, pp. 35–48.

[12] LiKamWa, P., A. Miller, M. Ogawa, and R. M. Park, "All-Optical Bistable Switching in an Active InGaAs Quantum-Well Waveguide," *IEEE Photonics Tech. Lett.*, Vol. 3, No. 6, June 1991, pp. 507–509.

[13] Mace, D. A. H., M. J. Adams, and C. Seltzer, "MQW Amplifier Optical Bistability," *Electron. Lett.*, Vol. 27, No. 15, July 1991, pp. 1363–1365.

[14] Kawaguchi, H., K. Inoue, T. Matsuoka, and K. Otsuka, "Bistable Output Characteristics in Semiconductor Laser Injection Locking," *IEEE J. Quantum Electronics*, Vol. QE-21, No. 9, September 1985, pp. 1314–1317.

[15] Adams, M. J., and R. Wyatt, "Optical Bistability in Distributed Feedback Semiconductor Laser Amplifiers," *IEE Proc. Pt. J*, Vol. 134, No. 1, February 1987, pp. 35–40.

[16] Hui, R., and A. Sapia, "Nonlinearity Difference in the Two Passbands of a Distributed-Feedback Semiconductor Laser Amplifier," *Opt. Lett.*, Vol. 15, No. 17, September 1990, pp. 956–958.

[17] Marshall, I. W., M. J. O'Mahony, D. M. Cooper, P. J. Fiddyment, J. C. Regnault, and W. J. Devlin, "Gain Characteristics of a 1.5 μm Nonlinear Split Contact Laser Amplifier," *Appl. Phys. Lett.*, Vol. 53, No. 17, October 1988, pp. 1577–1579.

[18] Adams, M. J., "Theory of Two-Section Laser Amplifiers," *Optical and Quantum Electronics*, Vol. 21, 1989, pp. S15–S31.

[19] Barnsley, P. E., I. W. Marshall, H. J. Wickes, P. J. Fiddyment, J. C. Regnault, and W. J. Devlin, "Absorptive and Dispersive Switching in a Three Region InGaAsP Semiconductor Laser Amplifier at 1.57 μm," *J. Modern Optics*, Vol. 37, No. 4, 1990, pp. 575–583.

[20] Adams, M. J., J. V. Collins, and I. D. Henning, "Analysis of Semiconductor Laser Optical Amplifiers," *IEE Proc. Pt. J*, Vol. 132, No 1, February 1985, pp. 58–63.

[21] Adams, M. J., "Physics and Applications of Optical Bistability in Semiconductor Laser Amplifiers," *Solid-State Electronics*, Vol. 30, No. 1, 1987, pp. 43–51.

[22] Kawaguchi, H., "Multiple Bistability and Multistability in a Fabry-Perot Laser Diode Amplifier," *IEEE J. Quantum Electronics*, Vol. QE-23, No. 9, September 1987, pp. 1429–1433.

[23] Kawaguchi, H., "Absorptive and Dispersive Bistability in Semiconductor Injection Lasers," *Optical and Quantum Electronics*, Vol. 19, 1987, pp. S1–S36.

[24] Adams, M. J., "Optical Amplifier Bistability on Reflection," *Optical and Quantum Electronics*, Vol. 19, 1987, pp. S37–S45.

[25] Calvani, R., and R. Caponi, "$\chi^{(3)}$ Measurements in a Bistable GaAlAs Fabry-Perot Amplifier," *Electron. Lett.*, Vol. 26, No. 18, August 1990, pp. 1513–1514.

[26] Pan, Z., H. Lin, and M. Dagenais, "Switching Power Dependence on Detuning and Current in Bistable Diode Laser Amplifier," *Appl. Phys. Lett.*, Vol. 58, No. 7, February 1991, pp. 687–689.

[27] Ogasawara, N., and R. Ito, "Static and Dynamic Properties of Nonlinear Semiconductor Laser Amplifier," *Japanese J. Appl. Phys.*, Vol. 25, No. 9, September 1986, pp. L739–L742.

[28] Pan, Z., and M. Dagenais, "Observation of Dynamic Frequency Chirp in Bistable Semiconductor Laser Amplifiers," *Nonlinear Optics: Materials, Fundamentals, and Applications, Tech. Dig.*, WE6, August 1992, pp. 358–360.

[29] Kuszelewicz, R., and J. L. Oudar, "Bistable Flip-Flop Operation of Passive Optical Nonlinear Devices," *Int. Conf. Quantum Electron., Tech. Dig.*, MP-68, 1988.

30] Tsuda, H., and T. Kurokawa, "Construction of an All-Optical Flip-Flop by Combination of Two Optical Triodes," *Appl. Phys. Lett.*, Vol. 57, No. 17, October 1990, pp. 1724–1726.

31] Kawaguchi, H., H. Tani, and K. Inoue, "Optical Bistability Using a Fabry-Perot Semiconductor-Laser Amplifier With Two Holding Beams," *Opt. Lett.*, Vol. 12, No. 7, July 1987, pp. 513–515.

32] Tani, H., J. Nishikido, and H. Kawaguchi, "Optical Switching by Another Optical Beam Using a Fabry-Perot Semiconductor Laser Amplifier," OQE 87-1 (in Japanese).

33] Okada, M., H. Kikuchi, K. Takizawa, and H. Fujikake, "Optical Bistability and Set-Reset Operation of a Fabry-Perot Semiconductor Laser Amplifier With Two Detuned Light Injection," *IEEE J. Quantum Electronics*, Vol. QE-27, No. 8, August 1991, pp. 2003–2015.

34] Mukai, T., Y., Yamamoto, and T. Kimura, "Optical Direct Amplification for Fiber Transmission," *NTT Rev. ECL*, Vol. 31, No. 3, 1983, pp. 340–348.

35] Walker, A. C., J. S. Aitchison, S. Ritchie, and P. M. Rodgers, "Intrinsic Optical Bistability and Multistability in a Passive GaAlAs Waveguide," *Electron. Lett.*, Vol. 22, No. 7, March 1986, pp. 366–367.

36] Osinski, M., and J. Buus, "Linewidth Broadering Factor in Semiconductor Lasers--An Overview," *IEEE J. Quantum Electronics*, Vol. QE-23, No. 1, January 1987, pp. 9–29.

37] Pan, Z., T.-N. Ding, and M. Dagenais, "Demonstration of Cascadability and Spectral Bistability in Bistable Diode Laser Amplifiers," *Dig. of Nonlinear Optics: Materials, Phenomena and Devices*, Kauai, New York: IEEE, p. 230.

38] Lang, R., "Injection Locking Properties of a Semiconductor Laser," *IEEE J. Quantum Electronics*, Vol. QE-18, No. 6, June 1982, pp. 976–983.

39] Kobayashi, K., H. Nishimoto, and R. Lang, "Experimental Observation of Asymmetric Detuning Characteristics in Semiconductor Laser Injection Locking," *Electron. Lett.*, Vol. 18, No. 2, January 1982, pp. 54–56.

40] Goldberg, L., H. F. Taylor, and J. F. Weller, "Locking Bandwidth Asymmetry in Injection-Locked GaAlAs Lasers," *Electron. Lett.*, Vol. 18, No. 23, November 1982, pp. 986–987.

41] Otsuka, K., and H. Kawaguchi, "Period-Doubling Bifurcations in Detuned Lasers With Injected Signals," *Phys. Rev. A*, Vol. 29, No. 5, May 1984, pp. 2953–2956.

42] Hui, R., S. Benedetto, and I. Montrosset, "Optical Bistability in Diode Laser Amplifiers and Injection-Locked Laser Diodes," *Opt. Lett.*, Vol. 18, No. 4, February 1993, pp. 287–289.

43] Hui, R., A. Paradisi, S. Benedetto, and I. Montrosset, "Dynamics of Optically Switched Bistable Laser Diodes in the Injection-Locked State," *Opt. Lett.*, Vol. 18, No. 20, October 1993, pp. 1733–1735.

44] Sacher, J., D. Baums, P. Panknin, W. Elsässer, and E. O. Göbel, "Intensity Instabilities of Semiconductor Lasers Under Current Modulation, External Light Injection, and Delayed Feedback," *Phys. Rev. A*, Vol. 45, No. 3, February 1992, pp. 1893–1905.

45] Lee, E.-K., H.-S. Pang, J.-D. Park, and H. Lee, "Bistability and Chaos in an Injection-Locked Semiconductor Laser," *Phys. Rev. A*, Vol. 47, No. 1, January 1993, pp. 736–739.

46] Lang, R., and K. Kobayashi, "External Optical Feedback Effects on Semiconductor Injection Laser Properties," *IEEE J. Quantum Electronics*, Vol. QE-16, No. 3, March 1980, pp. 347–355.

47] Glas, P., and R. Müller, "Different Kinds of Bistable Behaviour of a GaAlAs Diode Coupled to an External Cavity of Variable Length," *Optical and Quantum Electronics*, Vol. 19, 1987, pp. S61–S74.

48] Yoshikuni, Y., H. Kawaguchi, and T. Ikegami, "Intensity Fluctuation of 1.5 μm InGaAsP/InP Distributed Feedback Lasers Involving the Optical Feedback Effect," *IEE Proc. Pt. J*, Vol. 132, No. 1, February 1985, pp. 20–27.

49] Olsson, N. A., W. T. Tsang, R. A. Logan, I. P. Kaminow, and J.-S. Ko, "Spectral Bistability in Coupled Cavity Semiconductor Lasers," *Appl. Phys. Lett.*, Vol. 44, No. 4, February 1984, pp. 375–377.

50] Dutta, N. K., G. P. Agrawal, and M. W. Focht, "Bistability in Coupled Cavity Semiconductor Lasers," *Appl. Phys. Lett.*, Vol. 44, No. 1, January 1984, pp. 30–32.

51] Phelan, P., L. Reekie, D. J. Bradley, and W. A. Stallard, "Hysteretical and Spectral Behaviour of Bistable Cleaved Cavity Semiconductor Lasers," *Optical and Quantum Electronics*, Vol. 18, 1986, pp. 35–41.

2] Margalit, M., R. Nagar, N. Tessler, G. Eisenstein, M. Orenstein, U. Koren, and C. A. Burrus, "Bistability and Optical Control of a Distributed-Bragg-Reflector Laser," *Opt. Lett.*, Vol. 18, No. 8, April 1993, pp 610–612.

[53] Mori, Y., J. Shibata, and T. Kajiwara, "Optical Polarization Bistability in TM Wave Injected Semiconducto Lasers," *IEEE J. Quantum Electronics*, Vol. QE-25, No. 3, March 1989, pp. 265–272.

[54] Mori, Y., "Dynamic Properties of Transverse-Magnetic Wave Injected Semiconductor Lasers," *IEEE J Quantum Electronics*, Vol. QE-27, No. 11, November 1991, pp. 2415–2421.

[55] Mori, Y., J. Shibata, and T. Kajiwara, "Analysis of Optical Polarization Bistability in Transverse-Magnetic Wave Injected Semiconductor Lasers," *J. Appl. Phys.*, Vol. 67, No. 5, March 1990, pp. 2223–2228.

[56] Ogawa, T., Y. Ida, and K. Hayashi, "Analysis of Polarization Bistability of Phase-Shifted DFB Lase Due to TM Light Injection," *J. Lightwave Technology*, Vol. 10, No. 7, July 1992, pp. 913–917.

Chapter 5
Two-Mode Bistability via Gain Saturation

5.1 GENERAL CONSIDERATION

If a laser is oscillating in two modes, the cross-effect between the two modes arises via gain saturation. The possibility of achieving bistability in two-mode lasers via gain saturation, in general, has been demonstrated mathematically in terms of an idealized mode [1,2]. Yamada [3] and Tang et al. [4] have analyzed the required conditions for bistable operation of a two-mode LD based on mode competition through gain saturation. Tang et al. [4] also extended the theory to the case where the bistability is controlled by an injected optical signal.

To the lowest order in the nonlinearity, the saturated gain $g_i(I)$ of a particular mode i) of an idealized laser is related to the unsaturated gain g_{i0} through the mode intensity I_i and saturation parameter I_{isat} as follows.

$$g_i(I) = g_{i0}/(1 + \epsilon_{ii}I_i) \tag{5.1}$$

where $\epsilon_{ii} = 1/I_{isat}$. When $\epsilon_{ii}I_i \ll 1$, $g_i(I)$ can be rewritten as follows.

$$g_i(I) = g_i(1 - \epsilon_{ii}I_i) \tag{5.2}$$

In the case of two modes, mode 1 and mode 2, the rate of change of one-mode intensity also depends on the intensity of the other:

$$\frac{dI_1}{dt} = g_1I_1(1 - \epsilon_{11}I_1 - \epsilon_{12}I_2) \tag{5.3}$$

$$\frac{dI_2}{dt} = g_2I_2(1 - \epsilon_{21}I_1 - \epsilon_{22}I_2) \tag{5.4}$$

where ϵ_{11} and ϵ_{22} are self-saturation and ϵ_{12} and ϵ_{21} are cross-saturation coefficients. The stability of such a gain-coupled two-mode laser system can be analyzed in the phase plane (I_1, I_2). Equations (5.3) and (5.4) for the steady state give

$$I_1 = 0 \quad \text{or} \quad \epsilon_{11}I_1 + \epsilon_{12}I_2 = 1 \tag{5.5a}$$

$$I_2 = 0 \quad \text{or} \quad \epsilon_{21}I_1 + \epsilon_{22}I_2 = 1 \tag{5.5b}$$

since $\dot{I}_1 = 0$ and $\dot{I}_2 = 0$, where the dot denotes the time derivative.

We draw a graph with I_1 as the abscissa and I_2 as the ordinate, as in Figure 5.1. Then (5.5a) gives the I_2 axis and the solid slant line, while (5.5b) gives the I_1 axis and the broken line. It can be seen from (5.5a) that $\dot{I}_1 = 0$ on the solid slant line, while $\dot{I}_1 < 0$ above the line on the right and $\dot{I}_1 > 0$ below it on the left. The condition for $\dot{I}_2$ is similar with regard to the broken line. Therefore, the time variations of I_1 and I_2 in the four domains partitioned by the two slant lines are shown in Figure 5.1. It is thus found that the states satisfying the steady-state conditions simultaneously are given by the three black points in Figure 5.1, of which the stable point is $X(x_1, x_2)$:

$$x_1 = (\epsilon_{12} - \epsilon_{22})/(\epsilon_{21}\epsilon_{12} - \epsilon_{11}\epsilon_{22}) \tag{5.6a}$$

$$x_2 = (\epsilon_{21} - \epsilon_{11})/(\epsilon_{21}\epsilon_{12} - \epsilon_{11}\epsilon_{22}) \tag{5.6b}$$

It should be noted that this holds for the weak-coupling case.

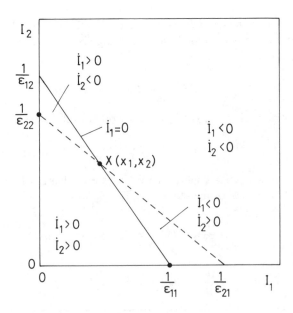

Figure 5.1 Diagram for light intensities I_1 and I_2 of two-mode oscillation.

Two possible characteristic situations, weak coupling and strong coupling, as shown in Figure 5.2 can occur. If $\epsilon_{11}\epsilon_{22} > \epsilon_{12}\epsilon_{21}$ (weak coupling), in general, there can be stable one- or two-mode operation as shown in Figures 5.2(a–c). If X is below the I_1 axis ($x_1 > 0$, $x_2 < 0$), only mode 1 will oscillate (Figure 5.2(b)). If $x_1 > 0$ and $x_2 > 0$, stable simultaneous two-mode oscillation will result (Figure 5.2(a)). If $x_1 < 0$ and $x_2 > 0$, the laser will oscillate only in mode 2 (Figure 5.2(c)).

Bistability will occur if $\epsilon_{21}\epsilon_{12} > \epsilon_{11}\epsilon_{22}$ (strong coupling). In this case, if X is in the first quadrant of the phase plane as shown in Figure 5.2(d), the laser is bistable. Also, in

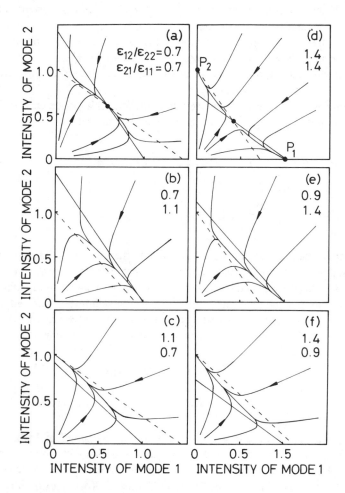

Figure 5.2 Examples of trajectories of the intensities of the two modes, I_1 and I_2 (abscissa and ordinates in units of $1/\epsilon_{11}$ and $1/\epsilon_{22}$, respectively), for various initial values in the phase plane: (a–c) $\epsilon_{11}\epsilon_{22} > \epsilon_{21}\epsilon_{12}$; (d–f) $\epsilon_{21}e_{12} > \epsilon_{11}e_{22}$; g_1 is assumed equal to g_2 for all cases; solid straight line: $dI_2/dI_1 = \infty$; dashed straight line: $dI_2/dI_1 = 0$; heavy dot: crossing point X. (From [4]. Reproduced by permission of AIP.)

this case, if $x_1 > 0$ and $x_2 < 0$, mode 2 will oscillate (Figure 5.2(f)); if $x_1 < 0$ and $x_2 > 0$, mode 1 will oscillate (Figure 5.2(e)). Thus, in the strong-coupling case, only one of the modes oscillates, the other mode being suppressed. In the resulting steady state, therefore, the laser does not oscillate simultaneously in two modes, but it oscillates in only one of the modes. At the stable point P_1 in Figure 5.2(d), it oscillates in mode 1, while at P_2 it oscillates in mode 2. Which of the bistable states is reached is dependent on the initial conditions, as can be seen from the curves in Figure 5.2(d).

The bistable operation of such a two-mode laser can be controlled either electrically or optically. Electrical control can obviously be achieved through current control of the gain or absorption and needs no further elaboration.

Consider the case of optical control of a two-wavelength laser through an injected signal $I_2^{(i)}$ at or near the wavelength of mode 2, for example. There are two possibilities, depending on whether the oscillation in mode 2 adds coherently or incoherently to the injected signal.

If the injected signal near the wavelength of mode 2 adds incoherently to the intensity of mode 2, the rate of change for each mode intensity becomes

$$\frac{dI_1}{dt} = g_1 I_1 [1 - \epsilon_{12}^{(i)} I_2^{(i)} - \epsilon_{11} I_1 - \epsilon_{12} I_2] \tag{5.7a}$$

$$\frac{dI_2}{dt} = g_2 I_2 [1 - \epsilon_{22}^{(i)} I_2^{(i)} - \epsilon_{21} I_1 - \epsilon_{22} I_2] \tag{5.7b}$$

where I_2 refers to the intensity of mode 2, excluding the injected signal. The effect of the injected signal is mainly to depress the population inversion and hence the gains for the two modes. Because the injected signal could have a mode volume and spatial distribution different from those of modes 1 and 2, its effect on the saturation coefficients of the two modes can be different and is indicated by a superscript (i). In the presence of the injected signal, the coordinates of the crossing point X become correspondingly

$$x_1(I_2^{(i)}) = x_1(0) - \{[\epsilon_{12}\epsilon_{22}^{(i)} - \epsilon_{22}\epsilon_{12}^{(i)}]/[\epsilon_{21}\epsilon_{12} - \epsilon_{11}\epsilon_{22}]\} I_2^{(i)} \tag{5.8a}$$

$$x_2(I_2^{(i)}) = x_2(0) - \{[\epsilon_{21}\epsilon_{12}^{(i)} - \epsilon_{11}\epsilon_{22}^{(i)}]/[\epsilon_{21}\epsilon_{12} - \epsilon_{11}\epsilon_{22}]\} I_2^{(i)} \tag{5.8b}$$

where $x_{1,2}(0)$ refers to the case in the absence of an injected signal and is given by (5.6a and (5.6b). It is clear from (5.8a) and (5.8b) that bistability with optical control can be achieved. For example, if $x_1(0) < 0$, $x_2(0) > 0$, and the laser initially oscillates in mode 1 (Figure 5.3(a)), the laser can be made bistable (Figure 5.3(b)) or to oscillate in mode 2 (Figure 5.3(c)) by making $x_1(I_2^{(i)}) > 0$ and $x_2(I_2^{(i)}) < 0$ through the injected signal, provided

$$(\epsilon_{12}\epsilon_{22}^{(i)} - \epsilon_{22}\epsilon_{12}^{(i)}) < 0 \tag{5.9a}$$

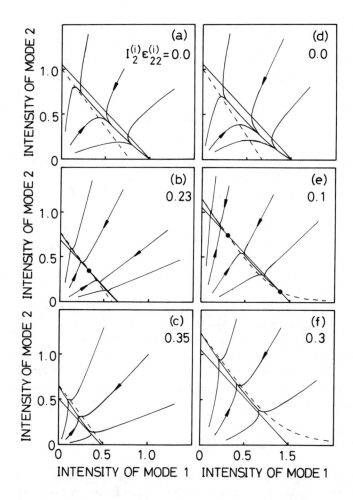

Figure 5.3 Examples of trajectories of the intensities of the two modes, I_1 and I_2 (abscissa and ordinates in units of $1/\epsilon_{11}$ and $1/\epsilon_{22}$, respectively), in the phase plane in the presence of injected signals: (a–c) incoherent injection, $\epsilon_{12}/\epsilon_{22} = 0.95$, $\epsilon_{21}/\epsilon_{11} = 1.3$, $\epsilon_{12}^{(i)}/\epsilon_{22}^{(i)} = 1.5$; (d–f) coherent injection, $\epsilon_{12}/\epsilon_{22} = 0.95$, $\epsilon_{21}/\epsilon_{11} = 1.5$, $\eta'/g_2 = 0.5$. (From [4]. Reproduced by permission of AIP.)

$$(\epsilon_{21}\epsilon_{12}^{(i)} - \epsilon_{11}\epsilon_{22}^{(i)}) > 0 \qquad (5.9b)$$

from (5.8a) and (5.8b). Note that the conditions in (5.9a) and (5.9b) do not conflict with the requirement of bistability $\epsilon_{21}\epsilon_{12} > \epsilon_{11}\epsilon_{22}$. The size of the hysteresis loop defining the bistable region can be determined easily from the conditions $x_1(I_{21}^{(i)}) = 0$ and $x_2(I_{12}^{(i)}) = 0$ using (5.8a) and (5.8b), where $I_{21}^{(i)}$ and $I_{12}^{(i)}$ refer to the injection levels for switching from mode 2 to mode 1 and mode 1 to mode 2, respectively.

In the coherent case, the rate equation for the intensity of mode 1 remains formally the same as (5.3a). For mode 2, the intensity equation must be determined from the rate equation for the field amplitude with an added source term proportional to the complex amplitude of the injected signal:

$$\frac{dE_2}{dt} = \left(\frac{g_2}{2}\right)[1 - \epsilon_{22}|E_2|^2 - \epsilon_{21}|E_1|^2]E_2 + \left(\frac{\eta}{2}\right)E^{(i)} \qquad (5.10)$$

where $\eta = \eta' + i\eta''$ is a proportionality constant. Assuming the emitted field is stimulated by the injected field, E_2 is, therefore, in phase with an includes $E^{(i)}$. Equation (5.10) leads to the rate equation for the intensity of mode 2:

$$\frac{dI_2}{dt} = g_2[1 - \epsilon_{22}I_2 - \epsilon_{21}I_1]I_2 + \eta'[I_2 I^{(i)}]^{1/2} \qquad (5.11)$$

in place of (5.4), where $I_2 > I^{(i)}$. The trajectories in the phase plane now depend on the injected signal $I^{(i)}$. In particular, with an injected signal there is no longer a stable state corresponding to oscillation in mode 1 only. As $I^{(i)}$ increases, the stable point initially on the I_1 axis moves into the first quadrant (Figure 5.3(e)). In addition, because the spatial distribution of the injected field can be slightly different from that of a freely oscillating mode 2, the values of the coefficients g_2, ϵ_{22}, ϵ_{21}, and ϵ_{12} are also dependent on the injected signal. Thus, with the injected signal, the laser can be made to switch from stable oscillation in mode 1 (Figure 5.3(d)) to oscillation in mode 2 (Figure 5.3(f)) and with a bistable region in between (Figure 5.3(e)).

5.2 POLARIZATION BISTABILITY

The bistability results from the competition in the LD cavity between its transverse electric field polarized parallel to the junction plane (TE) and its transverse magnetic field polarized parallel to the junction plane (TM) modes. Figure 5.4 shows some examples of the schematic structure of polarization BLD caused by gain saturation [5]. In the first report of polarization bistability by Chen and Liu [6], an InGaAsP/InP LD was cooled down to the polarization transition temperature (190K), which resulted in an enhancement in the TM mode gain due to the internal stress in the active region (Figure 5.4(a)). They suggested that nonlinear gain saturation is the origin of this bistability. Hysteresis in the polarization resolved output power was obtained by controlling the optical feedback power of the TE mode via an intracavity electro-optic modulator by Fujita et al. [7]. Tang et al. [4] predicted that theoretically LDs having two-mode bistabilities in polarization (TE/TM) or in wavelength (λ_1/λ_2) can be switched between the bistable states through coherent or incoherent optical control input, and demonstrated wavelength bistability experimentally. Linton et al. [8] reported optically triggered polarization bistability in GaInAsP twin-stripe LD

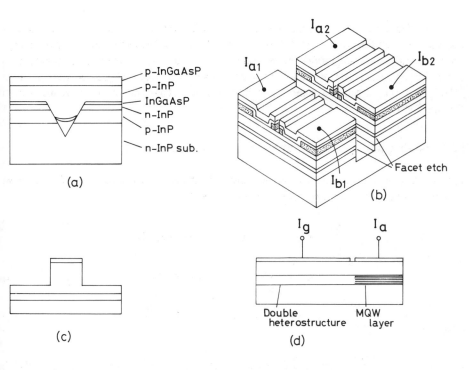

Figure 5.4 Polarization BLDs via gain saturation: (a) V-grooved substrate BH InGaAsP/InP LD operating at about 190K; (b) four-contact, twin-stripe LD; (c) ridge-waveguide LD; (d) LD consisting of DH gain and MQW absorber sections with separate electrical contacts. (From [5].)

(Figure 5.4(b)). Rheinländer et al. [9] have reported a polarization-bistable InGaAsP/InP ridge-waveguide LD, and claimed the hole transfer between valence bands, which may cause the difference in the gain saturation level of the TE and TM polarization modes, as the origin of the mutual-mode coupling (Figure 5.4(c)). Klehr et al. observed the switching time of about 50 ps for the transition between TE and TM polarization under 900-MHz current modulation in the ridge-waveguide LD [10]. Recently, it has also been reported that two-mode bistability could result from the mode-dependent saturable absorption in the LD cavity (Figure 5.4(d)) [11,12]. Although various forms of two-mode bistability in LDs have been realized, only hysteresis like bistability (S-shaped) has so far been observed, and the width of the hysteresis loop is usually small. Details of S-shaped polarization bistability is described in Section 5.2.4.

Kawaguchi et al. [13] predicted theoretically that there are two kinds of polarization bistability: conventional S-shaped bistability and pitchfork bifurcation bistability. In the pitchfork bifurcation polarization bistability, when the LD current is increased from zero, the output intensities of TE and TM bifurcate at a critical point, and one polarization mode oscillates. With the injection of an optical trigger input having a polarization opposite to the oscillation polarization, the oscillation polarization changes to that of the input. It was also predicted that pitchfork bifurcation bistability would have major speed advantage over conventional S-shaped bistability. They have observed this new type of polarization bistability by using a two-armed polarization-selective external cavity configuration [14,15]. All-optical flip-flop operation has been successfully demonstrated by using both TE and TM trigger inputs.

5.2.1 Rate Equation Analysis: Model and Basic Equation

We analyze the static behavior of polarization bistability in LDs using the following rate equations, taking into account gain saturation [13,16].

$$\frac{dn}{dt} = P - \frac{n}{\tau_s} - v_g g(n)(1 - \epsilon_{EE}S_E - \epsilon_{EM}S_M)\dot{S}_E$$

$$- v_g g(n)(1 - \epsilon_{ME}S_E - \epsilon_{MM}S_M)S_M \tag{5.12}$$

$$\frac{dS_E}{dt} = v_g \Gamma_E g(n)(1 - \epsilon_{EE}S_E - \epsilon_{EM}S_M)S_E - \frac{S_E}{\tau_{pE}} + \beta B n^2 \tag{5.13}$$

$$\frac{dS_M}{dt} = v_g \Gamma_M g(n)(1 - \epsilon_{ME}S_E - \epsilon_{MM}S_M)S_M - \frac{S_M}{\tau_{pM}} + \beta B n^2 \tag{5.14}$$

In these equations, S_E and S_M are the photon densities for the TE and TM modes, respectively. The carrier density in the active layer is n. The material gain is given by $g(n) = an - b$, P is the pump rate, τ_s is the carrier lifetime, and v_g is the group velocity. The optical confinement factors are Γ_E and Γ_M, where the subscripts E and M refer to the TE and TM modes, respectively. The coefficients for self-saturation are ϵ_{EE} and ϵ_{MM}, and ϵ_{EM} and ϵ_{ME} are those for cross-saturation from the nonlinear gain. τ_{pE} and τ_{pM} are the photon lifetimes in the laser cavity, β is the spontaneous emission factor, and B is the recombination coefficient.

Consider the case of optical control of a two-mode laser through an optical injected signal. There are two possibilities, depending on whether the injected signal adds coherently or incoherently to the laser field.

When a coherent optical signal $S_{iE(M)}$ (E(M) refers to the TE (TM) mode) at a wavelength within the injection-locking band is injected into the BLD, an additional term $\sqrt{S_{E(M)} \eta S_{iE(M)} / \tau_{pE(M)}}$ is included in (5.13) or (5.14), depending on its polarization [4]. Here, η is the coupling coefficient. The injection-locking half bandwidth of LDs, Δf_L, is expressed as [17] $\Delta f_L = \sqrt{S_i/S_{out}}/4\pi\tau_p$. When $S_i = S_{out}$ and $\tau_p = 3$ ps, Δf_L becomes 27 GHz.

If the injected signal with TM polarization, S_{iM}, adds incoherently to the intensity of the TM laser mode, rate equations (5.12) to (5.14) become [4]

$$\frac{dn}{dt} = P - \frac{n}{\tau_s} - v_g g(n)(1 - \epsilon_{EE}S_E - \epsilon_{EM}S_M - \epsilon_{EM}^{(i)}S_{iM})S_E$$

$$-v_g g(n)(1 - \epsilon_{ME}S_E - \epsilon_{MM}S_M - \epsilon_{MM}^{(i)}S_{iM})S_M \tag{5.15}$$

$$\frac{dS_E}{dt} = v_g \Gamma_E g(n)(1 - \epsilon_{EE}S_E - \epsilon_{EM}S_M - \epsilon_{EM}^{(i)}S_{iM})S_E$$

$$-\frac{S_E}{\tau_{pE}} + \beta B n^2 \tag{5.16}$$

$$\frac{dS_M}{dt} = v_g \Gamma_M g(n)(1 - \epsilon_{ME}S_E - \epsilon_{MM}S_M - \epsilon_{MM}^{(i)}S_{iM})S_M$$

$$-\frac{S_M}{\tau_{pM}} + \beta B n^2 \tag{5.17}$$

The effect of the injected signal is mainly to depress the population inversion and hence the gains for the two modes, as described in Section 5.1.

TE/TM polarization bistability is found under the following conditions: (1) coexistence probability for both the TE and the TM modes, and (2) mutual coupling of both modes via gain saturation. Bistability occurs [4] if $\epsilon_{ME}\epsilon_{EM} > \epsilon_{EE}\epsilon_{MM}$. Usually, an LD oscillates with the TE mode because facet reflectivities and the modal confinement are larger for the TE mode than for the TM mode. However, if, for example, the structure of an LD is optimized, the device becomes polarization-insensitive [18,19]. From the calculations, taking account of the intraband relaxation processes (lifetime about 0.1 ps), the following relations can be expected [3]: $\epsilon_{ME} = \epsilon_{EM} = k\epsilon_{EE} = k\epsilon_{MM}$ ($k = 0$ to 2). The value of k depends on the difference between the TE and TM oscillation wavelengths, $\lambda_E - \lambda_M$. The value reaches a maximum when both wavelengths are equal, and is greater than unity in the range of $|\lambda_E - \lambda_M| < 1$ nm. These relations are also applicable for the competition between two modes with different wavelengths and with the same polarization, and have been considered as the origin of the wavelength bistability in the longitudinal modes.

Yu and Liu [20] reported theoretical values for the self-($\epsilon_{EE}, \epsilon_{MM}$) and cross-saturation coefficients ($\epsilon_{EM}, \epsilon_{ME}$) calculated using the density-matrix formalism for both nonstrained and strained 1.3-μm InGaAsP LDs. They have shown that the value of k is 1.7 when $\lambda_E = \lambda_M$ and is greater than unity in the range of $|\lambda_E - \lambda_M| < 1$ nm for a nonstrained LD. Calculated self-saturation coefficients at various carrier densities are shown in Figure 5.5(a). It is found that $\epsilon_{EE} \sim \epsilon_{MM}$. The cross-saturation coefficients depend on the frequencies of both polarizations. Calculated results are shown for ϵ_{EM} and ϵ_{ME} in Figure 5.5(b) and (c). It can be seen from the figure that the cross-saturation coefficients are significant

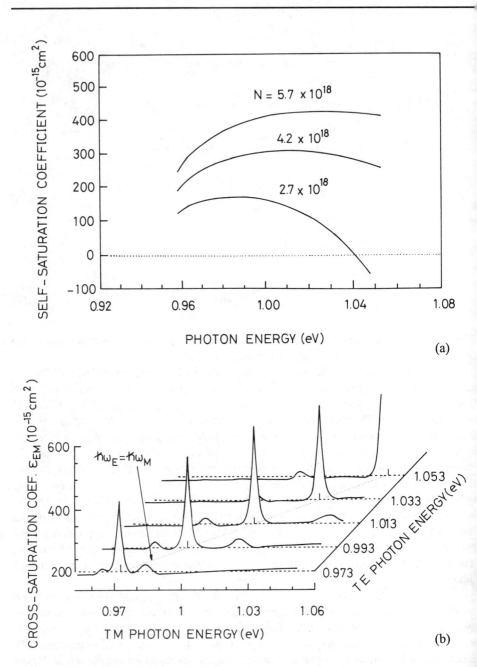

Figure 5.5 (a) Spectra of self-saturation coefficient at various carrier densities. ϵ_{EE} and ϵ_{MM} degenerate. (b,c) Cross-saturation coefficient as functions of $\hbar\omega_E$ and $\hbar\omega_M$ at $N = 4.2 \times 10^{18}$ cm^{-3}: (b) ϵ_{EM}; (c) ϵ_{ME}. The dashed baselines are for the values of the cross-saturation coefficient equal to 2×10^{-13} cm^2. (From [20]. Reproduced by permission of AIP.)

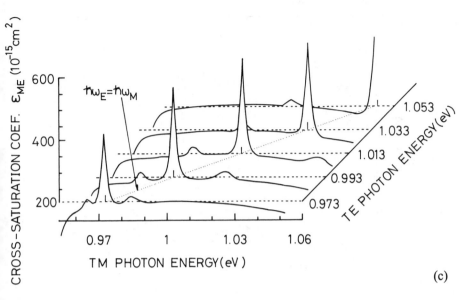

Figure 5.5 (continued).

only when $\omega_E \sim \omega_M$. The two small bumps of the cross-saturation coefficients on both sides of the line of $\omega_E = \omega_M$ in the figure are caused by beating the transition from the conduction band to the heavy-hole band and from the conduction band to the light-hole band.

They used in their calculation the value of intraband relaxation time of $\tau_{in} = 3$ ps. The spectral width of the holes is essentially determined by the carrier-carrier scattering rates, which in the time scale is on the order of 50 fs [21]. Henneberger et al. have recently reported the hole width of 35 nm for GaAs. Therefore, we can expect the values of $k > 1$ for a much wider wavelength region in the actual LDs.

5.2.2 Calculated Results

Figure 5.6(a) shows the computed variations in photon density versus the excitation rate P, which demonstrates the ordinary hysteresis (S-shaped) bistability. The stable and unstable regions are depicted by solid and dashed lines, respectively. The parameters used in the calculations are $\epsilon_{EE} = \epsilon_{MM} = 7 \times 10^{-18}$ cm^3, $\epsilon_{EM} = 6 \times 10^{-17}$ cm^3, $\epsilon_{ME} = 1.65 \times 10^{-17}$ cm^3, $\Gamma_E = 0.15$, $\Gamma_M = 0.135$, $\tau_{pE} = 2.71 \times 10^{-12}$ sec, $\tau_{pM} = 3 \times 10^{-12}$ sec, $\tau_s = 3 \times 10^{-9}$ sec, $v_g = 6.7 \times 10^9$ cm s^{-1}, $\beta = 1 \times 10^{-4}$, $B = 1.33 \times 10^{-10}$ cm s^{-1}, $a = 3.25 \times 10^{-16}$ cm^2, and $b = 325$ cm^{-1}. For hysteresis bistability to be obtained, an enhanced cross-saturation effect ($\epsilon_{EM} \gg \epsilon_{EE}, \epsilon_{MM}$), which does not inherently exist in LDs, was included. In a high-pump-rate region ($P > 1.22 \times 10^{27}$ s^{-1} cm^{-3}), other solutions appear for the outputs of TE and TM polarizations, which will be described later in detail.

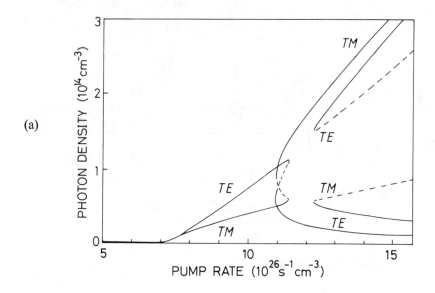

(a)

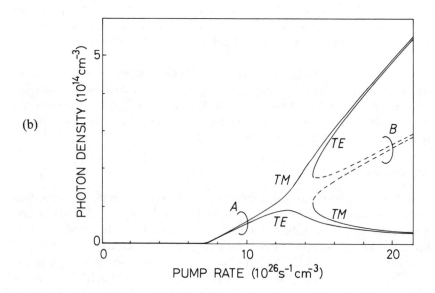

(b)

Figure 5.6 (a) Ordinary S-shaped polarization bistability arising from the enhancement of the cross-saturation effect. $\epsilon_{EE} = \epsilon_{MM} = 7 \times 10^{-18}$ cm^3, $\epsilon_{EM} = 6 \times 10^{-17}$ cm^3, and $\epsilon_{ME} = 1.65 \times 10^{-17}$ cm^3. (b) Pitchfork bifurcation bistability arising from the inherent intraband relaxation processes. $\epsilon_{EE} = \epsilon_{MM} = 1 \times 10^{-17}$ cm^3 and $\epsilon_{EM} = \epsilon_{ME} = 2 \times 10^{-17}$ cm^3. (From [15].)

Only the intraband relaxation processes, which intrinsically exist in LDs, need to be taken into account to allow the demonstration of pitchfork bifurcation bistability as shown in Figure 5.6(b). Here we assume $k = 2$; that is, $\epsilon_{EM} = \epsilon_{ME} = 2\epsilon_{EE} = 2\epsilon_{MM}$. The parameters are $\epsilon_{EE} = \epsilon_{MM} = 1 \times 10^{-17}$ cm³, $\epsilon_{EM} = \epsilon_{ME} = 2 \times 10^{-17}$ cm³, $\Gamma_E = 0.15$, $\Gamma_M = 0.135$, $\tau_{pE} = 2.7 \times 10^{-12}$ sec, and $\tau_{pM} = 3 \times 10^{-12}$ sec. The other parameters are the same as for Figure 5.6(a). In this case, as the pump rate is allowed to increase from zero, the solutions for the outputs of the TE and TM polarizations bifurcate at the critical point and form branch A. Branch B is obtained only by injection of an optical trigger input with TM polarization at the appropriate bias pump rate.

Hereafter we concentrate our interest on this pitchfork bifurcation bistability. The parameters used in the calculations are the same as for Figure 5.6(b).

Figure 5.7 shows the computed variations in the photon density versus the excitation rate for a different photon lifetime of the TE mode, τ_{pE}, which demonstrates the pitchfork bifurcation polarization bistability. Photon lifetime τ_p in a laser cavity is generally expressed by the following equation:

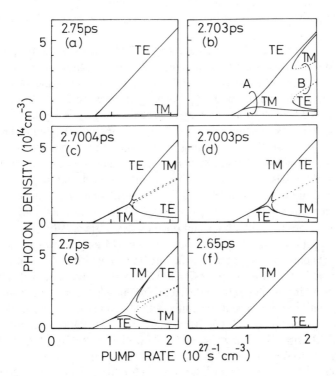

Figure 5.7 Computed static variations in photon density versus the excitation rate for a different photon lifetime of the TE mode, τ_{pE}. Photon lifetime of the TM mode, τ_{pM}, is 3 ps.

$$\tau_p = \frac{L}{v_g \{ \alpha L + \ln(1/R) + \ln(1/T) \}} \qquad (5.18$$

Here, L is the cavity length, v_g is the velocity of light in the cavity, α is the average distributed loss constant, T is the transmittance of the attenuator inserted into the cavity and R is the power reflectivity of the cavity. Therefore, when the transmittance of the attenuator is decreased, the photon lifetime becomes short. The stable and unstable regions are depicted by solid and dashed lines in the figure, respectively. Parameters used in the calculations are the same as for Figure 5.6(b) except the value of τ_{pE}. When $\tau_{pE} = 2.75$ ps (Figure 5.7(a)), the TE mode threshold is lower than the TM mode threshold, and the TE mode oscillates. When $\tau_{pE} = 2.703$ ps (Figure 5.7(b)) as the pump rate is increased from zero, the solutions for the outputs of TE and TM polarizations bifurcate at the critical point and form branch A. Branch B is obtained only by injection of an optical trigger input with TM polarization at the appropriate bias pump rate. Between $\tau_{pE} = 2.7004$ ps (Figure 5.7(c)) and $\tau_{pE} = 2.7003$ ps (Figure 5.7(d)), the dominant oscillation mode at a high pump rate changes from TE to TM. On a condition of $\tau_{pE} = 2.65$ ps (Figure 5.7(f)) the TM mode threshold becomes lower than the TE mode threshold, and the TM mode oscillates.

Figure 5.8 shows the calculated static TE and TM outputs versus the photon lifetime τ_{pM}, of the TM mode for different pump rates P. The photon lifetime, τ_{pE}, of the TE mode is chosen to be 2.7 ps. When P is small ($P = 1.3 \times 10^{27}$ s^{-1} cm^{-3}), the switching between the TE and TM modes occurs when τ_{pM} is changed. However, the TE- and TM-output-versus-τ_{pM} curves do not show any hysteresis. When $P = 1.4 \times 10^{27}$ s^{-1} cm^{-3}, the polarization switching with hysteresis occurs. This corresponds to the polarization-bistable switching observed by Fujita et al. [7] by changing the voltage of the modulator inserted into the laser cavity. As shown in Figure 5.8, if P is further increased and is set in the regime where the branch B exists ($P > 1.45 \times 10^{27}$ s^{-1} cm^{-3}), with decreasing τ_{pM} from 3 ps, the output state is abruptly switched from TM mode oscillation to TE mode oscillation. TE mode oscillation continues even if τ_{pM} returns to 3 ps. The initial and final states are marked in the figure by solid circles and open circles, respectively.

Figure 5.9 shows the computed static TE and TM outputs versus TE optical input for different bias pump rates for the case in which the injection light couples to the lasing light coherently. Here, $\tau_{pE} = 2.7$ ps, $\tau_{pM} = 3$ ps, and $\eta = 1$ are assumed. Three types of polarization switching are possible for different bias pump rates: switching without hysteresis, switching with hysteresis, and switching with a memory effect. Typical examples are shown in the figure. For $P = 1.3 \times 10^{27} s^{-1}$ cm^{-3}, the TM output decreased and the TE output increased with increasing S_{iE}, and polarization switching without hysteresis is observed. We observe polarization-bistable switching when $P = 1.4 \times 10^{27}$ s^{-1} cm^{-3}. At a bias pump rate of 1.5×10^{27} s^{-1} cm^{-3}, when the TE optical input is increased, the output state is abruptly switched from the TM mode oscillation (corresponding to branch A of Figure 5.7(e)) to TE mode oscillation (branch B of Figure 5.7(e)) at an injection photon

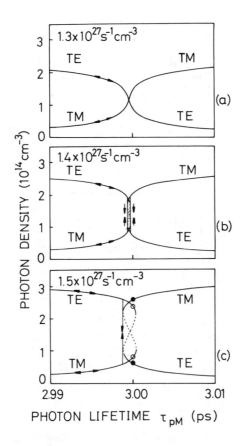

Figure 5.8 Computed static TE and TM outputs versus photon lifetime of the TM mode for different bias pump rate. $\tau_{pE} = 2.7$ ps.

density of 1.6×10^7 cm^{-3}. The laser oscillation with TE polarization continues even if the TE optical input intensity decreases to zero.

ON ($S_w\uparrow$) and OFF ($S_w\downarrow$) switching photon densities of the TE mode are shown in Figure 5.10 versus bias pump rates for the case of coherent optical injection. The parameters used in the calculation are the same as those used in Figure 5.9. When $P < 1.37 \times 10^{27}$ s^{-1} cm^{-3}, the laser shows the polarization switching without hysteresis. Therefore, the ON switching photon density equals the OFF switching photon density in this case. For the pump rate of 1.37×10^{27} s^{-1} cm$^{-3} < P < 1.45 \times 10^{27}$ s^{-1} cm^{-3}, the switching photon density becomes different for ON switching and OFF switching, and the laser shows bistability. The bistable region increases with increasing pump rate. For the pump rate of $P > 1.45 \times 10^{27}$ s^{-1} cm^{-3}, the LD has only ON switching. This means that the laser oscillation with TE polarization continues even if the TE optical input intensity decreases to zero.

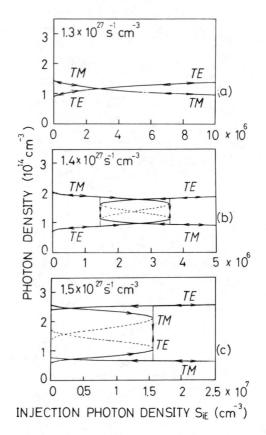

Figure 5.9 Computed static TE and TM outputs versus TE optical input adding coherently to the output for different bias pump rate. $\tau_{pE} = 2.7$ ps and $\tau_{pM} = 3$ ps. (From [15].)

If the injection light is coupled incoherently, much higher input power is needed for the switching. The switching characteristics are qualitatively the same as for the coherent case shown in Figure 5.9.

The all-optical flip-flop is realized with the help of trigger optical pulses. Figure 5.11 shows the calculated transient response of TE and TM output. The BLD is biased in the regime of the bifurcated solution, and trigger pulses, with either TE or TM polarization, are injected. Pump rate P is set at 1.5×10^{28} cm^{-3} s^{-1}, $\eta = 1$, and $S_{iE} = S_{iM} = S_E(P = 1.5 \times 10^{28}$ cm^{-3} s$^{-1})$. Ultrafast flip-flop operation is predicted at a rate of 50 Gbps as shown in Figure 5.11.

5.2.3 Experimental Results on Pitchfork Bifurcation Bistability

The experimental setup used to demonstrate the pitchfork bifurcation polarization bistability is shown in Figure 5.12. A T-shaped polarization-selective external cavity was con-

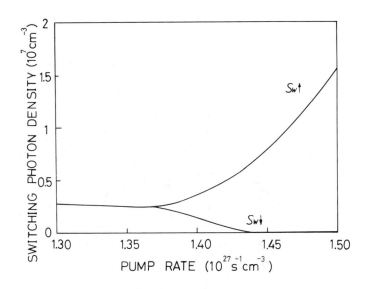

Figure 5.10 ON ($S_w\uparrow$) and OFF ($S_w\downarrow$) switching photon density versus bias pump rate.

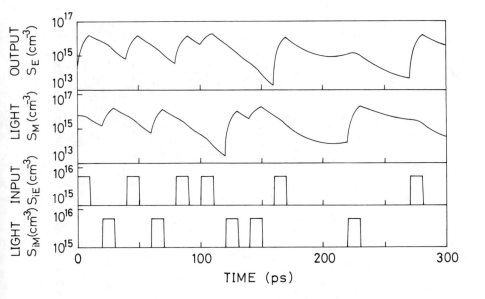

Figure 5.11 Flip-flop operation in pitchfork bifurcation polarization bistability.

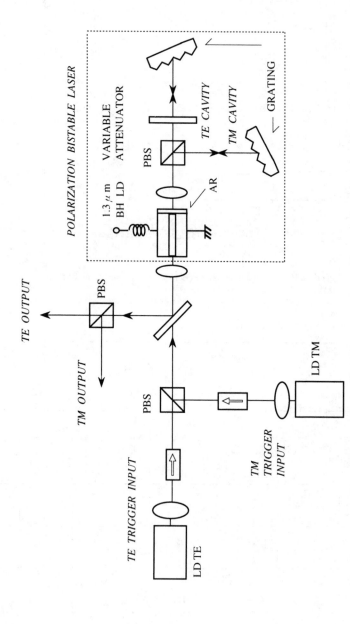

Figure 5.12 Experimental setup demonstrating polarization bistability. (From [14].)

tructed for the TE and TM modes by separating the beam with a polarization beam splitter (PBS). A 1.3-μm InGaAsP BH LD was used in the experiment. One facet was coated to achieve <1 % reflectivity for better coupling to external cavities. Gratings were used as external cavity mirrors, so that the oscillation wavelengths for TE and TM modes could be adjusted independently. The cavity lengths for the TE and TM modes were about 26 cm. The threshold current was reduced from 97 to 54 mA by the construction of the external cavity. The optical loss of the TE mode cavity was changed by the variable attenuator to increase the TE mode threshold current. By controlling the attenuation, the TE mode threshold became much the same as the TM mode laser threshold. TE and TM trigger optical inputs were injected into the BLD through optical isolators.

When one of the TE and TM cavities was blocked, light-output-versus-current (L-I) curves were approximately straight-line with small ripples caused by the interference between the LD chip cavity and the external cavity above the laser threshold. As shown in Figure 5.13, when both cavities were coupled simultaneously, the L-I curves were drastically changed. When transmittance of the variable attenuator T was set to 0.6, the TE mode threshold was lower than the TM mode threshold. Therefore, the TE mode oscillated. When T was decreased to 0.56, polarization-bistable switching occurred. Although this laser system did not include any additional enhanced cross-gain saturation effect, the L-I curve showed clear bistable characteristics. This relates to the interference between the LD chip cavity and the external cavity. The point at which polarization switching from TE to TM mode occurred coincided with the antiphase condition for the LD chip cavity and the external cavity of the TE mode. With further decrease in T, the laser oscillated with the TM mode in the high pump region. Very complex polarization bistability was seen in the low pump region.

One arm was blocked and then recoupled. The L-I curves for TE and TM modes are shown in Figure 5.14. The operation conditions are the same as that of T = 0.52 in Figure 5.13. In the low pump region (I = 72 to 95 mA), when the TM arm was blocked, the TE mode output power increased (the arrow in the upper direction at A). When the TM arm was recoupled, the TE mode output decreased (the arrow in the downward direction at A). Therefore, bistability was not observed. However, in the high pump region of I = 95 to 120 mA, the laser system showed a new type of polarization bistability, as theoretically predicted. The TE output changed to the upper branch when the TM arm was blocked. Then the TE output stayed on the upper branch even when the TM arm was recoupled. When the current was gradually increased, this continued until the current reached the next antiphase condition for the LD chip cavity and the external cavity (B in the figure). Moreover, when the TE output stayed on the upper branch and then the TE arm was cut off, TM oscillation occurred. Even when the TE arm was recoupled, TM oscillation continued. These results suggest that the LD system shown in Figure 5.12 shows pitchfork polarization bistability.

From the simulation of rate equations, it has been shown that when the optical input is injected into the BLD, which is biased in the region where branch B exists, the polarization of the laser changes to the same polarization as that of the incident light.

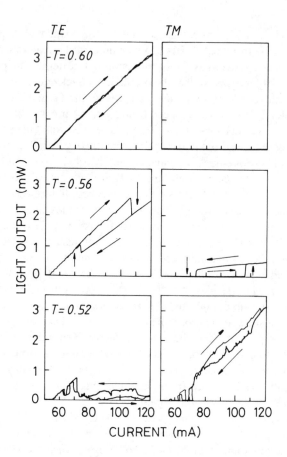

Figure 5.13 TE and TM light-output-versus-LD-current characteristics when both arms are coupled simultaneously. T is the transmittance of the variable attenuator in the TE arm. (From [14].)

Even when the incident light is cut off, the polarization of the BLD remains unchanged. This has been verified as shown in Figure 5.15. A TE light was injected into the BLD, which was oscillating with the TM mode (Figure 5.15). When the TE optical input was increased, the TE output was seen to abruptly increase and the TM oscillation completely stopped. The laser oscillation with TE polarization continued even when the TE optical input intensity was decreased to zero. The opposite switching (switching from TE to TM by TM input) has also been observed. To obtain the TE oscillation as the initial condition, the TM arm was temporally blocked in this case. The typical switching optical power was about 0.3 mW, which was measured just before the coupling lens. The lowest switching power measured was about 45 μW. If we assume the coupling loss to be 10 dB, the lowest switching power was as low as 4.5 μW.

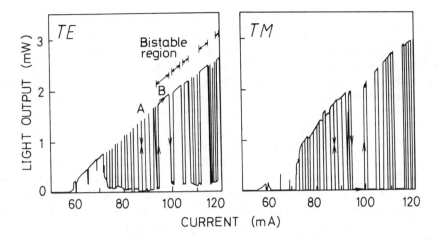

Figure 5.14 TE and TM light output versus LD current for $T = 0.52$. In the high pump region, the TE output changed to the upper branch by the cutoff of the TM arm, and the TE output stayed on the upper branch when the TM arm was recoupled. (From [14].)

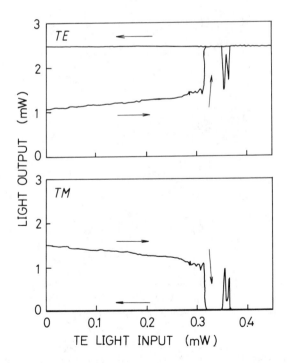

Figure 5.15 Switching from TM to TE by TE light input.

The wavelength characteristics of the optical switching power are a very important factor for the practical use of bistable devices. Figure 5.16 shows the optical switching power as functions of the TE oscillation wavelength (Figure 5.16(a)) and TM oscillation wavelength (Figure 5.16(b)) for switching from TM to TE. The TE and TM output wavelengths were changed by changing the angle of the gratings used as the external cavity. The threshold currents for TE and TM modes, when one cavity was coupled at a time, are also shown in the figure. In Figure 5.16(a), the mark ↑ expresses the TM oscillation wavelength and the mark ⇓ expresses the TE input wavelength with a FWHM of 3 nm. The LD was biased at 90 mA. The switching from TM to TE was obtained by the optical input of approximately 0.2 mW for the wavelength of 1.284 to 1.293 μm. For the wavelength of $\lambda_{TE} < 1.284$ μm and $\lambda_{TE} > 1.293$ μm, the switching was not obtained with a 1.6-mW TE input, because of the small difference between the TE and TM threshold current. As shown in Figure 5.16(b), when the bias current was set at 90 mA, the switching from TM to TE occurred in the wide-wavelength region of 1.283 to 1.298 μm by the injection of 0.2- to 0.3-mW TE input. For the wavelength of $\lambda_{TM} < 1.283$ μm and $\lambda_{TM} > 1.298$ μm, the LD oscillated with the TE mode without the TE light injection, because

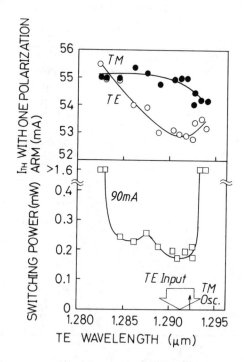

Figure 5.16(a) Switching power as a function of the oscillation wavelength for switching from TM to TE. Threshold currents for TE and TM mode are shown in the upper figure: TE oscillation wavelength dependence: ↑ is TM oscillation wavelength and ⇓ is TE input wavelength.

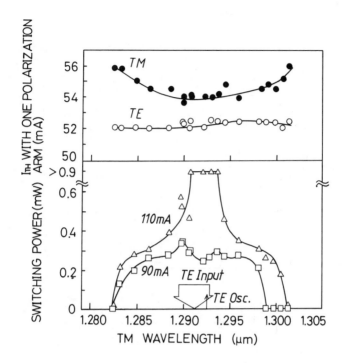

Figure 5.16(b) TM oscillation wavelength dependence: ↑ is TE oscillation wavelength and ⇓ is TE input wavelength.

the TM threshold was much higher than the TE threshold. For the condition of $I_b = 110$ mA, the switching was not obtained in the wavelength region of 1.291 to 1.293 μm. This is because the difference between the TE and TM thresholds was very small.

The optical switching power as functions of the TE oscillation wavelength (Figure 5.16(a)) and TM oscillation wavelength (Figure 5.16(b)) for switching from TE to TM were also examined. To obtain the TE oscillation as the initial condition, the TM arm was temporally cut off. The tendencies are very similar to that of Figure 5.16. As clearly observed in these figures, a precise wavelength matching is not needed for the polarization switching. The optical switching power, rather, depends on the difference in the threshold currents.

The switching between the two polarization states was accomplished by injecting the trigger optical inputs with the TE and TM modes as shown in Figure 5.17. The bias current of the BLD was set at about 125 mA. The TE and TM oscillation wavelengths were set at 1.2928 and 1.2925 μm, respectively. Single-wavelength operation DFB LDs with $\lambda = 1.2925$ μm and $\lambda = 1.2909$ μm were used as the sources for the TE and TM trigger inputs, respectively. When a TE trigger pulse was injected, the BLD changed its polarization from TM to TE. The laser oscillation with TE polarization continued even

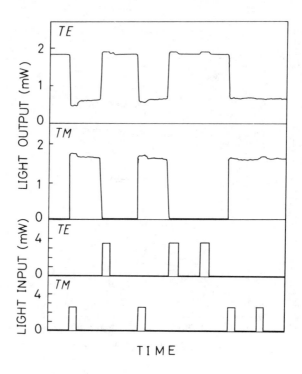

Figure 5.17 Time variability for all-optical flip-flop operation. (From [15].)

when the injection of the TE trigger pulse was stopped. And at the TE state, when a TM trigger pulse was injected, the polarization of the BLD changed from TE to TM. If the optical trigger input having the same polarization as the BLD is injected, the polarization state of the BLD is not changed. All-optical flip-flop operation has been successfully demonstrated as shown in the figure. The light input power was measured just before the coupling lens. If we assume that the coupling loss is 10 dB, the BLD provides a gain of about 7 dB.

TE-TM polarization bistability is found under the following two conditions as described above: (1) Probability of coexistence for both the TE and TM modes; in other words, the threshold currents and the differential gains for TE and TM modes are very close to each other. (2) Existence of strong mutual coupling between the modes through gain saturation. If we take into account the intraband relaxation processes that inherently exist in LDs, the second condition is satisfied. To satisfy the first condition, a narrow-stripe BH LD [19] or a strained MQW LD [22] can be used. Ultrafast flip-flop operation, which was theoretically predicted, may be achieved in such solitary LDs because of their short cavity length.

5.2.4 S-Shaped Bistability Caused by Stress

When the cross-saturation effect is enhanced and the modal gain for the TE mode is about the same as that for the TM mode, we can obtain S-shaped polarization bistability as described in Section 5.2.2.

Under normal stress-free conditions, the laser output is TE-polarized because the TE mode has a larger mode-confinement factor in the waveguide and higher reflectivity at the facets. However, the TM mode can be promoted to compete with the TE mode by introducing a small amount of lattice deformation, on the order of 10^{-4}, into the active layer that changes the band structure and thereby enhances the optical gain of the TM mode relative to that of the TE mode. In InGaAsP LDs, the lattice deformation can be created by lattice mismatch between the epitaxial layers. Polarization bistability is observed in some InGaAsP/InP LDs whose TM net gain is slightly larger than the TE net gain at the threshold.

Figure 5.18(a) shows the polarization-resolved-power-versus-current characteristics of a polarization-bistable InGaAsP/InP LD at various temperatures of interest [23]. The LDs have a V-grooved substrate BH and emit laser light at 1.3 μm. The measurement is done by recording the laser power transmitted through a polarizer. Special care is taken to eliminate any optical feedback that may complicate the system. At 195K and above, the laser operates in a pure TE mode and the power-versus-current characteristics are kink-free. Below this temperature, the laser output starts to show the TM emission above the threshold current. The TM stimulated emission becomes dominant at lower temperatures. At 194.2K, the laser operates in a pure TM mode at a low injection current and abruptly switches operation to a pure TE mode at a higher current. With a further decrease in the temperature, hysteresis can be observed in the polarization-resolved-power-versus-current curves. The width of the hysteresis loop varies from almost zero at 193K to 23 mA at 186.7K. Below 186.7K, the laser operates in a TM mode and switches into a mixture of TE and TM modes at a higher current and the hysteresis disappears.

The temperature-dependent polarization flip is attributed to the buildup of thermal stress created by the difference in the thermal expansion coefficients of the epitaxial layers. As the temperature decreases, the increase in the TM-mode gain relative to the TE-mode gain, estimated from the difference in thermal expansion coefficients of InP and InGaAsP, is on the order 0.1 cm^{-1} K. Figure 5.18(b) shows the net-gains-versus-current relation for the TE and TM modes measured at 295K and 191K. The net gains are calculated using the measured contrast ratios of the longitudinal mode spectra. The net gain is zero above the threshold. It can be seen that the temperature change causes a relative shift of the gain curves. (The slope for the TE mode is always larger than that for the TM mode because the confinement factor is larger for the TE mode.) From Figure 5.18(b), the rate of increase of the TM net gain relative to the TE net gain is 0.1 cm^{-1} K. The measured value varies from device to device, because the thermal stress at the active layer is also affected by external conditions.

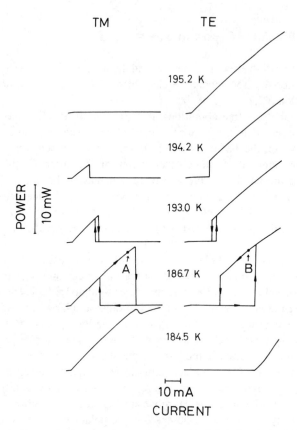

Figure 5.18(a) Polarization-resolved-power-versus-current characteristics of a polarization BLD at various temperatures. (Reprinted with permission from Chapman & Hall Ltd. From [23].)

At room temperature, polarization bistability is very rare in V-grooved substrate BH LDs, and various kinds of TE-TM mixed emission characteristics are observed, as shown in Figure 5.19 [20]. The behavior of emission characteristics can be described as follows (as in Fig. 5.19):

(a) Single-TE emission.
(b) Single-TM emission.
(c) TE-TM coexisting: the TE emission appears first as the current reaches a threshold. As the current is increased, both TE and TM modes increase monotonically in power.
(d) TM-TE coexisting: the characteristics are similar to that of type (c), but with the roles of the two polarizations exchanged.
(e) TE-TM switching: the TE mode reaches the lasing threshold first. It is totally suppressed eventually after the TM emission appears at a higher current. Switching

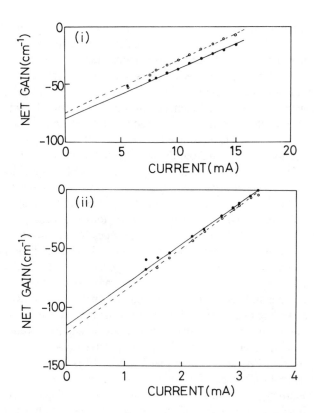

Figure 5.18(b) Net-gains-versus-current characteristics of a laser measured at (i) 295K and (ii) 191K. TM: solid lines/filled circles; TE: broken line/open circles. (Reprinted with permission from Chapman and Hall Ltd. From [23].)

from TE to TM may take place gradually over a wide current range or dramatically in a very narrow current range.

(f) TM-TE switching: the characteristics are similar to that of type (e), but with the roles of the two polarizations exchanged.

(g) TE-TM incomplete switching: the TE mode reaches the lasing threshold first. The TM emission appears and suppresses the TE emission in a manner similar to the TE-TM switching case, but the TE emission never disappears completely, even at a very high current. Eventually the emission characteristics behave similar to the coexisting characteristics.

(h) TM-TE incomplete switching: the characteristics are similar to that of type (g) but with the roles of the two polarizations exchanged.

(i) TM-TE bistability: the TM emission appears first like the TM-TE switching case, but a bistable region exists before the TE mode takes over completely at a high current.

Each type of polarization-dependent emission characteristic can be simulated by properly choosing the stress level, refractive indexes of the active and cladding layers, and polarization-dependent internal loss coefficients as free parameters. By looking into the polarization-dependent gain, gain nonlinearities, and losses related to these parameters, the criteria for various types of emission characteristics can be understood. The polarization-dependent linear gain and gain saturation coefficients as functions of carrier density are determined by the stress, while the refractive indexes of the active and cladding layers and polarization-dependent internal loss coefficients decide the total loss of the two polarizations. Though the confinement factors calculated from the refractive indexes of the active and cladding layers can also affect the emission of the two polarizations, they are not as significant as the gain and loss coefficients in determining the emission characteristics.

The polarization-dependent emission characteristics are essentially determined by the gain and loss differences between the two polarizations and the polarization-dependent gain nonlinearities. Based on these factors, various types of polarization-dependent emission characteristics can be analyzed, shown in Figure 5.19. This diagram was obtained by running simulations with numerous sets of stress levels, refractive indexes of the active and cladding layers, and polarization-dependent internal loss coefficients. For each set of these parameters, the simulated characteristics can be qualitatively classified and located on the diagram according to the value of the internal stress and the loss difference between the two polarizations. After the results of numerous simulations were mapped on the diagram, the boundaries dividing different types of polarization-dependent emission characteristics were drawn. One should note that Figure 5.19 is obtained for the V-grooved substrate BH InGaAsP/InP lasers. For lasers made of the other semiconductor material or with different structures and dimensions, each area on the diagram will be shifted and deformed. Since $\alpha_e < \alpha_m$ and $X \geq 0$ are the most common conditions for the InGaAsP lasers, most observed emission characteristics should be in the first quadrant of Figure 5.19. Here, α_e is the loss coefficient for the TE mode, α_m is that for the TM mode, and X is a value of the internal stress. This is in agreement with an experimental observation.

5.2.5 S-Shaped Bistability Caused by Saturable Absorber

In the QW structure, the selection rules for conduction to heavy-hole transitions yield a large anisotropy in the absorption cross sections for TE and TM polarized light near the band edge (see Section 2.2.1). A room-temperature polarization bistability based on this effect has been demonstrated in an LD with an intracavity MQW saturable absorber. The structure of the S-shaped polarization BLD proposed by Ozeki et al. is shown in Figure 5.4(d) [11]. It consists of a conventional DH gain section coupled to an MQW section, which acts as a polarization-sensitive saturable absorber. Each region has separate electrical contacts so that they may be pumped independently. For conduction to heavy-hole transitions, the TE mode is more heavily absorbed in the MQW absorber than the TM mode.

The mechanism by which optical bistability occurs in the polarization BLD can be qualitatively understood as follows, with the aid of Figure 5.20. In this figure, the sum

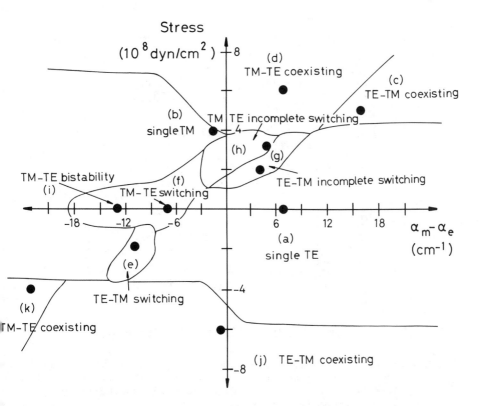

Figure 5.19 The loss difference of the two polarizations and stress for various types of polarization-dependent emission characteristics. The compressive stress has a positive value, whereas a negative value of stress indicates tensile stress. (See pages 166 and 167 for definitions.) (From [20]. Reproduced by permission of AIP.)

of the mirror, cavity, and saturable MQW absorber losses for the TE and TM modes are plotted as a function of intracavity intensity, along with the saturable gain of the DH section. Since the loss in the MQW section is saturated faster by TE than by TM, there are two loss curves for each polarization (labeled $X(Y)$ for mode X loss saturated by mode Y intensity). The gain curves are plotted for different pumping levels in the gain section. The steady-state solution is given by the intersection of the gain and loss curves. In the polarization BLD with the intracavity MQW saturable absorber, the net TM loss is reduced relative to the TE, and lasing is initially in TM. The locus of the solution starts on the TM (TM) curve, and moves to the right with increasing pump current J_g until it reaches the point where the TM (TM) loss is equal to the TE (TM) loss, at $J_g = J_{th}^{high}$. An infinitesimal increase in J_g causes the TE mode to grow at the expense of TM, and since the TE saturation is stronger, the MQW becomes further saturated, providing positive feedback for the growth of the TE mode and extinction of the TM mode. The locus moves

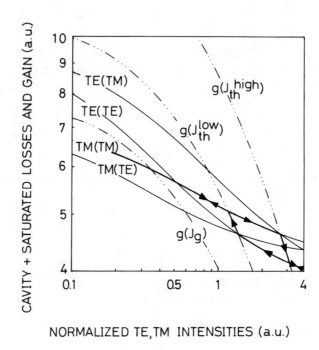

Figure 5.20(a) Polarization-dependent loss and polarization-independent gain as a function of intracavity intensity, showing the evolution of the two-mode bistability. The gain curves are for constant injection current $g(J_g)$. Loss curves are labeled $X(Y)$ for loss of mode X saturated by mode Y. (From [11]. Reproduced by permission of AIP.)

abruptly along the gain curve at $J_g = J_{th}^{high}$, until the TE (TE) loss curve is reached. Decreasing J_g causes the TE intensity to decrease until J_g reaches J_{th}^{low}, where the TE (TE) and TM (TE) losses are equal. An infinitesimal decrease in J_g causes the TM mode to grow at the expense of TE, reducing the saturation of the MQW section, further enhancing the growth of TM and extinguishing TE until only the TM mode is present. Thus, the two-mode bistability is seen to be the result of the complementary processes of gain competition in the DH section, which is essentially polarization-independent, and strong anisotropic absorption in the MQW section. The existence of polarization bistability in the structure was verified by using a numerical analysis of the two-mode rate equations [11].

Ozeki and Tang observed polarization switching and hysteresis in an LD with a two-armed polarization-sensitive external cavity [12]. An intracavity polarization-dependent saturable absorber (an AR-coated BH LD) was placed asymmetrically in one of the two arms of the external cavity so that one polarization mode is saturated more strongly than the other.

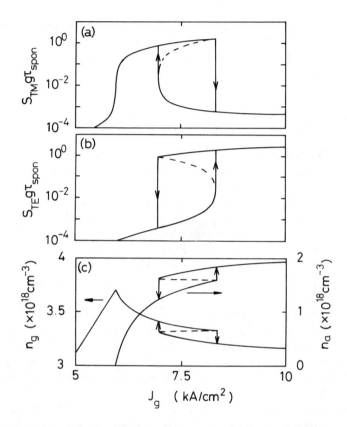

Figure 5.20(b) Polarization-dependent loss and polarization-independent gain as a function of intracavity intensity, showing the evolution of the two-mode bistability. The gain curves are for constant injection current $g(J_g)$. Loss curves are labeled $X(Y)$ for loss of mode X saturated by mode Y. (From [11]. Reproduced by permission of AIP.)

However, it is not very easy to achieve S-shaped polarization bistability in an LD monolithically integrating a polarization-dependent saturable absorber at this time. Tanaka and Shimada demonstrated TE-TM mode switching in GaAsP/AlGaAs tensile strained QW LDs with multiple electrodes [24]. For a 250-μm-long cavity, the threshold current density of LD wafers at room temperature is about 1.6 kA/cm². With a long cavity, these LDs operate in the TM mode, but the TE dominates when the cavity is shorter than 200 μm. TE-TM mode switching is obtained in a two-electrode laser with electrodes 150 and 30 μm long. When current is injected into both electrodes, the LD oscillates at 790 nm in the TM mode and with a threshold current of 40 mA. When current is injected only into the longer electrode, however, it oscillates at 800 nm in the TE mode, and absorption at the region under the shorter electrode increases the threshold current to 80 mA. For both kinds of oscillation, the suppression ratio is greater than 10 dB. This LD operates in the fundamental mode over an injection-current range of several milliwatts. Bistability has not been seen in this LD.

5.3 WAVELENGTH BISTABILITY

Many LDs exhibit hysteresis in the characteristics of lasing longitudinal (axial) mode versus temperature and lasing longitudinal mode versus injection current, especially at higher powers [25–27]. Mode competition phenomena have been analyzed using a density matrix formalism, and a strong gain suppression among the longitudinal modes has been shown to be an intrinsic property of LDs and has been considered the origin of wavelength bistability [3,28]. Sasaki et al. demonstrated the electro-optical switch, logic, and memory operations using a mode-hopping phenomenon of an LD [29]. Tang et al. observed the bistability in output versus injected optical power for injected light signals at wavelengths corresponding to slave laser resonances up to 5 nm higher or lower than the oscillation wavelength without light injection [4].

Bistable wavelength switching has been realized using a two-electrode DFB LD with strongly inhomogeneous excitation [30]. For this laser, it has been considered that the change in the distribution of gain and refractive index due to the axial spatial hole burning causes the wavelength switching, and the nonlinear effect due to gain saturation is responsible for the bistable characteristics. Jinno et al. also reported similar results for a DFB LD divided into three sections [31]. A fast wavelength switching operation of less than 200 ps between two modes separated by 1.7 nm was observed. A high-speed repetitive wavelength switching operation up to 1 GHz was also achieved [32]. Kuznetsov has observed the bistable-wavelength-latch operation of a two-segment DFB LD. Mode power switching in less than 200 ps with 450-ps delay between set and reset pulses has been demonstrated using short electrical trigger pulses [33]. Gray and Roy have studied the two-mode bistable operation, and investigated the dwell times of the two modes in relation to the average power ratio between the two modes, taking into account Langevin noise sources [34]. Ozeki and Tang [35] have proposed and analyzed a new type of two-mode BLD with an asymmetric cavity configuration, which consists of a gain region, a saturable absorber region, and a set of mode-selective mirrors (or filters). These elements are arranged such that the saturable absorber acts as an intracavity loss for one mode and a bleachable absorber external to the cavity for the other mode. Both modes share the same gain medium and compete for the same gain. Bleaching of the absorber leads to a switching of the modes and possible bistability.

The laser-power change associated with longitudinal mode jumping in AlGaAs CSP lasers was examined by Ogasawara and Ito [27]. Figure 5.21 shows the results obtained for one of the CSP lasers at three different injection levels. The nominal reflectivity R of

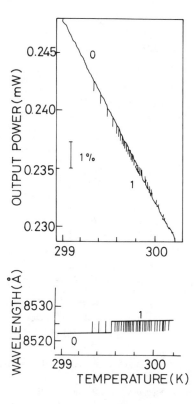

Figure 5.21(a) Output-power-versus-temperature characteristics in the random mode-switching regime. Lasing-mode-versus-temperature characteristics are illustrated in the lower parts. The LD used was a CSP laser with the nominal facet reflectivities of 0.9. (From [27]. Reproduced by permission of JJAP.)

the end facet was 0.9 and the threshold current was 40 mA. Figure 5.21(a) shows the temperature dependence of the output power P at a fixed injection current of 44 mA. Note that the external efficiency is low because of the high facet reflectivity; the photon density inside the cavity for $P = 0.24$ mW in this sample is estimated to be comparable to that for $P \sim 2.8$ mW in samples with $R = 0.3$. The heat-sink temperature was varied at a rate of about 1°/min. As the temperature was raised, the output power gradually decreased because of the increase in the threshold current. In addition, the output power was observed to switch between two distinct levels over the temperature region of about 1°. A simultaneous observation of the dispersed light power revealed that the two power levels corresponded to oscillation in the different longitudinal modes; the power was higher when the laser oscillated in a shorter-wavelength mode (0-mode) and the power was lower when the laser oscillated in a longer wavelength mode (1-mode), as schematically illustrated in the lower part of Figure 5.21(a).

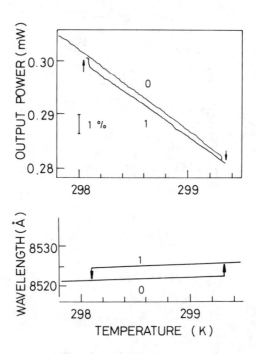

Figure 5.21(b) Output-power-versus-temperature characteristics in the hysteresis regime (upper parts). Lasing-mode-versus-temperature characteristics are illustrated in the lower parts. The LD used was a CSP laser with the nominal facet reflectivities of 0.9. (From [27]. Reproduced by permission of JJAP.)

In Figure 5.21(a), it can also be seen that the oscillation in the 0-mode is predominant in the lower temperature region. With an increase in the temperature, the probability of the 1-mode oscillation gradually increases. As a result, the 1-mode oscillation is predominant in the higher temperature region. In the middle-temperature region, the probabilities of the oscillation of the two modes are nearly equal and random mode switching takes place most frequently.

At higher injection levels, a clear hysteresis was observed in the oscillation-mode-versus-temperature characteristics, as illustrated in Figure 5.21(b,c). Figure 5.21(b) shows the dependence of the output power P (upper part) and lasing wavelength (lower part) on the heat-sink temperature at an injection current of 44.5 mA. A clear hysteresis can be seen between the 0-mode and 1-mode. Note that the output power is decreased upon mode jumping from the 0-mode to the 1-mode, while the output power jumps up upon mode jumping from the 1-mode to the 0-mode. A further increase in the injection current gave rise to hysteresis loops that involved more than two longitudinal modes.

Hysteresis in the lasing mode has been explained in terms of *strong coupling* among longitudinal modes by Yamada and Suematsu [36] and by Kazarinov et al. [37]; there the gain of the nonlasing mode was shown to be suppressed on both shorter and longer wavelength sides of the lasing mode. This type of mode coupling will hereafter be called *symmetric mode coupling*, since the nonlinear gain that accounts for the gain suppression was shown to be symmetric with respect to the lasing mode to a good approximation. Although the presence of hysteresis can be explained by this model, the power reduction upon mode jumping towards a longer wavelength mode cannot be understood by this model, as described in the following.

Figure 5.22 schematically illustrates how the gain spectra vary in a mode-versus-temperature hysteresis loop according to the symmetric mode-coupling model [27]. A simple hysteresis loop formed by two adjacent modes, s (the shorter wavelength mode) and l (the longer wavelength mode), is treated.

In the lower part of the figure, the gain spectra at the lower end temperature T_1, the middle temperature T_2, and the higher end temperature T_3 of the hysteresis loop are illustrated. The lasing mode in Figure 5.22(a–c) is the s-mode, while that in 5.22(a'–c') is the l-mode. The solid lines represent the spectra of linear gain $g(I)$ for the injection current I fixed above the threshold. The gain of the lasing mode is saturated at the threshold gain g_{th}. The quantity $g(I)$-g_{th} is proportional to the ability of each mode in emanating laser power. The broken lines represent the dispersion of saturated gain $\tilde{g}$, while the dotted lines represent that of linear gain at threshold $g(I_{th})$. If the gain spectrum were *homogeneously broadened*, $\tilde{g}$ would coincide with $g(I_{th})$. Here, $\tilde{g}$ is assumed to be appreciably lower than $g(I_{th})$ on both sides of the lasing mode owing to the symmetric gain suppression.

In the upper part of the figure, $g(I)$, $g(I_{th})$, and $\tilde{g}$ for the s- and l-mode are shown as functions of temperature. As temperature rises, the linear gain of the s-mode $g_s(I)$ decreases while that of the l-mode $g_l(I)$ increases, because the linear gain peak is shifted towards longer wavelengths owing to the decrease in bandgap energy as shown in Figure 5.22(a–c). The $s \rightarrow l$ mode jumping occurs at T_3, where the saturated gain of the l-mode g_l reaches g_{th}. Likewise, the $l \rightarrow s$ mode jumping takes place at T_1, where the saturated gain of the s-mode g_s reaches g_{th}.

In the symmetric mode-coupling model, illustrated in Figure 5.22, the output power should increase upon both $s \rightarrow l$ and $l \rightarrow s$ mode jumping, since the mode jumping occurs towards a mode with higher linear gain. The magnitudes of the power increase at $s \rightarrow l$ and $l \rightarrow s$ mode jumping are equal because of the symmetry in gain suppression. Therefore, this model is not consistent with the power reduction upon mode jumping towards longer wavelengths shown in Figure 5.21(b,c).

Instead, Ogasawara and Ito proposed an asymmetric mode-coupling model as illustrated in Figure 5.23. In this model, $\tilde{g}$ is assumed to be slightly higher than $g(I_{th})$ on the longer wavelength side of the lasing mode and considerably lower than $g(I_{th})$ on the shorter wavelength side, which is in accordance with the experimental observation on the gain saturation. Then g_l reaches g_{th} at T_3 even though $g_l(I)$ is less than $g_s(I)$, as shown in Figure 5.23(c). Therefore, output power decreases at the $s \rightarrow l$ mode jumping. This is consistent with the experimental observation of the power reduction at the mode jumpings towards

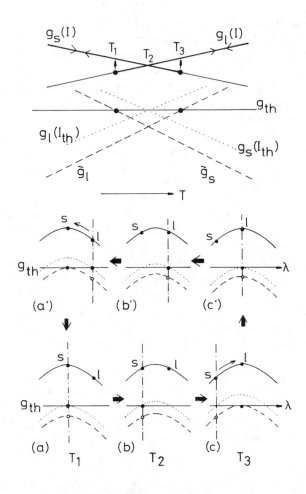

Figure 5.22 Schematic diagram of longitudinal-mode competition in the symmetric mode-coupling model. Gain spectra at (a,a') the lower end temperature T_1, (b,b') the middle temperature T_2, and (c,c') the higher end temperature T_3 of a hysteresis loop are shown in the lower part. Solid lines: linear gain spectra $g(I)$; broken lines: saturated gain spectra $\tilde{g}$; dotted lines: linear gain spectra at threshold $g(I_{th})$. Temperature dependence of $g(I)$, $\tilde{g}$, and $g(I_{th})$ for the shorter wavelength mode (s-mode) and the longer wavelength mode (l-mode) are shown in the upper part. (From [27]. Reproduced by permission of JJAP.)

longer wavelengths shown in Figure 5.21(b,c). When the $l \rightarrow s$ mode jumping is induced at T_1, on the other hand, the linear gain increases, as shown in Figure 5.23. (a'). This is also consistent with the experimental observation regarding the power increase upon the mode jumping towards shorter wavelengths, shown in Figure 5.21(b,c).

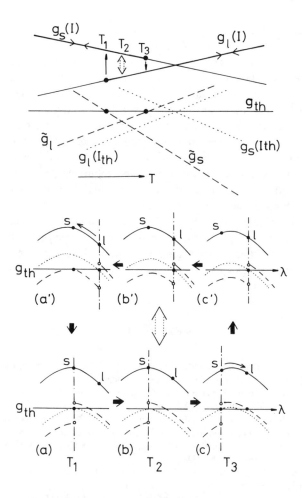

Figure 5.23 Schematic diagram of longitudinal-mode competition in the asymmetric mode-coupling model. Gain spectra at (a,a′) the lower end temperature T_1, (b,b′), the middle temperature T_2, and (c,c′) the higher end temperature T_3 of a hysteresis loop are shown in the lower part. Solid lines: linear gain spectra $g(I)$; broken lines: saturated gain spectra $\tilde{g}$; dotted lines: linear gain spectra at threshold $g(I_{th})$. Temperature dependence of $g(I)$, $\tilde{g}$, and $g(I_{th})$ for the shorter wavelength mode (s-mode) and the longer wavelength mode (l-mode) are shown in the upper part. (From [27]. Reproduced by permission of JJAP.)

It should be noted here that the mode transition in this asymmetric mode-coupling model is hysteretic; the $l \rightarrow s$ mode-jumping temperature (T_1) is lower than the $s \rightarrow l$ mode-jumping temperature (T_3), since the magnitude of the gain suppression on the shorter wavelength side is larger than the magnitude of the gain enhancement on the longer wavelength side. Therefore, the asymmetric mode coupling, described here, should be included within the category of strong coupling.

Random mode-switching phenomena can also be understood intuitively as follows. At lower laser power, where the hysteresis loop is small, g_s and g_l in the hysteresis loop are very close to g_{th}. Then mode jumping can readily be induced by intrinsic quantum noise, as depicted by the dotted arrow in Figure 5.23. At the temperatures where $g_s \approx g_l$, the probability of the oscillation of each mode is nearly equal, and mode switching takes place most frequently. In the lower temperature region, where $g_s > g_l$, the relative probability of the s-mode oscillation is greater than that of the l-mode oscillation. The l-mode oscillation becomes predominant in the higher temperature region, where $g_l > g_s$. However, the output power for the s-mode P_s is consistently higher than that for the l-mode P_l, since mode switching is induced in the temperature region, where $g_s(I) > g_l(I)$. Thus, random mode-switching phenomena involving the output-power change illustrated in Figure 5.21(a) is reasonably explained in terms of this asymmetric mode-coupling model.

In this way, the laser-power change upon mode jumping in both random mode-switching and hysteresis regimes can be reasonably explained in terms of the asymmetric feature in gain saturation.

REFERENCES

[1] Lamb, W. E., "Theory of an Optical Maser," *Phys. Rev.*, Vol. 134, No. 6A, June 1964, pp. A1429–A1450.

[2] Shimoda, K., *Introduction to Laser Physics*, Berlin: Springer-Verlag, 1984.

[3] Yamada, M., "Transverse and Longitudinal Mode Control in Semiconductor Injection Lasers," *IEEE J. Quantum Electronics*, Vol. QE-19, No. 9, September 1983, pp. 1365–1380.

[4] Tang, C. L., A. Schremer, and T. Fujita, "Bistability in Two-Mode Semiconductor Lasers via Gain Saturation," *Appl. Phys. Lett.*, Vol. 51, No. 18, November 1987, pp. 1392–1394.

[5] Kawaguchi, H., "Bistable Laser Diodes and Their Applications for Photonic Switching," *Int. J. Optoelectronics*, Vol. 7, No. 3, 1992, pp. 301–348.

[6] Chen, Y. C., and J. M. Liu, "Polarization Bistability in Semiconductor Lasers," *Appl. Phys. Lett.*, Vol. 46, No. 1, January 1985, pp. 16–18.

[7] Fujita, T., A. Schremer, and C. Tang, "Polarization Bistability in External Cavity Semiconductor Lasers," *Appl. Phys. Lett.*, Vol. 51, No. 6, August 1987, pp. 392–394.

[8] Linton, R. S., I. H. White, J. E. Carroll, J. Singh, M. J. Adams, and I. D. Henning, "Optically Triggered Polarisation Bistability in GaInAsP Twin Stripe Injection Lasers Using Integrated Devices," *Electron. Lett.*, Vol. 24, No. 19, September 1988, pp. 1232–1234.

[9] Rheinländer, B., A. Klehr, O. Ziemann, V. Gottschalch, and G. Oelgart, "Room Temperature Polarization Bistability in 1.3 μm InGaAsP/InP Ridge Waveguide Lasers," *Opt. Comm.*, Vol. 8, No. 3,4, January 1991, pp. 259–261.

[10] Klehr, A., A. Bärwolff, R. Müller, M. Voß, J. Sacher, W. Elsasser, and E. O. Gobel, "Ultrafast Polarisation Switching in Ridge-Waveguide Laser Diodes," *Electron. Lett.*, Vol. 27, No. 18, August 1991, pp. 1680–1682.

[11] Ozeki, Y., J. E. Johnson, and C. L. Tang, "Polarization Bistability in Semiconductor Lasers With Intracavity Multiple Quantum Well Saturable Absorbers," *Appl. Phys. Lett.*, Vol. 58, No. 18, May 1991, pp. 1958–1960.

[12] Ozeki, Y., and C. L. Tang, "Polarization Switching and Bistability in an External Cavity Laser With a Polarization-Sensitive Saturable Absorber," *Appl. Phys. Lett.*, Vol. 58, No. 20, May 1991, pp. 2214–2216.

[13] Kawaguchi, H., I. H. White, M. J. Offside, and J. E. Carroll, "Ultrafast Switching in Polarization-Bistable Laser Diodes," *Opt. Lett.*, Vol. 17, No. 2, January 1992, pp. 130–132.

[14] Kawaguchi, H., and T. Irie, "Optically Triggered Polarisation Bistable Switching in Laser Diodes With External Cavities," *Electron. Lett.*, Vol. 28, No. 17, August 1992, pp. 1645–1647.

[15] Kawaguchi, H., "Polarization Bistable Laser Diodes," *Int. J. Nonlinear Optical Physics*, Vol. 2, No. 3, 1993, pp. 367–389.

[16] Chen, Y. C., and J. M. Liu, "Polarization Bistability in Semiconductor Laser: Rate Equation Analysis," *Appl. Phys. Lett.*, Vol. 50, No. 20, May 1987 pp. 1406–1408.

[17] Mukai, T., Y. Yamamoto, and T. Kimura,"Optical Amplification by Semiconductor Lasers," in *Semiconductors and Semimetals*, R. K. Willardson and A. C. Beer, eds. New York: Academic, 1985, p. 291.

[18] Saitoh, T., and T. Mukai, "Structural Design for Polarization-Insensitive Travelling-Wave Semiconductor Laser Amplifiers," *Optical and Quantum Electronics*, Vol. 21, 1989, pp. S47–S58.

[19] Yamada, K., T. Ushikubo, M. Joma, H. Horikawa, T. Kunii, Y. Ogawa, M. Kawahara, and T. Nonaka, "Fabrication of Narrow Stripe Laser Diodes," *Extended Abstracts, 38th Spring Meeting, The Japan Society of Applied Physics and Related Societies*, No. 3, p. 966 (in Japanese).

[20] Yu, B. M., and J. M. Liu, "Polarization-Dependent Gain, Gain Nonlinearities, and Emission Characteristics of Internally Strained InGaAsP/InP Semiconductor Lasers," *J. Appl. Phys.*, Vol. 69, No. 11, June 1991, pp. 7444–7459.

[21] Henneberger, K., F. Herzel, S. W. Koch, R. Binder, A. E. Paul, and D. Scott, "Spectral Hole Burning and Gain Saturation in Short-Cavity Semiconductor Lasers," *Phys. Rev. A*, Vol. 45, No. 3, February 1992, pp. 1853–1859.

[22] Magari, K., M. Okamoto, H. Yasaka, K. Sato, Y. Noguchi, and O. Mikami, "Polarization Insensitive Travelling Wave Type Amplifier Using Strained Multiple Quantum Well Structure," *IEEE Photonics Tech. Lett.*, Vol. 2, No. 8, August 1990, pp. 556–558.

[23] Chen, Y. C., and J. M. Liu, "Switching Mechanism in Polarization-Bistable Semiconductor Lasers," *Optical and Quantum Electronics*, Vol. 19, 1987, pp. S93–S102.

[24] Tanaka, H., and J. Shimada, "TE/TM Mode Switching of GaAsP Strained Quantum-Well Laser Diode," *OE/LASER'93, Laser Diode Technology & Application V*, 1993, 1850-17.

[25] Nakamura, M., K. Aiki, N. Chinone, R. Ito, and J. Umeda, "Longitudinal-Mode Behaviors of Mode-Stabilized $Al_xGa_{1-x}As$ Injection Lasers," *J. Appl. Phys.*, Vol. 49, No. 9, September 1987, pp. 4644–4648.

[26] Ogasawara, N., and R. Ito, "Output Power Change Associated With Longitudinal Mode Jumping in Semiconductor Injection Lasers," *Japanese J. Appl. Phys.*, Vol. 25, No. 7, July 1986, pp. L617–L619.

[27] Ogasawara, N., and R. Ito, "Longitudinal Mode Competition and Asymmetric Gain Saturation in Semiconductor Injection Lasers. I. Experiment," *Japanese J. Appl. Phys.*, Vol. 27, No. 4, April 1988, pp. 607–614.

[28] Ogasawara, N., and R. Ito, "Longitudinal Mode Competition and Asymmetric Gain Saturation in Semiconductor Injection Lasers. II. Theory," *Japanese J. Appl. Phys.*, Vol. 27, No. 4, April 1988, pp. 615–626.

[29] Sasaki, W., H. Nakayama, M. Mitsuda, and T. Ohta, "Fast Optical Wavelength Bistability Under the Mode-Hopping Phenomenon in Semiconductor Lasers," *SPIE*, Vol. 1280, 1990, pp. 245–256.

[30] Shoji, H., Y. Arakawa, and Y. Fujii, "New Bistable Wavelength Switching Device Using a Two-Electrode Distributed Feedback Laser," *Electron. Lett.*, Vol. 24, No. 14, July 1988, pp. 888–889.

[31] Jinno, M., M. Koga, and T. Matsumoto, "Optical Tristability Including Spectral Bistability Using an Inhomogeneously Excited Multielectrode DFB LD," *Electron. Lett.*, Vol. 24, No. 16, August 1998, pp. 1030–1031.

[32] Shoji, H., Y. Arakawa, and Y. Fujii, "Fast Bistable Wavelength Switching Characteristics in Two-Electrode Distributed Feedback Laser," *IEEE Photonics Tech. Lett.*, February 1990, Vol. 2, No. 2, pp. 109–110.

[33] Kuznetsov, M., "Picosecond Switching Dynamics of a Bistable-Wavelength-Latch Two-Segment Distributed Feedback Laser," *IEEE Photonics Tech. Lett.*, Vol. 2, No. 9, September 1990, pp. 623–625.

[34] Gray, G., and R. Roy, "Bistability and Mode Hopping in a Semiconductor Laser," *J. Opt. Soc. Am. B*, Vol. 8, No. 3, March 1991, pp. 632–638.

[35] Ozeki, Y., and C. L. Tang, "New Two-Mode Bistability in an Asymmetric Absorptive Multisection Laser Diode: Theoretical Analysis," *IEEE J. Quantum Electronics*, Vol. 27, No. 5, May 1991, pp. 1160–1170.

[36] Yamada, M., and Y. Suematsu, "Analysis of Gain Suppression in Undoped Injection Lasers," *J. Appl. Phys.*, Vol. 52, No. 4, April 1981, pp. 2653–2664.

[37] Kazarinov, R. F., C. H. Henry, and R. A. Logan, "Longitudinal Mode Self-Stabilization in Semiconductor Lasers," *J. Appl. Phys.*, Vol. 53, No. 7, July 1982, pp. 4631–4644.

Chapter 6
Waveguiding Bistability

wo-mode intensity bistability between spatially distinct laser outputs may be preferred r some applications because of its compatibility with intensity-modulated optical signals. this chapter, we look at two kinds of BLDs for this purpose: twin-stripe laser and ross-coupled laser.

.1 GENERAL CONSIDERATION

he use of twin-stripe laser structures can give rise to a number of interesting effects ssociated with spatial instability [1–3]. Observed phenomena for these devices include eam steering [4] and bistability [5,6]. The bistable switching is believed to be associated ith a change in the dominant waveguiding mechanism from self-focusing to gain guiding s the mode is observed to switch from a central self-focused mode to a gain-guided one, rossing from one side of the laser at one facet to the other side at the opposite facet igure 6.1(a)) [8]. Modal bistability in asymmetric twin ridge LDs has been investigated y Davies et al. (Figure 6.1(b)) [9]. A bistable switch has been observed when the switching ccurs between the first and zero-order modes. The switching speed was limited to 5 ns, ecause of the charge carrier redistribution that inevitably accompanies the modes crossing e structure. McInerney et al. have analyzed the case of two laterally coupled active uides and obtained the result that an optical signal can be injected into one of the guides nd switched into the other by nonlinear guiding [10]. They have claimed that diffusion-mited switching times of about 10 ps should be possible [10]. An experimental demonstra-on of such ultrafast switching is awaited.

Bistability has also been observed between a cross-coupled optical mode and a traight-pass optical mode in four-contact twin-stripe injection lasers (Figure 6.1(c)) [11]. Vatanabe et al. have analyzed cross-coupled resonant lateral modes in a twin-stripe four-ontact laser in which the current injection into each diagonal pair of stripes is equal

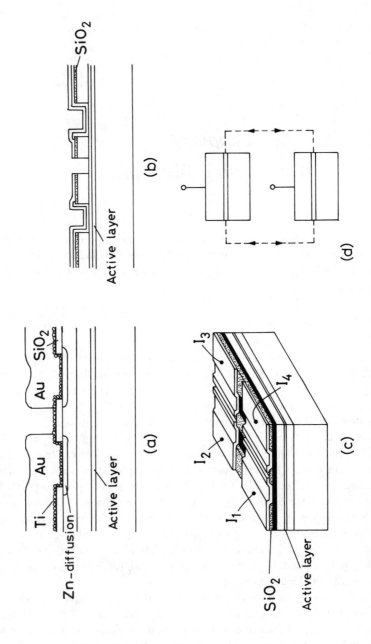

Figure 6.1 Structures of waveguiding BLDs: (a) closely coupled twin-stripe LD; (b) asymmetric twin ridge LD; (c) four-contact twin-stripe LD; (d) self-focused coupled cavity LD. (From [7].)

12,13]. When the carrier density difference between the adjacent waveguides is small nd the laser is shorter than the coupling length, the highest optical power appears in the waveguide with the lowest carrier density. This shows that a cross-coupled mode can be supported even in a two-contact twin-stripe laser, and bistable switching between the mode and its mirror image is possible.

Optical bistability has also been demonstrated using the self-focused coupled cavity aser, consisting of two wide-stripe GaAs/GaAlAs diodes strongly coupled in an external ing resonator (Figure 6.1(d)) [14].

.2 TWIN-STRIPE LASER

The structure of a twin-stripe laser is illustrated in Figure 6.2, which shows a laser with win BH waveguides made of an AlGaAs material [13]. The dielectric constants of the active regions, the lateral cladding, and the transverse cladding are denoted by ϵ_a, ϵ_l, and $_t$, respectively. The two active regions with cross sections of 0.1×2.0 μm^2 are separated by 1 μm. The parameters have been selected so that the coupled-mode approximation 15] is valid and the coupling length of the twin waveguide, L_c (510 μm long for the tructure shown in Figure 6.2), is comparable to the cavity length of practical lasers. Each waveguide (length L) is segmented at the center between the facets so that four active waveguide regions WG_1 to WG_4 are formed as labeled in Figure 6.2(a). This allows investigation of cross coupling. The relationship between the output light distribution and the injected current distribution is determined.

First, it is supposed that the carrier densities N_j ($j = 1$ to 4) in each diagonal pair of active regions are fixed to an identical value (i.e., so that $N_4 = N_1$ and $N_3 = N_2$) and the density difference between the adjacent waveguides is represented by Δ_c as shown in Figure 6.2, where N_0 is the average carrier density. Based on the effective index approximation [16], a three-layer slab equation is solved for the fundamental mode along the transverse (y) direction, along the lateral (x) direction for each waveguide, and, finally, the fields of the modes in each waveguide are coupled. For this calculation, the change of the complex dielectric constant with the carrier density N in each active region is calculated as $(A_r + iA_i)N$, where the values of A_r and A_i are as follows. $A_r + iA_i = (-5.4 + 1.5i) \times 10^{-20}$ cm^3. Let us define $\mathbf{E}_j(x)$ as the normalized electric field of the uncoupled fundamental mode of each region WG_j ($j = 1$ to 4) such that $\int_{-\infty}^{\infty} |\mathbf{E}_j(x)|^2 dx = 1$. Also, a column vector $\mathbf{a} = t(a_1, a_2)$ is defined, which expresses a field by $a_1\mathbf{E}_1(x) + a_2\mathbf{E}_2(x)$ and $a_1\mathbf{E}_3(x) + a_2\mathbf{E}_4(x)$ in the lower and the upper half of the laser cavity, respectively. Due to symmetry, $\mathbf{E}_3(x) = \mathbf{E}_2(x + d)$ and $\mathbf{E}_4(x) = \mathbf{E}_1(x - d)$, where d is the distance between the centers of the adjacent waveguides. By using this formalism, the wave propagation is expressed by the change of the vector $\mathbf{a}$ along the z-direction, and a field pattern that does not change after a round-trip is obtained by solution of the eigenvalue equation:

$$M'M\mathbf{a} = G^2\mathbf{a} \qquad (6.1)$$

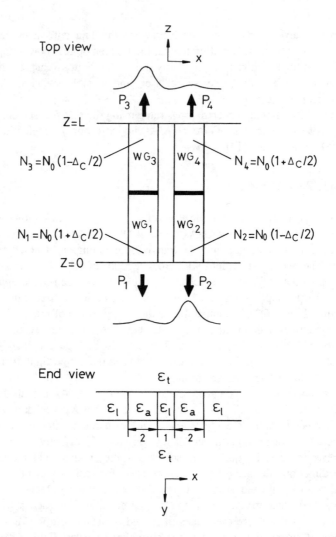

Figure 6.2 Schematic structure of a twin-stripe laser with diagonal carrier distribution. $N_i(i = 1$ to 4) is the carrier density for each waveguide region and N_0 is the average of them. The dielectric constant without carriers are $\epsilon_a = 13.1 - 0.023i$, $\epsilon_i = 12.9$, and $\epsilon_t = 11.4$. (©1992 IEEE. From [13].)

where the matrices M and M' [17] represent the coupled wave propagation in the forward and backward direction, respectively. The eigenvalue G^2 in (6.1) is generally a complex value, which means that the phase of the eigenmode field deviates from $2m\pi$ (where m is an integer) after a round-trip, although its power distribution does not change. However this phase offset is of little consequence, since it corresponds to a slight wavelength offset from that which would be generated by the laser. Each optical output power P_i emitted from WG$_j$ ($j = 1$ to 4) is therefore calculated from the field distribution of the uncoupled modes and the solution of amplitude values from (6.1).

In the next stage of the model, the current distribution required to maintain the carrier and field distribution is calculated using the electron rate equation assuming that the carriers are well confined and uniformly distributed throughout each active region. This approximation reduces the rate equation to

$$-BN_j^2 + \frac{J_j}{ew_t} - \frac{2\pi\Gamma_y(\epsilon_{ai} + A_iN_j)}{n_ghcw_tw_l} \int_{\mathrm{WG}_j} \left| \sum_{n=1+k}^{2+k} a_nE_n(x) \right|^2 dx = 0$$

$$(j = 1 \text{ to } 4) \quad (6.2)$$

where $k = 0$ and 2 for the lower and the upper half of the cavity, respectively, B is the bimolecular recombination constant, J_j is the current density in each region, WG$_j$, e is the electronic change, w_t is the transverse thickness of the active layer, Γ_y is the transverse confinement factor, ϵ_{ai} is the imaginary part of ϵ_a, $\epsilon_{ai} + A_iN_j$ is the carrier-dependent imaginary part of the dielectric constant in the active region, n_g is the group index in the material, h is Planck's constant, c is the light velocity in the vacuum, w_l is the lateral width of each active waveguide, and $\sum_{1+k}^{2+k} a_nE_n(x)$ is the electric field of the supported lateral field, where $E_n(x)$ is the uncoupled-mode field of WG$_n$ ($n = 1$ to 4). The injection current Cu$_j$ ($j = 1, 4$) required in WG$_j$ is therefore calculated by integrating the current density over each region.

Using the simulated model of the twin-stripe laser, two solutions are normally found that maintain the shape of the field after the propagation of one round-trip. These are evenlike and oddlike resonant lateral modes whose phase difference between the waveguides are near 0° and 180°, respectively. Of these modes, the one with higher gain is observed as it becomes much larger than the other after several round-trips. With a diagonal carrier distribution as shown in Figure 6.2, each modal field always has a cross-coupled distribution whose field pattern at a facet is the mirror image of that at the other facet ($P_4 = P_1$ and $P_3 = P_2$). Since both the carrier and the light distributions are diagonal, the current distribution calculated by (6.2) is also diagonal (Cu$_4$ = Cu$_1$ and Cu$_3$ = Cu$_2$).

A laser with $L < L_c$ (the coupling length) is much more interesting than one with $L > L_c$ because only the former gives bistability between cross-coupled modes (even with uniform current injection). Thus, a typical case with $L = 0.95L_c$ is investigated here. Figure 6.3 shows the near-field output light patterns at Cu$_1$/Cu$_2$ = 1 (uniform current injection) for $P_0 < 0.5$ mW and $P_0 = 0.55$, 0.88, and 2.2 mW, where Figure 6.3(b–d) are one of the bistable states. The patterns of the other bistable state are the mirror images of them. The output patterns (solid curve) for $P_0 = 0.88$ and 2.2 mW are composed of both the oddlike (broken curve) and the evenlike (dotted curve) resonant modes. Since the wavelength of these modes are generally different from each other, they may be observed independently from each other by resolving spectrally.

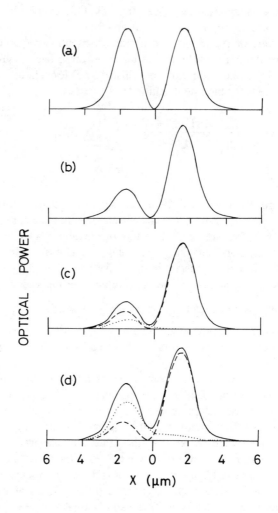

Figure 6.3 Near-field output patterns with $Cu_1/Cu_2 = 1$ and (a) $P_0 < 0.5$ mW, (b) $P_0 = 0.55$ mW, (c) $P_0 = 0.8$ mW, and (d) $P_0 = 2.2$ mW are shown in solid curves, where (b–d) are of one of the bistable states. The broken and the dotted curves in (c,d) are the oddlike and evenlike modes that compose the whole pattern. (©1992 IEEE. From [13].)

Two cross-coupled lateral modes in gain-guided twin-stripe LDs and switching between these modes by changing the injection current have been experimentally observed [18]. Near- and far-field patterns at both facets show a skewed light field coupling from under one stripe to under the other.

6.3 CROSS-COUPLED LASER

The cross-coupled bistable laser diode (XCBLD), in which lasing can be toggled between two orthogonal outputs, has been developed [19], which is based on a hybrid device proposed by Lasher and Fowler [20]. Figure 6.4 shows a diagram of the XCBLD. It uses the folded-cavity geometry, which has a much lower threshold current density and higher coupling efficiency into single-mode waveguides than an equivalent square laser with planar mirrors. The XCBLD has a common gain region, where the two modes compete for stimulated emission, and separate saturable absorbers, which cause latching in the individual modes. The XCBLD is operated with the gain region forward-biased and the saturable absorbers biased below transparency. At high enough injection current in the gain region, noise or a small asymmetry in the device will cause one mode to start lasing, saturating its saturable absorber. The gain is then clamped at a level high enough to sustain lasing in this direction, but not in the cross mode with its unsaturated saturable absorber. By momentarily applying a short electrical pulse to this saturable absorber or injecting light into the cross mode, the loss in the cross mode is reduced, allowing it to lase at a lower gain region carrier density, which turns off the other mode. When the

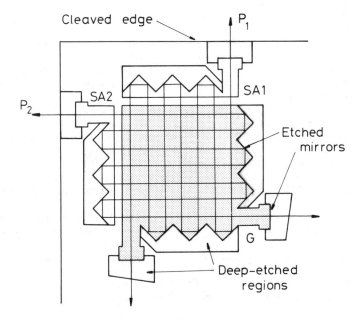

Figure 6.4 Diagram of a folded-cavity cross-coupled BLD with seven folds. The dashed lines show the idealized beam paths in the folded cavity. Bistability between the cross modes results from competition for gain in region G and saturable absorption in regions SA1 and SA2. (From [19].)

stimulus is removed, lasing in the new configuration is maintained. Since the XCBLD is symmetrical, lasing can be switched back to the original direction by the same method.

Electrically triggered set-reset optical memory operation has been achieved in a folded-cavity device under pulsed bias at 77K.

REFERENCES

[1] Shore, K. A., and T. E. Rozzi, "Near-Field Control in Multistripe Geometry Injection Lasers," *IEEE J. Quantum Electronics*, Vol. QE-17, No. 5, May 1981, pp. 718–722.

[2] Shore, K. A., "Optically Induced Spatial Instability in Twin-Stripe-Geometry Lasers," *Optical and Quantum Electronics*, Vol. 14, 1982, pp. 177–181.

[3] Shore, K. A., "Semiconductor Laser Bistable Operation With an Adjustable Trigger," *Optical and Quantum Electronics*, Vol. 14, 1982, pp. 321–326.

[4] Scifres, D. R., W. Streifer, and R. D. Burnham, "Beam Scanning With Twin-Stripe Injection Lasers," *Appl. Phys. Lett.*, Vol. 33, No. 8, October 1987, pp. 702–704.

[5] White, I. H., and J. E. Carroll, "New Mechanism for Bistable Operation of Closely Coupled Twin Stripe Lasers," *Electron. Lett.*, Vol. 19, No. 9, April 1983, pp. 337–339.

[6] White, I. H., J. E. Carroll, and R. G. Plumb, "Room-Temperature Optically Triggered Bistability in Twin-Stripe Lasers," *Electron. Lett.*, Vol. 19, No. 14, July 1983, pp. 558–560.

[7] Kawaguchi, H., "Bistable Laser Diodes and Their Applications for Photonic Switching," *Int. J. Optoelectronics*, Vol. 7, No. 3, 1992, pp. 301–348.

[8] White, I. H., and J. E. Carroll, "Optical Bistability in Twin-Stripe Lasers," *IEE Proc. Pt. H*, Vol. 131, No. 5, October 1984, pp. 309–321.

[9] Davies, D. A. O., I. H. White, and J. E. Carroll, "Modal Bistability in Asymmetric Twin Ridge Semiconductor Lasers," *IEE Proc. Pt. J*, Vol. 136, No. 1, February 1989, pp. 6–13.

[10] McInerney, J. G., L. Yuan, T. C. Salvi, and D. M. Heffernan, "Optical Bistability and Switching by Nonlinear Waveguiding in Semiconductor Laser Amplifiers," *Dig. of Nonlinear Optics: Materials, Phenomena and Devices*, Hawaii, pp. 232–233.

[11] White, I. H., R. S. Linton, and J. E. Carroll, "Directional Coupling and Bistability in Four Contact Twin-Stripe Injection Lasers," *IEE Proc. Pt. J*, Vol. 135, No. 1, February 1988, pp. 25–30.

[12] Watanabe, M., I. H. White, and J. E. Carroll, "Analysis of the Cross-Coupled Lateral Mode in a Twin-Stripe Four-Contact Laser With Diagonal Current Injection," *IEEE J. Quantum Electronics*, Vol. QE-26, No. 11, November 1990, pp. 1942–1953.

[13] Watanabe, M., I. H. White, and J. E. Carroll, "Two Cross-Coupled-Mode Operation of a Twin-Stripe Laser: Bistability for Wide Ranges of Output Power Level and Current Ratio," *IEEE J. Quantum Electronics*, Vol. QE-28, No. 2, February 1992, pp. 395–399.

[14] Heffernan, D. M., J. McInerney, L. Reekie, and D. J. Bradley, "Bistability by Induced Waveguiding in Coupled Semiconductor Lasers," *IEEE J. Quantum Electronics*, Vol. QE-21, No. 9, September 1985, pp. 1505–1512.

[15] Marcuse, D., *Light Transmission Optics*, New York: Van Nostrand, 1992, p. 420.

[16] Streifer, W., D. R. Scifres, and R. D. Burnham, "Analysis of Gain Guided Waveguiding in Stripe Geometry Lasers," *IEEE J. Quantum Electronics*, Vol. QE-14, No. 6, June 1978, pp. 418–427.

[17] Setterlind, C. J., and L. Thylen, "Directional Coupler Switches With Optical Gain," *IEEE J. Quantum Electronics*, Vol. QE-22, No. 5, May 1986, pp. 595–602.

[18] Watanabe, M., H. Fujiki, S. Mukai, M. Ogura, H. Yajima, K. Shimoyama, and H. Gotoh, "Observation of Bistability Between Two Crosscoupled Lateral Modes in Twin-Stripe Lasers," *Technical Report of IEICE*, OQE 92-136, December 1992, pp. 1–6 (in Japanese).

[19] Johnson, J. E., C. L. Tang, and W. J. Grande, "Optical Flip-Flop Based on Two-Mode Intensity Bistability

in a Cross-Coupled Bistable Laser Diode,'' *Appl. Phys. Lett.*, Vol. 63, No. 24, December 1993, pp. 3273–3275.

[20] Lasher, G. J., and A. B. Fowler,''Mutually Quenched Injection Lasers as Bistable Devices,'' *IBM J.*, Vol. 8, September 1964, pp. 471–475.

Chapter 7

Self-Pulsation and Ultrashort-Optical-Pulse Generation

7.1 PERIOD DOUBLING AND CHAOS IN DIRECTLY MODULATED LASER DIODES

7.1.1 Introduction

A laser can have different forms of emission as far as the time dependence is concerned [1]. The output power and laser frequency may be constant or variable in time. For the dynamic behavior of a laser, it is often important that the laser can have two reservoirs of energy. The energy may be contained in the laser field or in the medium. Thus, the laser resembles a pendulum or LC (inductance and capacitance) circuit having a pronounced resonance frequency. The resonance may be more or less strongly damped. If it is relatively weakly damped, the resonance manifests itself (e.g., when the laser is rapidly switched on).

The solution of the rate equations may also be unstable, which means that small perturbations of the laser will be amplified. The laser can then attain another emission form (e.g., corresponding to a time-periodic solution reflected in regular periodic pulsing of the laser output). The time dependence (pulsing) of the laser output power can be accompanied by a corresponding time dependence of the laser frequency.

Another emission form that is possible when the time-independent solution is unstable is nonperiodic pulsing, also termed *chaotic* pulsing. In this case, although there is still some regularity or periodicity in the output, the exact form of pulsing never repeats and many characteristics of the laser output show stochastic properties. The spectrum of the laser field or intensity, for example, is continuous and noiselike. Evidently, with the intensity the laser frequency may vary correspondingly nonperiodically.

Multiperiodic pulsing is also often encountered (Figure 7.1). The laser output shows a periodic pulsing with period T. When one parameter (for example, a frequency of the modulation current) is changed, it closes on itself not after one, but after two round-trips, so that the period changes from T to $2T$. This is known as a *period-doubling* bifurcation. When the parameter is further changed, a new period-doubling bifurcation appears: the curve closes on itself only after four round-trips, so that the period suddenly changes to $4T$.

7.1.2 Period Doubling in Directly Modulated Laser Diodes

When the relaxation time of the population inversion is much larger than the memory time of the induced dipoles, the familiar rate equations can be used for analyzing the laser output power. In this situation, it was found that chaotic behavior arises in a single mode laser when its losses are periodically modulated with a frequency comparable to the relaxation oscillation frequency [2,3]. To demonstrate the existence of chaotic behavior in a laser with a modulated pump, the LD provides a good example. Kawaguchi showed, as described in the next section in detail, that the self-pulsating LD with a sinusoidal current modulation shows period-doubling bifurcation and chaos [4].

Soon after, Lee et al. showed that the directly modulated LD, which does not have self-pulsation, with the modulation frequency of the injection current comparable to the

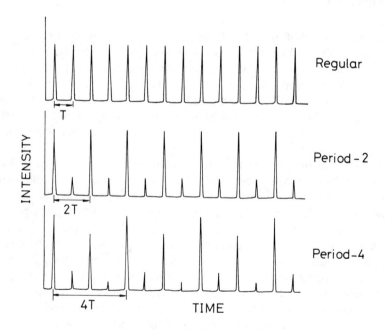

Figure 7.1 Waveforms of multiperiodic pulsing.

relaxation oscillation frequency, exhibits a period-doubling route to chaos as the modulation index of current is increased [5]. The LD has no additional disturbance (for example, external feedback) except sinusoidal current modulation. This phenomenon simply relies on the photon-electron resonances and their interaction with the modulation frequency.

As the charge is increased above the laser threshold, the photon density begins to build up exponentially. Starting from a very low level, this building process can take 100 ps or so. Consequently, when the optical density does become significant, a very high gain exists in the laser cavity, resulting in further increase of photon density. The resulting generation of the stimulated emission light in turn causes depletion of the gain, and thus the pulse is rapidly turned off as the charge is depleted to below the threshold.

In normal operation, the carriers depleted through the generation of the picosecond laser pulse are replaced by the bias current so that in the next modulation cycle the charge rises to above the threshold to allow the generation of another picosecond optical pulse. However, if the bias current to the laser is not sufficient to provide this charge within the required time, then the charge carrier concentration is not able to pass the threshold sufficiently to amplify the photon density in the cavity enough to generate a significant picosecond optical pulse in the next modulation cycle. Under certain conditions, therefore, the laser may be caused to operate so that only one substantial optical pulse is generated in every two cycles.

By varying the dc current bias level to the laser, one may obtain other gain-switching regimes. For example, apparently chaotic or nonrepeating fluctuations of pulse amplitudes may be generated, or, alternatively, pulse trains consisting of repetitive cycles of pulses of varying amplitude may be generated.

Kao et al. [6] examined the influence of noise on the period doubling by employing the single-mode rate equations with Langevin noise. It is concluded that the noise bump near half the driving frequency is an important precursor of period doubling. The mode gain in LDs decreases with an increase in power due to nonlinear process such as spectral hole burning. When these small nonlinearities are included in the single-mode rate equations, it is found that they can eliminate the sequence of period-doubling bifurcations leading to chaos in directly modulated LDs [7]. For InGaAsP lasers used in optical communication systems, the nonlinear effects are strong enough that the possibility of chaotic behavior should be of no concern. Using a smaller gain compression factor, Hemery et al. showed in their rate equation simulations the existence of period doubling and achieved good agreement with their experiment [8]. Tang and Wang [9] found that Auger recombination tends to suppress chaos for $\omega_m < \omega_r$. Here ω_m is the modulation frequency and ω_r is the relaxation frequency. However, for $\omega_m > \omega_r$, chaotic behavior becomes prominent.

Ngai and Liu [10] presented a detailed study of the dynamic characteristics of a directly modulated 1.55-μm DFB LD. They investigated the large-signal modulation characteristics up to a modulation index of 12. At a high modulation index, period doubling, period tripling, and period quadrupling were observed. By examining these irregular behaviors at various bias current levels and modulation frequencies, the route to chaos in directly modulated DFB LDs was found to be through period doubling.

The device used in the experiment was a 1.55-μm InGaAsP DFB laser with a threshold current of 16 mA. The laser oscillated at a single longitudinal mode with side-mode suppression of about 15 dB, even under strong modulation. The small-signal modulation bandwidth of the device was measured to be higher than 8 GHz.

When the laser was biased just above the threshold, it exhibited a weak resonance peak with a broad pedestal in the power spectrum. As the RF modulation was applied, the resonance frequency f_r shifted to lower frequencies. This resonance frequency shift stopped at half of the drive frequency when the input power reached a certain level, and a further increase in input power only resulted in the sharpening of the resonance peak. This was the onset of period doubling. For example, when $I_{dc} = 22$ mA, f_r was initially at 2.3 GHz. When a 4-GHz modulation signal was applied, f_r shifted continuously to a lower frequency with the increase in input power. Since the resonance peak was lower than the signal component at 4 GHz by more than 20 dB, the laser at this state essentially followed the modulation signal. However, once f_r reached 2 GHz, period doubling occurred and the intensity of the resonance peak began to increase with the increase in injection level. The power level at 2 GHz may become so intense that it is even higher than that at 4 GHz by 3 dB. The intensity at 2 GHz was reduced at even higher input power levels and new components at one-fourth and three-fourths of the modulation frequency began to grow, which corresponded to period quadrupling.

Figure 7.2 summarizes all the observations, where (a,b) correspond to the results obtained at bias currents of 22 and 18 mA, respectively. When biased at 22 mA, the laser, as shown in Figure 7.2(a), behaved normally without any bifurcation for modulation frequencies up to 2 GHz at all the injection levels. However, when the frequency was increased to 2.5 GHz, period doubling began to occur at $P_{in} = 2$ dBm and persisted until $P_{in} = 20$ dBm, after which the period doubling was suppressed and the laser began to follow modulation again. As the modulation frequency was increased to 3.0 and 3.5 GHz, the period doubling tended to persist up to higher levels of P_{in}. When the modulation was increased to 4 GHz, further bifurcation to period quadrupling was found at $P_{in} > 16$ dBm. Similar bifurcation processes were observed at lower bias currents, where, as shown in Figure 7.2(b), both period doubling and period quadrupling started at slightly lower frequencies. However, when large-signal high-frequency modulations were applied, more complicated bifurcations occurred. At low frequencies (2.5 and 3 GHz), the laser changed its state from period doubling to period quadrupling and then back to period doubling again when P_{in} was increased. However, at high frequencies (3.5 GHz), the period quadrupling was found to proceed to period tripling instead of period doubling. In some cases, the period tripling would change to a state where only a broad continuous noise spectrum with a sharp component at modulation frequency existed. This corresponds to chaos.

7.1.3 Optical-Pulse Generation Through Harmonic-Frequency Modulation

One example of applications of the period doubling in directly modulated LDs is the generation of single-longitudinal-mode gigabit-rate optical pulses from LDs through har-

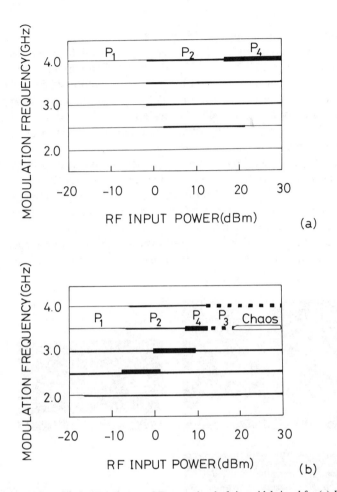

Figure 7.2 Modulation frequency against the input RF power level of sinusoidal signal for (a) I_{dc} = 22 mA and (b) I_{dc} = 18 mA in 1.55-μm InGaAsP DFB LDs. ⸻, ⸻, ▬▬ ▬▬, ▬▬▬▬, and ⬜⬜ represent a single period (P_1), period doubling (P_2), period tripling (P_3), period quadrupling (P_4), and chaos, respectively. Note in (b) a clear route to chaos is shown to be through period doubling, period quadrupling, and period tripling at 3.5-GHz modulation. (From [10]. Reproduced by permission of AIP.)

monic-frequency sinusoidal modulation [11]. The LD used in the experiment was a 1.3-μm InGaAsP/InP self-aligned stripe Fabry-Perot laser with a 7-μm waveguide width and 200-μm cavity length. The threshold current I_{th} was 74.0 mA. The laser showed typical damped oscillations for step driving-current pulses and single-longitudinal-mode oscillation under CW operation. The laser was directly modulated by a sinusoidal current (current value from peak-to-peak was I_p) superimposed on a dc bias current I_b.

Typical experimental results for generation of single-longitudinal-mode gigabit-rate optical pulses through harmonic-frequency, sinusoidal-injection-current modulation are shown in Figure 7.3(a). The upper and lower traces in the left-hand-side figure of Figure 7.3(a) show the waveforms of the laser output and the modulation current, respectively. The bias current I_b was 82.9 mA ($1.12I_{th}$), the modulation current I_p was about 30 mA, and the modulation frequency f_m was 3.1 GHz. The waveform of the light is limited by the detector frequency response. The time delay notable between the upper and lower trace does not signify any important information. The right-hand-side figure of Figure 7.3(a) shows the corresponding spectrum. The center wavelength was 1.2945 μm. The mode spacing is about 8 Å.

We can obviously see from Figure 7.3(a) that the laser is oscillating in the form of a spikelike shape, and that the frequency of the spiking is $f_m/2$. We can also see that the laser oscillated in a single longitudinal mode in spite of the high bit rate (1.65 GHz) and the very deep modulation. About 90% of the photon energy was concentrated in the main oscillating mode at this condition.

An $f_m/2$ spikelike oscillation occurred over a wide operating range. For example, under the operating condition of $I_b = 79.66$ mA ($1.08I_{th}$) and $I_p = 30$ mA, $f_m/2$ spiking

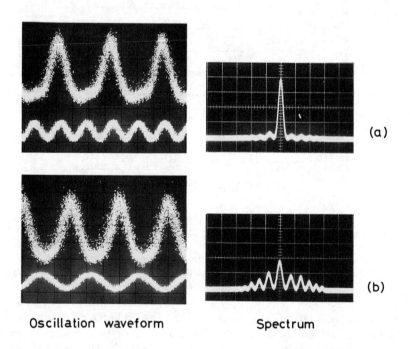

(a)

(b)

Oscillation waveform   Spectrum

Figure 7.3 Oscillation waveform and spectrum: (a) under harmonic-frequency modulation: (left) waveforms of laser output (upper) and modulation current (lower) with 200-ps/horizontal division, (right) lasing spectrum; longitudinal-mode spacing is about 8 Å, $I_b = 82.9$ mA ($1.12I_{th}$), $I_p = 30$ mA, $f_m = 3.1$ GHz; (b) under usual fundamental-frequency modulation; $I_b = 85.56$ mA ($1.16I_{th}$), $I_p = 30$ mA, $f_m = 1.65$ GHz. (From [11].)

oscillation was observed in the frequency range from $f_m = 2.0$ to 3.1 GHz. However, the single-longitudinal-mode optical-pulse generation was found to occur only in the higher frequency side. The single-longitudinal-mode component in Figure 7.3(a) was observed to become more dominant as the modulation frequency f_m approached the highest-frequency edge for subharmonic pulse generation, although the modulation efficiency decreased accordingly. In contrast, when f_m approached the lowest-frequency edge, the modulation efficiency became higher, whereas the number of oscillating modes increased.

The mechanisms of these phenomena can be considered as follows. During the first period of a modulation current, the carrier density in the active region increases slightly over the CW threshold. It does not reach a value sufficient to produce an optical pulse, but a very weak light begins to oscillate continuously just like during dc operation (see bottom half of the temporal plot in Figure 7.3(a)). In the succeeding period, the carrier density becomes sufficient to produce an optical pulse. After radiation of one optical pulse, the carrier density returns to the initial value. This process repeats, resulting in $f_m/2$ spiking oscillation. When a weak light produced in the first period has a single-longitudinal-mode component resulting from staying as the constant carrier density, it affects the following optical pulse as a coherent external injection light, and the spectrum of the optical pulse becomes very narrow.

Generation of optical pulses with repetition frequency f_m was examined for the same LD shown in Figure 7.3(a). This was done to compare it with the experimental results mentioned above. Figure 7.3(b) shows a typical observed waveform for the laser output and spectrum. $I_b = 85.56$ mA ($1.16I_{th}$), $I_p = 30$ mA, and $f_m = 1.65$ GHz. Though the modulation index and intensity of the modulated optical output were about the same as for Figure 7.3(a), the laser showed multimode operations under all experimental conditions.

7.1.4 Pulse Position Bistability

Bistable devices based on a particular nonlinear regime of gain switching can be constructed, where short optical pulses can be generated at one of two stable temporal positions relative to a reference clock pulse [12]. The temporal position of the optical generated pulse constitutes the state for the optical bistable device. The position of the pulse and hence the state of the bistable element can be changed by an optical trigger.

The bistability is based on period doubling, a distinct regime of gain switching, where a picosecond optical pulse is generated only every other modulation period. To investigate the bistable mechanism, a conventional rate equation analysis is carried out. A typical example of period doubling is shown in Figure 7.4(a). The top trace of Figure 7.4(a) shows the charge carrier concentration at the laser junction that is sustained during excitation by a 1-GHz sinusoidal current superimposed on a dc current level.

It can be seen that a period-doubling regime is inherently bistable, as shown in Figure 7.4(b). If the modulation cycles are labeled 1, 2, 3, ..., then there are two possible states for the laser to operate: (1) when optical pulses are generated in cycles 1, 3, 5, ...,

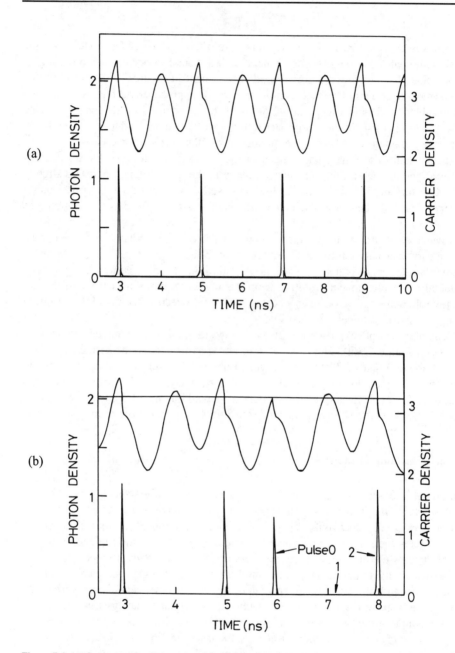

Figure 7.4 (a) Computer simulation of a gain-switched GaAs laser driven with 1-GHz RF + $1.15I_{th}$ bias: the top trace shows the carrier concentration and the lower trace the resulting photon density profile in the laser cavity, which shows period-doubling operation; (b) simulation showing switching between stable states by injection of a small trigger optical pulse at $t = 6$ ns. (©1987 IEEE. From [12].)

and (2) when generated in cycles 2, 4, 6, ... To be useful, the bistable mechanism must allow switching between its two states. This can be achieved by the injection of light during those cycles when pulses are not generated. Although during these cycles the carrier concentration does not reach the gain-switching threshold, it is in fact close to, if not above, the lasing threshold. The laser cavity thus has very high optical gain, so that if light is injected into the cavity, it will be greatly amplified. The resulting depletion of carriers may be large enough to prevent the buildup of charge sufficient to cause the generation of a gain-switched pulse during the next cycle. The next gain-switched pulse may not therefore occur until two cycles after the injected pulse, so a change in step of the alternate pulsing sequence has thus been caused.

Figure 7.4(b) is the simulation of the bistable switching of a GaAs LD similar to that of Figure 7.4(a). Here a 30-ps trigger pulse of energy 18% of that of the output from one facet of the bistable laser successfully triggers a bistable switch. The required trigger energy is a typical figure.

The effective switching speed of the BLD is one modulation period and is thus limited by the frequency response of the gain-switching mechanism. It is related to the relaxation oscillation frequency of the laser and thus increases with bias. Although the modulation frequency and hence speed of switching may be increased by increasing the bias, a limit is eventually reached that is determined by thermal dissipation and spontaneous lifetime. However, by using a 50-μm cavity and InGaAsP lasers that have a shorter carrier lifetime, switching speeds below 100 ps have been predicted at a bias of $2.3I_{th}$.

Some experiments were carried out to demonstrate the operating principle of this bistable mechanism. Figure 7.5 shows streak camera traces of the trigger and bistable laser outputs. In the left half of the lower trace, the bistable laser can be seen to be operating in a stable period-doubling mode, generating optical pulses at a 1-GHz rate from a 2-GHz electrical modulation. The top trace shows the trigger pulse to be injected into the bistable laser to obtain switching. The lower trace shows the trigger pulse's effect of interrupting the sequence of period-doubled pulsing. A switch of the bistable laser pulse from one stable set of temporal positions to the other is apparent.

In the case of pulse position bistability, the levels 1 and 0 must be represented by optical pulses occurring at different temporal positions. However, a complete logic system would require AND, OR, and NOT gates as well as bistable elements. It is not clear yet how AND and OR gates might be constructed, although a NOT gate could be readily constructed using an optical delay line.

7.2 SELF-PULSATION AND OPTICAL CHAOS

Self-pulsation (i.e., repetitive, strong intensity change in output power with a repetition rate of hundreds of megahertz to a few gigahertz) strongly connects with bistability as described in Section 3.2. The output stability depends on the carrier lifetime distribution along the laser cavity [4]. When the carrier lifetime is roughly the same over the whole

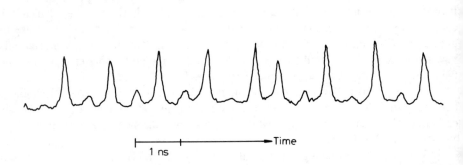

Figure 7.5 Streak camera traces of laser output. The optical trigger pulse can be seen to cause a shift in output pulse of the bistable laser. Top trace: trigger laser; bottom trace: bistable laser. (©1987 IEEE. From [12].)

laser cavity, inhomogeneously excited LDs show bistability. When the carrier lifetime of the unpumped or low-pumped region is shorter than the gain region, self-pulsation occurs in spite of applying only a dc bias current.

Let us consider the case where self-pulsation occurs and the current of the gain region is modulated by a sinusoidal RF signal superimposed on the dc current; that is,

$$I = I_b + I_m \sin(2\pi f_m t) \tag{7.1}$$

where f_m is the frequency of the modulation current. The pulse repetition frequency will be locked to the modulation frequency f_m if f_m is near the self-pulsating frequency f_R on the free-running condition (i.e., without sinusoidal RF modulation). When the frequency difference $|f_R - f_m|$ becomes large, this nonlinear laser system shows chaotic pulsation through period-doubling bifurcation as the frequency difference is increased [4]. Such locking in repetition frequency can also be obtained by the injection of optical pulses into the saturable absorber, and was used for all-optical timing extraction, described further in Section 10.2.1.

Calculated responses of a self-pulsating laser for sinusoidal current modulation are shown in Figure 7.6 [4,13]. Waveforms of photon density and corresponding RF spectra are shown on the left side and the right side, respectively. The dc bias current I_b and current modulation depth I_m are set at $2.2 \times P_{th}$ and $1.1 \times P_{th}$, respectively, where P_{th} is the threshold pump rate. A nonradiative carrier lifetime of 0.2 ns is assumed for the saturable absorption region. At 1.50 GHz, the laser shows sustained pulse oscillation with

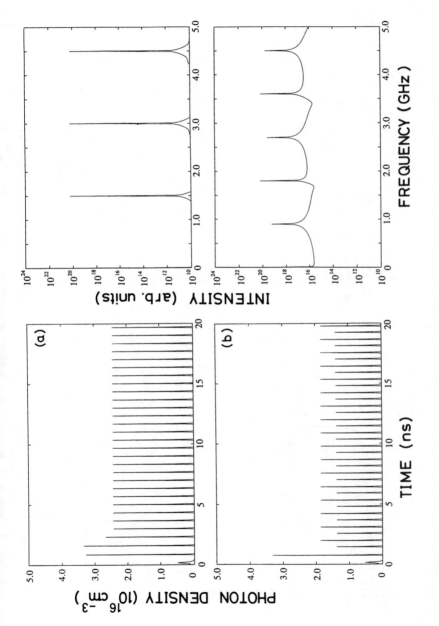

Figure 7.6(a,b) Output response and spectra of self-pulsating LDs for sinusoidal current modulation: (a) period 1; (b) period 2.

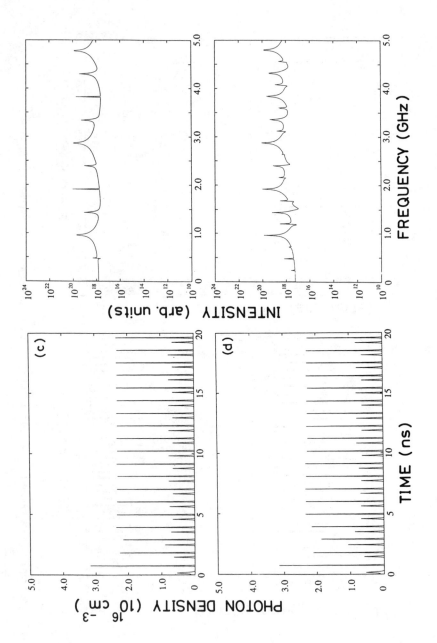

Figure 7.6(c,d) Output response and spectra of self-pulsating LDs for sinusoidal current modulation: (c) period 4; (d) period 8.

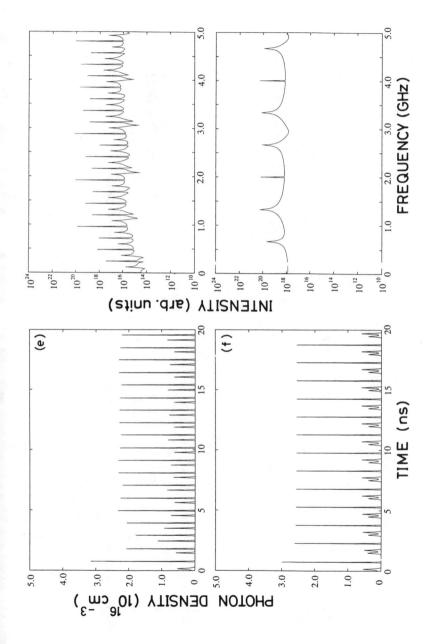

Figure 7.6(e,f) Output response and spectra of self-pulsating LDs for sinusoidal current modulation: (e) period 16; (f) period 3.

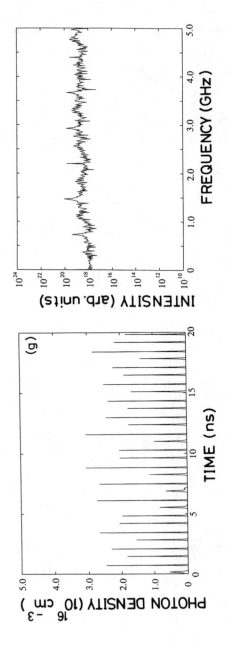

Figure 7.6(g) Output response and spectra of self-pulsating LDs for sinusoidal current modulation: (g) chaotic.

the modulation frequency. The first subharmonic appears at 1.80 GHz. At 1.91 GHz, period 4 oscillation is obtained. Period 8 and period 16 oscillations are observed at 1.918 and 1.92 GHz, respectively. We also observe period 3 oscillation at 2.0 GHz. Furthermore, chaotic outputs are seen at above 2.2 GHz.

A bifurcation diagram, showing the variation of the sampled peak photon density with the modulation frequency f_m of injection current (Figure 7.7), shows a period-doubling route to chaos.

These characteristics were observed in the experiment [14]. Experimental waveforms and RF spectra from a GaAlAs LD are shown in Figure 7.8. In the experiment, f_m was fixed to 1 GHz and f_R was changed by varying the dc bias current. By increasing the difference between the self-pulsation frequency f_R and the modulation frequency f_m, period 2 oscillation was observed with $f_R = 875$ MHz. A clear subharmonic resonance peak was observed at 500 MHz in the RF spectrum. Chaotic oscillation was also observed with f_R = 820 MHz.

A transition to chaos via quasiperiodicity was also observed by Winful et al. [15] in the output of a directly modulated self-pulsing LD. By sweeping the frequency and amplitude of the current modulation, several frequency-locked states (*Arnol'd tongues*) are mapped out. Good agreement with the predictions of a rate equation model was obtained. For the

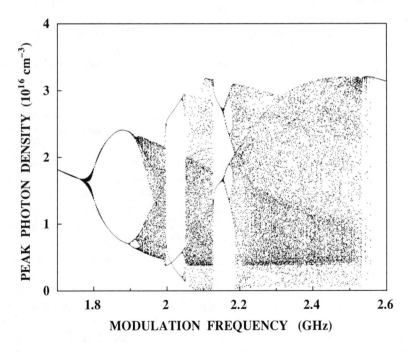

Figure 7.7 Bifurcation diagram of the sampled peak photon density versus the modulation frequency of injected current.

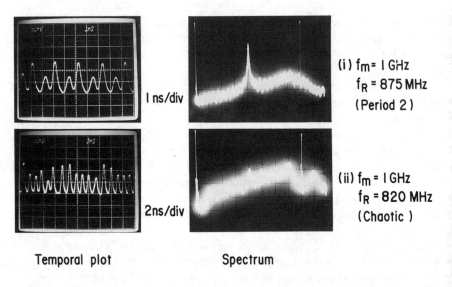

Temporal plot   **Spectrum**

Figure 7.8 Observed output response of self-pulsating laser for sinusoidal current modulation: (i) period 2; (ii) chaotic. (From [14].)

AlGaAs/GaAs lasers used in the experiments, the fundamental pulsation frequency f_0 can be tuned between 0.5 and 3 GHz by varying the dc bias current.

The second frequency f_{ext} is imposed by an RF generator whose output modulates the laser pumping current at rates between 0.3 and 2.0 GHz. In the presence of the external modulation, the intrinsic resonance frequency is shifted by an amount that depends on the amplitude of the modulation. We thus speak of a "dressed" intrinsic frequency f_0' and define a winding number $\rho = f_0'/f_{ext}$. If the winding number takes on a rational value p/q, where p and q are integers, the output pulsation frequency locks to a harmonic or subharmonic of the modulation. There is a range of frequency detunings within which the external modulation can effectively entrain the self-pulsation frequency. This locking range increases with the amplitude of the external modulation. By sweeping the frequency and amplitude of the modulation, we have mapped out the structure of the frequency locked states. On a plot of modulation amplitude versus frequency ratio, the locked states form regions known as Arnol'd tongues [16], whose boundaries separate the periodic motions from the quasiperiodic and aperiodic oscillations. Figure 7.9 shows several of these frequency-locked regions for small integer values of p and q. The organization of the locking regions follows the Farey tree structure [17]. Between any two locked bands with winding numbers p_1/q_1 and p_2/q_2, there exists another locked band whose winding number is given by the Farey sum $p_3/q_3 = (p_1 + p_2)/(q_1 + q_2)$. For example, the locked band with winding number 2/3 is the Farey composition of the bands 1/1 and 1/2. Locked states with large denominators have very narrow widths, are easily destabilized by noise and are thus difficult to resolve. There are other locked states (such as $\rho = 4/5$), which

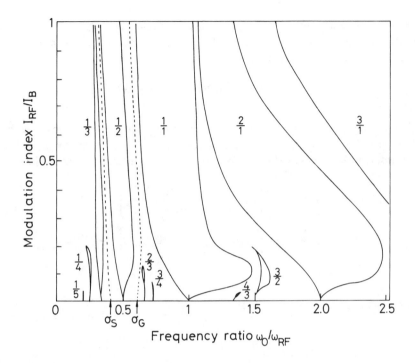

Figure 7.9 Frequency-locked regions (Arnol'd tongues) for a modulated self-pulsing laser. The plot shows the modulation depth versus the ratio of the intrinsic pulsation frequency at zero drive (ω_0) to the external modulation frequency. The dotted lines show paths of fixed winding numbers at the golden mean σ_g and at the silver mean σ_s. Because the intrinsic frequency shifts with the amplitude of the modulation, paths of fixed winding numbers are not straight lines. (From [15].)

by virtue of their proximity to the strong fundamental resonance are pulled by that resonance and tend to merge with it. We note that the measured Arnol'd tongue structure is independent of the direction in which the frequency or amplitude of the modulation is varied. No sign of hysteresis is observed.

7.3 MODE LOCKING IN LASER DIODES

7.3.1 Basic Theory

In an inhomogeneously broadened laser, oscillation can take place at a number of frequencies, which are separated by [18]

$$\omega_q - \omega_{q-1} = \frac{\pi c}{l} = \omega \tag{7.2}$$

Now consider the total optical electric field resulting from such multimode oscillation at some arbitrary point, say, next to one of the mirrors, in the optical resonator. It can be taken, using complex notation, as

$$e(t) = \sum_n E_n \exp\{i[(\omega_0 + n\omega)t + \phi_n]\} \tag{7.3}$$

where the summation is extended over all the oscillating modes and ω_0 is chosen, arbitrarily as a reference frequency. ϕ_n is the phase of the nth mode. One property of (7.3) is that $e(t)$ is periodic in $T = 2\pi/\omega = 2l/c$, which is the round-trip transit time inside the resonator.

$$
\begin{aligned}
e(t + T) &= \sum_n E_n \exp\left\{i\left[(\omega_0 + n\omega)\left(t + \frac{2\pi}{\omega}\right) + \phi_n\right]\right\} \\
&= \sum_n E_n \exp\left\{i\left[(\omega_0 + n\omega)t + \phi_n\right]\right\} \exp\left\{i\left[2\pi\left(\frac{\omega_0}{\omega} + n\right)\right]\right\} \\
&= e(t)
\end{aligned} \tag{7.4}
$$

Since ω_0/ω is an integer ($\omega_0 = m\pi c/l$),

$$\exp\left[2\pi i\left(\frac{\omega_0}{\omega} + n\right)\right] = 1 \tag{7.5}$$

Note that the periodic property of $e(t)$ depends on the fact that the phases ϕ_n are fixed. In typical lasers, the phases ϕ_n are likely to vary randomly with time. This causes the intensity of the laser output to fluctuate randomly and greatly reduces its usefulness for many applications where temporal coherence is important.

Two ways in which the laser can be made coherent are (1) to make it possible for the laser to oscillate at a single frequency only, so that mode interference is eliminated (2) to force the mode phases ϕ_n to maintain their relative values. This is the so-called *mode-locking* technique, which causes the oscillation intensity to consist of a periodic train with a period of $T = 2l/c = 2\pi/\omega$.

One of the most useful forms of mode locking results when all the phases are made the same. To simplify the analysis of this case, assume that there are N oscillating modes with equal amplitudes. Taking $E_n = 1$ and $\phi_n = 0$ in (7.3) gives

$$
\begin{aligned}
e(t) &= \sum_{-(N-1)/2}^{(N-1)/2} e^{i(\omega_0 + n\omega)t} \\
&= e^{i\omega_0 t} \frac{\sin(N\omega t/2)}{\sin(\omega t/2)}
\end{aligned} \tag{7.6}
$$

The average laser power output is proportional to $e(t)e^*(t)$ and is given by

$$P(t) \propto \frac{\sin^2(N\omega t/2)}{\sin^2(\omega t/2)} \qquad (7.7)$$

Some of the analytic properties of $P(t)$ are immediately apparent.

1. The power is emitted in the form of a train of pulses with a period $T = 2\pi/\omega = 2l/c$ (i.e., the round-trip delay time).
2. The peak power, $P(sT)$ (for $s = 1, 2, 3, \ldots$), is equal to N times the average power, where N is the number of modes locked together.
3. The peak field amplitude is equal to N times the amplitude of a single mode.
4. The individual pulsewidth, defined as the time from the peak to the first zero is $\tau_0 = T/N$. The number of oscillating modes can be estimated by $N \simeq \Delta\omega/\omega$—that is, the ratio of the transition lineshape width $\Delta\omega$ to the frequency spacing ω between modes. Using this relation, as well as $T = 2\pi/\omega$ in $\tau_0 = T/N$, we obtain

$$\tau_0 \sim \frac{2\pi}{\Delta\omega} = \frac{1}{\Delta\nu} \qquad (7.8)$$

Thus, the length of the mode-locked pulses is approximately the inverse of the gain linewidth.

Mode locking is accomplished by using one or more nonlinear elements in the cavity. In the frequency domain, the process can be described in terms of side-band generation of frequencies separated by $f_m = 1/T$. An alternative is to follow the pulse as it travels in the cavity to analyze the pulse development in time. For LDs, there are two ways to achieve mode locking. The first, active mode locking, is to impose a periodic change in the gain by modulating the pumping current at frequency f_m. The second, passive mode locking, uses in the cavity a saturable absorber whose absorption decreases with higher incident light intensity. The active method is easier to implement, because the modulation is forced upon the laser from an external source. But just as the modulation is predetermined, its highest frequency is limited to the width of the pulse generated. The modulation due to saturating a transition, as in passive mode locking, is as fast as the width of the saturating pulse. In fact, it is the self-created modulation that makes passive mode locking so efficient: the shorter and more intense the pulse, the faster and deeper the modulation becomes, which shortens the pulse even further, and so on until the Fourier transform of the pulse reaches the bandwidth of the system. But passive mode locking has to satisfy some conditions that restrict its operating range.

7.3.2 Active Mode Locking [19]

The gain is varied at the same frequency as the light passing through it (i.e., at the round-trip frequency of the cavity or a multiple of it) and the peak of the gain exceeds the threshold slightly. At each passage, the pulse center sees more gain than the wings and is sharpened (Figure 7.10). For sinusoidal modulation of a homogeneously broadened gain, the pulses are Gaussian with width [20]

$$\Delta t \sim 1/\left(\sqrt[4]{M} \sqrt{f_m \Delta \nu_{BW}} \right) \tag{7.9}$$

where M is the modulation depth, f_m is the modulation frequency, and $\Delta \nu_{BW}$ is the bandwidth of the system; f_m need not be exactly the cavity round-trip frequency. In fact, when instantaneous gain saturation due to the pulse is taken into account, a slightly lower f_m yields the shortest pulse. The pulsewidth is a compromise between the shortening and the broadening processes as quantified by f_m and $\Delta \nu_{BW}$. At higher f_m, the gain changes faster, and therefore the net gain (shaded areas in Figure 7.10) will appear for a shorter time. On the other hand, the pulse will be broadened each time it passes through the element that limits the bandwidth of the system. The pulsewidth is inversely proportional to the square root of the bandwidth and is therefore less efficient in using the available bandwidth. Passive mode locking is more efficient, since the pulsewidth is inversely proportional to the bandwidth (see (7.10) and (7.13)).

When the gain is modulated by current pulses, the laser is more like the synchronously mode-locked dye laser if the modulation pulsewidth is shorter than the gain relaxation time. The optical pulse saturates the gain to below the threshold and a net gain window opens up, similar to the situation shown in Figure 7.10. The pulsewidth can still be estimated from (7.9) with $1/f_m$ replaced by the width of the modulation pulse. Since f_m is already in the gigahertz range, pulse modulation will not reduce the pulsewidth significantly.

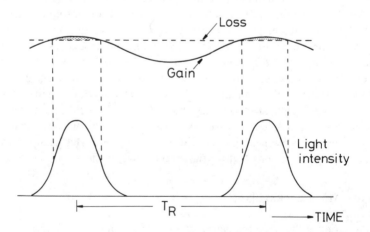

Figure 7.10 Active mode locking by gain modulation. The shaded areas represent excess gain over loss occurring for a short time in every cavity round-trip. (From [19].)

Implicit in (7.9) is the assumption that $\Delta t \gg 1/\Delta \nu_{BW}$, which is usually satisfied. It is also possible to modulate the loss of the laser or the phase of the light field instead.

Active mode locking is realized in a typical system shown in Figure 7.11. Modulating the dc pump of the diode modulates the gain. Since the round-trip time in the diode is only a few picoseconds, it is practical to extend the cavity with the lens and mirror to lower the modulation frequency. The radiative lifetime of the carriers is about 1 ns, which limits the modulation frequency to a few gigahertz. The AR coating on the diode eliminates internal reflection; otherwise, the three-reflector cavity will have a spectrum having clusters of modes. If clusters of modes are excited near the loss minima, the modes will be locked in clusters of center frequencies separated by f_d. Here, f_d is the round-trip frequency in the diode between the two cleaved faces. Noisy pulse bursts will appear if the clusters are not locked with one another; and if they are, a train of several pulses separated in time by $1/f_d$ will emerge. When the AR coating is imperfect, an etalon may be used to limit the laser oscillation to one cluster. If the gain curvature in frequency is steep enough, or excitation low enough, only the center cluster may oscillate without the bandwidth limiter.

7.3.3 Passive Mode Locking

In passive mode locking, the pulse itself creates its own modulation. The most common passive mode-locking device uses a saturable absorber. A saturable absorber may be called *fast* if its recovery time is shorter than the pulse it produces, so the change in absorption follows the shape of the saturating pulse. It is *slow* if its recovery time is longer than the pulse. The speed of the saturable absorber affects the mode-locking process.

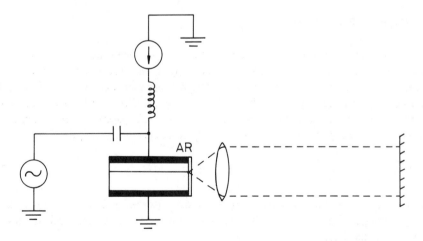

Figure 7.11 A typical LD mode locked by gain modulation. The AR coating eliminates internal reflection in the cavity. Sometimes an etalon is required in the cavity to generate coherent pulses.

The goal of combined mode locking is to receive the benefit of synchronization supplied by an RF generator (active mode locking) and yet take advantage of the pulse shortening due to the action of a saturable absorber (passive mode locking). This technique is called *hybrid mode locking*.

Fast Saturable Absorber Case

The gain is assumed to be slow and saturated by the average power only, and remains constant in time. As the light pulse in the cavity passes through the absorber, the absorption is saturated, the total loss (saturable and linear) of the system decreases, and so, near the peak of the pulse, where saturation is strongest, the gain exceeds the loss, and the total loss recovers after the pulse (Figure 7.12(a)). Again, a gain window opens during the presence of the pulse, and the opening time is determined by the pulse itself. For homogeneously broadened systems, the pulse intensity is [21] $\text{sech}^2(t/\Delta t)$ with

$$\Delta t = \frac{1}{[\Delta \nu_{BW} \sqrt{(q/2g)(P_p/P_A)}]} \tag{7.10}$$

where P_p is the peak power of the pulse, P_A is the saturation power of the absorber, q is the unsaturated loss due to the absorber, and g is the saturated gain; P_p and g depend on system parameters in more complicated ways, but (7.10) shows that Δt is now inversely proportional to the bandwidth $\Delta \nu_{BW}$.

Slow Saturable Absorber Case

When the relaxation time of the absorber is much longer than the pulsewidth, then an explanation of mode locking has to take into account the dynamics of both gain and absorber saturation. When the pulse passes through either medium, the gain or absorption is reduced and only recovers long after the pulse is gone (Figure 7.12(b)). If the system parameters are adjusted correctly, a net gain window again appears, which sharpens the pulse. For this to happen, the absorber must be saturated before the gain. That is,

$$E_A < E_G \tag{7.11}$$

where E_A and E_G are the saturation energies of the absorber and the gain, respectively. After the passage of the pulse, the absorber must recover faster than the gain to close the gain window; that is,

$$T_A < T_G \tag{7.12}$$

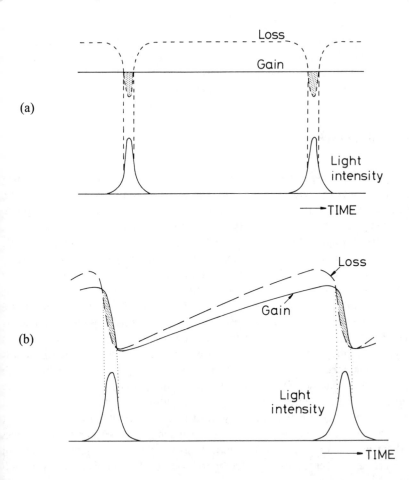

Figure 7.12 (a) Mode locking by a fast saturable absorber. The shaded areas represent excess gain over loss occurring in every round-trip when the pulse saturates the loss due to the absorber. The absorber has a response time fast enough to follow the pulse. (b) Mode locking by a slow saturable absorber. The shaded areas represent positive net gain. The response times of the gain and absorber are both longer than the pulsewidth. For a net gain to occur at the center of the pulse, the absorber must be saturated first when the pulse comes (the broken line bends down before the solid line), and recover first after the passage of the pulse (the broken line goes up before the solid line). (From [19].)

where T_A and T_G are the recovery times of the absorber and gain. If the gain recovers completely before the next pulse, the system will be unstable, since there is a net gain, so $T_G < T_R$. When mode locking is successful, the pulse intensity is again [22] $\mathrm{sech}^2(t/\Delta t)$, with

$$\Delta t \sim 4/(\Delta \nu_{BW}\sqrt{qE_0/E_A}) \qquad (7.13)$$

where E_0 is the energy of the pulse, E_A is the saturation energy of the absorber, and q is the unsaturated absorption normalized to the linear loss of the laser. The pulsewidth is inversely proportional to the pulse energy, a direct consequence of the fact that an energetic pulse induces deeper modulation by stronger saturation.

Colliding-Pulse Mode Locking [23]

The advent of the colliding-pulse mode-locked (CPM) dye laser [24] pushed pulse-generation methods into the femtosecond time domain with the report of the first pulses of less than 0.1 ps. The central idea is to use the interaction, or "collision," of two pulses in an optical cavity to enhance the effectiveness of the saturable absorber. Figure 7.13 illustrates this general process for two optical pulses in the laser cavity. With two optical pulses in a simple optical cavity, the saturable absorber must be placed precisely in the center of the cavity so that the two oppositely directed pulses will be able to interact in the saturable absorber at the same time. Since both pulses are coherent, they interfere with each other, creating a standing wave. At the antinodes of the wave, the intensity is greatest, more completely saturating the absorber and minimizing loss. At the nodes of the field, the absorber is unsaturated, but then, of course, the intensity of the field is a minimum, again minimizing the loss. The net effect of using the standing-wave field to saturate the absorber rather than the fields of the two pulses separately is a reduction in the energy required to saturate the absorber by a factor of approximately 1.5 [25]. Since the gain medium is being pumped continuously, each pulse reaches the gain medium at a point in time when the gain medium is fully recovered. The result is that when the two pulses meet in the saturable absorber, there is twice as much energy to saturate the absorber than when there is only one pulse in the optical cavity. Thus, the effective saturation parameter is increased by approximately a factor of 3 over that for a conventional passively mode-locked laser.

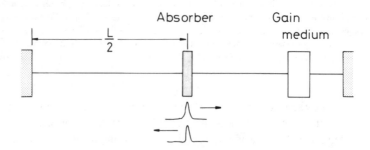

Figure 7.13 Diagram illustrating the colliding-pulse mode-locking configuration for two optical pulses in the cavity.

7.3.4 Short-Optical-Pulse Generation by Mode Locking

Much effort has been devoted to short-optical-pulse generation using LDs. The techniques fall into two general categories: relaxation oscillation and mode locking, although at times they both occur in the same system. Lee and Roldan [26] and Basov et al. [27] used two-section LDs to produce pulses 10 to 100 ps long. The repetition rate of the pulses was determined by the relaxation times of the system and was not directly connected with the round-trip time of a pulse within the system. Morozov et al. [28] used a two-section LD, one for gain, the other as an absorber in an external resonator. They only observed ripples at the time.

Schrans et al. have reported 320-fs pulse generation from external-cavity, two-section GaAs MQW LDs [29]. The absorber section is grounded through a 100Ω resistor, and the gain section is pumped with a dc current source. Typical durations of pulses generated by passive mode locking without compression were 4 to 10 ps. The pulses were compressed using a single-pass grating telescope compressor.

Actively mode-locked external-cavity LDs with a single gain segment are very susceptible to multiple-pulse formation. The undesired secondary pulses are initiated by reflections from imperfect AR coatings on the LD facet. The reflected pulse is then amplified because the main pulse does not fully deplete the gain, and the current drive to the segment may still be creating new carriers. Schell et al. [30] have calculated that even a very good antireflection coating, 10^{-5}, will cause multiple-pulse generation at the round-trip time of the LD.

Derickson et al. [31] showed that incorporation of reverse-biased intrawaveguide saturable absorber segments into the cavity suppresses the multiple-pulse generation problem. Single-pulse outputs of less than 2.8 ps with 0.7 pJ of energy are obtained using these techniques. To illustrate how the saturable absorber can suppress multiple-pulse formation, the propagation of a pulse around the mode-locked cavity is outlined in Figure 7.14. The modeled device is a two-section LD with a 500-μm overall length. Figure 7.14(a) shows rate equation simulations of the pulse energy versus distance for the main pulse as the pulse enters from the external cavity and propagates through the gain and absorber segments on the forward and reverse transits. The energy versus distance for the secondary pulse is shown in Figure 7.14(b,c) for the case of no saturable absorber recovery and complete saturable absorber recovery, respectively. If the absorber completely recovers, as seen in Figure 7.14(c), the secondary pulse is partially absorbed.

7.3.5 Mode Locking at Very High Frequencies

The possibility of mode locking an LD at millimeter-wave frequencies approaching and beyond 100 GHz was investigated by Lau [32]. Conventional thinking has it that since the electron density in a semiconductor device cannot fluctuate at a rate much faster than the spontaneous lifetime, or in the case of a laser, the geometric mean of the stimulated

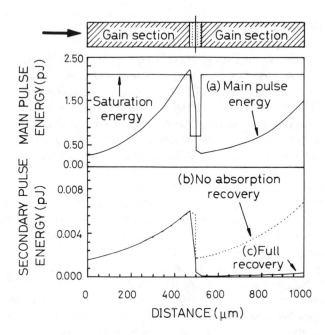

Figure 7.14 Simulated pulse energy versus distance along the LD for a 500-μm-long device: (a) the energy versus distance for the main pulse; (b) energy versus distance for a secondary pulse initiated by an imperfect AR coating without absorption recovery; (c) energy versus distance for a secondary pulse initiated by an imperfect AR coating with full absorption recovery. The saturation energy versus distance is also shown. (©1992 IEEE. From [31].)

lifetime and the photon lifetime (which is the direct modulation bandwidth of a laser), it would be improbable if not impossible to mode lock a laser at millimeter-wave frequencies of about 100 GHz. Analysis shows that this is not the case. The approach is based on the self-consistent model. First, a modulation in the electron density at approximately the intercavity mode spacing is assumed. The resulting optical modulation is then computed using a formalism similar to that of conventional active mode locking. The response of the electron density to the injection current modulation and the optical modulation is then computed and is self-consistently required to regenerate the electron density modulation assumed at the beginning of the analysis. For a fixed injection current modulation, the optical modulation can then be computed as a function of the cavity round-trip frequency.

The overall modulation response of a laser, with modulation frequency covering the range through the intercavity mode spacing frequency, is shown in Figure 7.15(a). At low modulation frequencies, the response follows the usual behavior of direct modulation of injection lasers, exhibiting the standard relaxation oscillation resonance. When modulated exactly at a cavity round-trip frequency $\Delta\omega$, the optical response depends on the value of $\Delta\omega$, represented by the dotted curve in Figure 7.15(a). The corner frequency for

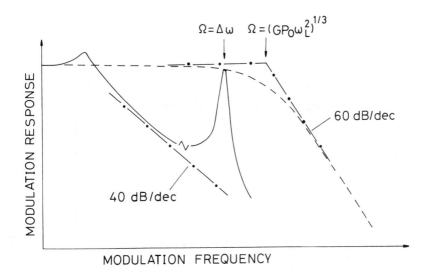

Figure 7.15(a) Overall modulation response of an injection laser over the entire frequency range beyond the intracavity modal frequency.

The roll-off occurs at $(GP_0\omega_L^2)^{1/3}$. Here, G is the optical gain constant of the active medium, P_0 is the average photon density, and ω_L is the half width of the gain spectrum. The parameter ω_L has an intrinsic value of approximately 2,500 GHz, and GP_0 is the inverse stimulated lifetime equal to 1/(0.5 ns); the corner frequency becomes 108 GHz. When the modulation frequency is detuned away from the intercavity mode spacing $\Delta\omega$, the response drops as a Lorentzian.

The analysis shows that there is basically no fundamental difficulty in trying to produce optical modulation at a cavity round-trip frequency of up to 100 GHz if one were to keep a small optical modulation depth. There is considerable difficulty, of course, in overcoming electrical parasitics and effectively injecting modulation current into the diode at those frequencies. This situation would be significantly alleviated if the modulation were internally applied (i.e., using an intracavity absorber) passive mode locking. But even in this case, the presence of a large parallel parasitic capacitance across the junction can cause clamping of the voltage of the junction and hence inhibit electron density modulation at such high frequencies. The effect is not serious, though, because the voltage is only a logarithmic function of the electron density, so that a small voltage fluctuation results in a large electron density fluctuation. Furthermore, at time scales of a few picoseconds, transit time effects of injection across the heterojunction might become significant. These effects, which are expected to take place at above 100 GHz, will be ignored.

We specifically have in mind a solitary LD with a tandem contact, one as the gain section and the other as absorber. We can use a self-consistent approach parallel to that of active mode locking to describe passive mode locking and to obtain the self-sustained

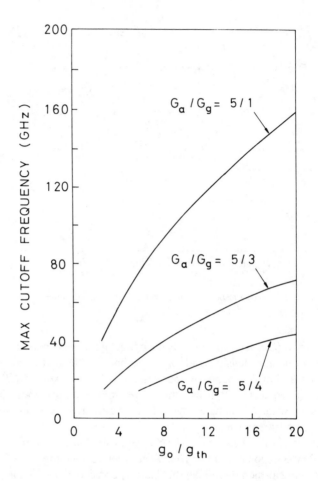

Figure 7.15(b) Maximum cutoff frequency as a function of g_0 for various values of G_a/G_g. (©1990 IEEE. From [32].)

optical modulation amplitude as a function of $\Delta\omega$ for different combinations of gain and absorber characteristics. The presence of a modulation in the net gain (gain minus absorber loss) at the intercavity modal frequency produces an optical modulation, which acts on the gain and absorber media separately to produce modulation in each of them. The difference in the gain and absorber modulation must self-consistently reproduce the net gain modulation assumed in the beginning. This self-consistency requirement imposes some restrictions on the parameters of the gain and absorber media, which we will examine below.

Figure 7.15(b) shows the highest attainable cutoff frequency (under the optimum value of a_0) as a function of g_0 for several values of absorber-gain coefficient ratios. The dvantage of a large ratio (G_a/G_g) is evident. An inhomogeneously pumped device with strong absorber actually lases at the wavelength off to one side of the gain peak, and ence it is possible to have a slightly larger differential absorption cross section. This is orroborated by the actual observation of passive mode locking, albeit a fairly weak one, n an inhomogeneously pumped device. Absorption-gain factors as large as 5:1, which re necessary for mode locking at above 100 GHz, will be attainable in an inhomogeneously umped laser diode made from single-QW materials.

The first observation of ultrashort light pulse generation by a mode-locked LD vithout an external cavity was made by Vasil'ev and Sergeev [33,34]. A modification of ne self-Q-switched AlGaAs/GaAs DH LD, which consisted of several sections amplifying nd absorbing the light, was used in the experiment. The length of the absorber and gain ection was 40 and 200 μm, respectively. To pump the LDs, rectangular current pulses f 10 to 20 ns in duration with 0- to 3-MHz repetition rate were applied. The reverse dc ias on the saturable absorber is varied within the 3V to 6V range. Generation of bandwidth-imited 2-ps pulses at 100-GHz repetition rate was demonstrated.

Theories of mode locking predict the recovery of gain and absorption in a laser efore an output pulse is emitted. From this point of view, it is unclear how the gain and bsorption can recover in the LD within such a short period of time (<10 ps). A possible xplanation is connected to the coherent effects of the light pulse interaction with the nverted laser population under the suppression of intraband relaxation processes by the trong electromagnetic field of the pulse [35] and the generation of steady-state $n\pi$ pulses.

Sanders et al. have confirmed these results [36,37]. A two-section MQW laser is assively mode locked without an external cavity at about 108 GHz [36]. The laser tructure is a 5-μm stripe, with the gain section about 245 μm and the absorber 90 μm. he sections are separated by a 5-μm-wide etch through the highly conductive cap layer nd have 5 kΩ electrical isolation. The pulsewidth average 2.4 ps and have a time-andwidth product of 1.1 when the absorber has a reverse bias of 0.67V to 0.80V. Self-ulsations at frequencies up to 8 GHz are also observed.

7.4 MODE LOCKING AND SELF-PULSATION

t was long recognized that the presence of a saturable absorber in a laser cavity causes elf-pulsation, which hinders or even inhibits mode locking [38].

Paslaski and Lau have obtained criteria for mode locking to take place and have hown that lasers with high-reflectivity coatings can most reliably achieve mode locking vithout being adversely affected by self-pulsation [39]. They experimentally map out the arameter ranges for mode locking (at 65 GHz) and self-pulsation (a few gigahertz) using GaAs MQW LD with a graded-index separate confinement. The structure has well and arrier widths of 100Å and 75Å, respectively, and an aluminum content of 0.3 in the arriers. Two separate top-contact sections define the gain and absorber regions with engths of 400 and 120 μm, respectively, and the gap between them is 50 μm. DC forward/

reverse biases were applied to the gain/absorber contacts and the mode-locked/self-pulsation output was observed.

We represent the bias conditions with the gain current (i_g) and the absorber reverse-bias voltage (V_b) and map the ranges in the i_g-V_b plane. The result is shown in Figure 7.16(a). The threshold line defines the lasing threshold. In principle, the mode-locking regime should extend almost all the way to the threshold line, but as seen in Figure 7.16(a), there is hardly a region of mode locking that is not affected by relaxation oscillation. The region labeled "strongly self-pulsing" is where self-pulsation completely quenches mode locking, and in the "weakly self-pulsing" region, self-pulsation with <100% modulation depth coexists with mode locking where they overlap.

The experiment was repeated with an identical laser coated with high-reflectivity coating of 70% and 90% on the front and back facets, respectively. The result is shown in Figure 7.16(b). The striking contrast with Figure 7.16(a) is that a mode-locked region is now clearly defined, which is free of self-pulsation.

To understand why high-reflection coatings help to minimize the simultaneous occurrence of mode locking and self-pulsation, we employ the phasor diagram analysis originally used to predict the feasibility of ultrahigh-frequency mode locking. In that approach, the photon intensity oscillation is represented by $S \exp(i\omega t)$, where S is the real amplitude of the oscillation. When considering mode locking, $\omega = \Omega$ = longitudinal-mode spacing (>50 GHz). When considering self-pulsation, $\omega = \omega_{sp}$ = self-pulsing frequency, in the 1- to 5-GHz range for typical lasers. The optical modulation drives the gain and absorber media to produce gain and loss modulation, whose difference is the net gain modulation given by

$$g_{net} = \left\{ \frac{-G_g g_0}{i\omega + 1/T_g} - \frac{-G_a a_0}{i\omega + 1/T_a} \right\} S \, e^{i\omega t} = \hat{g}_{net} \, e^{i\omega t} \qquad (7.14)$$

$\hat{g}_{net}$ is a phasor quantity, represented as a vector in the complex plane, which is responsible for driving the optical modulation. In (7.14), $G_{g/a}$ and $1/T_{g/a} = 1/\tau_{g/a} + G_{g/a}S_0$ are the differential gain and effective lifetimes for the gain and absorber, respectively, $\tau_{g/a}$ and S_0 being the spontaneous lifetime and average photon density, respectively; g_0 and α_0 are the steady-state gain and absorption. The first quantity in the braces ({}) in (7.14) is the gain modulation, and the second quantity is the loss modulation.

For self-consistency, $\hat{g}_{net}$ must lie within certain quadrants of the complex plane. For mode locking, it was shown that $\hat{g}_{net}$ must lie in the right half plane. For self-pulsation, the (relatively) low-frequency net gain modulation drives the average photon density in the laser cavity through the small signal photon rate equation:

$$\dot{S} = \hat{g}_{net} S_0 \qquad (7.15)$$

If self-pulsation were to take place, S increases with time, and therefore, from (7.15), $\hat{g}_{net}$ must also lie in the right half plane. These are illustrated in Figure 7.17. Thus, the

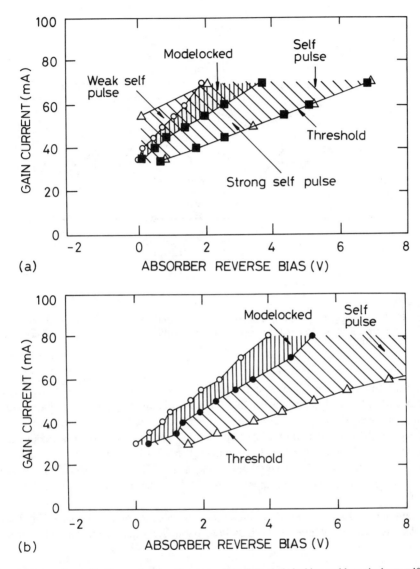

Figure 7.16 Experimentally observed parameter ranges of 65-GHz mode locking and low-gigahertz self-pulsation of (a) an uncoated (reflectivities of 0.3 on both facets) and (b) high-reflectivity coated (reflectivities of 0.7 and 0.9) QW laser. (From [39]. Reproduced by permission of AIP.)

conditions for self-pulsation and mode locking are actually very similar, given respectively by

$$\mathrm{Re}(\hat{g}_{\mathrm{net}}^{\mathrm{SP}}) > 0 \quad \text{and} \quad \mathrm{Re}(\hat{g}_{\mathrm{net}}^{\mathrm{ML}}) > 0 \tag{7.16}$$

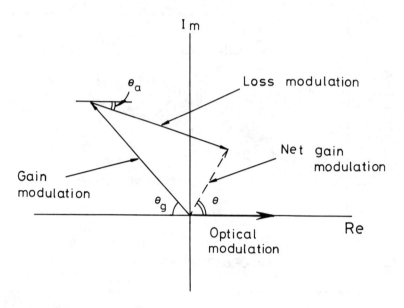

Figure 7.17 Phasor diagram representation of regimes of mode locking and self-pulsation. θ_a and θ_g are the phase lag of the loss and gain modulation, respectively, behind the optical modulation. (From [39].)

where $\hat{g}_{net}^{SP}$ and $\hat{g}_{net}^{ML}$ are $\hat{g}_{net}$ evaluated at $\omega = \omega_{sp}$ and Ω, respectively, and Re( ) denotes the real part. Physically, this means that the net gain modulation is in phase with the photon density modulation and hence tends to reinforce the latter. In order for $\hat{g}_{net}$ to be in the right half plane, it is preferable that the gain modulation lags in phase behind the loss modulation (Figure 7.17). The single-pole roll-off nature of the gain and loss modulation (7.14) dictates that if gain lags behind loss at $\omega = \Omega$, then the same will happen at $\omega = \omega_{sp}$. This is very unfortunate, since it implies self-pulsation will very likely accompany mode locking.

Can one find a situation where mode locking can take place without self-pulsation? Consider the situation where the various laser parameters satisfy the following criteria.

$$G_g S_0 \gg 1/\tau_g; \; G_a S_0 \gg 1/\tau_a \tag{7.17a}$$

$$\omega_{sp} \ll G_a S_0, \; G_g S_0 \ll \Omega \tag{7.17b}$$

These two conditions basically state that the stimulated lifetimes are short--in the subnanosecond range--which of course is much longer than $1/\Omega$, the cavity round-trip time. Both (7.17a) and (7.17b) can be satisfied at high photon densities. Under these conditions, (7.14) becomes

$$\text{Re}(\hat{g}_{net}^{SP}) \sim (-g_0 + a_0)S/S_0 \tag{7.18a}$$

and

$$\text{Re}(\hat{g}_{net}^{ML}) \sim (-G_g^2 g_0 + G_a^2 a_0)SS_0/\Omega^2 \tag{7.18b}$$

But the average gain and loss must satisfy the threshold condition:

$$g_0 - a_0 = \hat{g}_{th} > 0 \tag{7.19}$$

where $\hat{g}_{th}$ is the threshold gain. Thus, if conditions given by (7.17a) and (7.17b) are satisfied, then from (7.16) and (7.18a), self-pulsation will not take place. On the other hand, from (7.16), (7.18b), and (7.19), mode locking will take place if

$$\left(1 - \frac{g_{th}}{g_0}\right) > \left(\frac{G_g}{G_a}\right)^2 \tag{7.20}$$

It is obvious that this last condition can be satisfied more easily if the threshold gain g_{th} is small. Equations (7.17a), (7.17b), and (7.20) together serve as a set of prescriptions for achieving mode locking without self-pulsation, and can be summarized as having a large (intracavity) photon density and a low threshold. These are readily achieved with high-reflectivity coatings, as the experimental results described above have shown.

These results can be physically interpreted as follows. Both self-pulsation and mode locking represent dynamically unstable situations, whose occurrence requires that the loss saturates "more easily" than the gain as the photon density increases. However, the unsaturated gain is always higher than the unsaturated loss in a laser above the threshold, the difference being determined by the passive cavity Q. If the gain and loss saturate instantaneously with rising photon density (as in self-pulsation where the frequency is relatively low), then it is inevitable that the reduction in gain is always larger than the reduction in loss. If, however, the saturation is not instantaneous (as in high-frequency mode locking), in that the response time of the gain is very long, then over a brief instant the absorber may be reduced by a large amount while the gain has not yet responded to the rising photon density. This situation is particularly favored in a high-Q cavity, since in this case the unsaturated loss is only slightly smaller than the unsaturated gain, and hence the reduction in loss due to increasing photon density is large compared to the corresponding reduction in gain.

According to Lau and Paslaski's results, it is difficult to generate very short pulses (about 1 ps) without the simultaneous occurrence of a low-frequency (a few gigahertz) self-pulsation envelope modulating the pulse train [40]. However, it is not clear at present whether the same conclusion applies to other passively mode-locked configurations [41].

Large-signal dynamics of mode locking of LDs is studied in the time domain using the traveling-wave rate equations [42,43]. Here we consider a two-section structure. The

rate equations take into account the spatial variations of photon and carrier densities b
dividing the laser cavity into small segments along the propagating direction. The rat
equations therefore have the following traveling-wave form.

$$\frac{\partial S_{tr}^{\pm}(x,t)}{\partial t} \pm v_g \frac{\partial S^{\pm}(x,t)}{\partial x}$$

$$= \Gamma G S^{\pm}(x,t) - \alpha v_g S^{\pm}(x,t) + \Gamma \beta R_{sp} \tag{7.21}$$

$$\frac{\partial S^{\pm}(x,t)}{\partial t} = \frac{S_{tr}^{\pm}(x,t) - S^{\pm}(x,t)}{t_d} \tag{7.22}$$

$$\frac{\partial N(x,t)}{\partial t} = -G S^{\pm}(x,t) + \frac{J(x,t)}{d} - R_{sp} - R_{nr} - \frac{N(x,t)}{\tau_a} \tag{7.23}$$

$$G = \frac{g_0(N - N_{th})}{1 + \epsilon(N)S} \tag{7.24}$$

where S and N are the photon and carrier densities, respectively, Γ is the confinemen
factor, v_g is the group velocity, α is the distributed loss, β is the spontaneous emissio
coupling factor, τ_a is the absorber recovery time, g_0 is the differential gain coefficient
R_{sp} and R_{nr} are spontaneous and nonradiative recombination rates, respectively, N_{th} is the
threshold carrier density, $\pm$ denotes the waves in + and − directions. S_{tr} is the transien
photon density, which takes account of the signal delay effect, and t_d is the characteristic
delay time.

The results show simultaneous occurrence of mode locking and self-pulsation ove
a large parameter range. Pulsewidth is found to be mainly limited by the saturation
induced gain modulation.

7.5 MONOLITHIC MODE-LOCKED LASER DIODES

Good high-frequency mode locking requires (1) a gain with a wide modulation bandwidth.
(2) efficient and stable coupling between the active and passive sections, and (3) a low-
loss extended cavity. In usual experiments using actively mode-locked LDs, the extended
cavity is fabricated using bulk optics or hybrid integration techniques. These structures
are bulky and may suffer from excessive losses and poor stability.

Tucker et al. [44] report a monolithically integrated extended-cavity laser that
overcomes these problems. Using this laser, active mode-locking has been achieved at
repetition rates as high as 40 GHz. Morton et al. [45] report the results of hybrid mode
locking combining both active and passive mode locking of an LD. These functions
were integrated into a monolithic device with a 1.3-μm GaInAsP gain region, an active

waveguide, and a saturable absorber. The devices have low threshold currents and exhibit hysteresis in their light/current characteristics. The long integrated waveguides allow mode locking at a repetition rate of 15 GHz without the need for an external cavity. Pulsewidths as short as 1.4 ps have been demonstrated.

Wu et al. [41] report the generation of short optical pulses from monolithic CPM QW lasers. A schematic diagram of the monolithic CPM laser is shown in Figure 7.18. A BH laser is fabricated with a two-step organometallic vapor-phase epitaxy (OMVPE) growth technique. First, graded-index separate-confinement heterostructure InGaAsP/InGaAs MQW layers are grown on an n^+ InP substrate. Then iron-doped semi-insulating InP is selectively grown around the patterned 2-μm-wide active region to provide electric isolation and reduce parasitic capacitance. Standard lithography and wet chemical etching are used to produce the final structure. As shown in Figure 7.18, the active QW region extends throughout the entire cavity to remove undesired waveguide mismatch and to simplify the fabrication process. The uncoated cleaved facets are used as symmetric mirrors. The cavity is divided into five sections: the two end sections near the facets are modulators, the center section is a saturable absorber, and the remaining two sections between the modulators and the absorber are active waveguides.

In this monolithic CPM laser, an external RF source is used to actively mode lock two counterpropagating pulses in the linear cavity. Instead of relying on the two counterpropagating pulses to adjust themselves to collide in the saturable absorber, here the two end modulators force the pulses to collide in the center saturable absorber. The transient grating in the saturable absorber produced by these two colliding wave packets stabilizes and shortens the pulses. The timing accuracy between the two modulators is then determined by the equal path length of the integrated microstrip transmission lines, which is guaranteed by the lithographic process. The cavity length is 2.54 mm, which

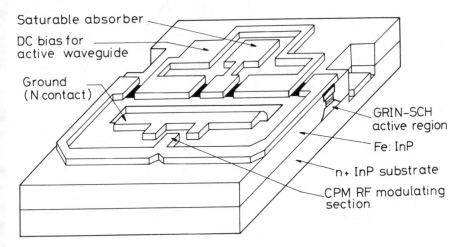

Figure 7.18 The monolithic CPM QW laser. (From [41].)

corresponds to a round-trip frequency of 16.3 GHz. Both the modulators and the saturable absorber are 70 μm long, and the gaps between sections are 10 μm long. The fabricated laser has a typical threshold current of 90 mA with the active waveguides biased only.

In the mode-locked experiment, 32.6-GHz sinusoidal signals are fed from a low-phase-noise synthesizer. This driving frequency is selected to match one-half of the cavity round-trip time. The shortest pulse is obtained when the active waveguide is biased just above the threshold (94 mA), with the end modulators and the center saturable absorber tied together through a bias tee. The second-harmonic autocorrelation trace exhibits a clean pulse with an FWHM of 2.2 ps. The signal-to-noise ratio is much better than that obtained with the non-CPM active mode-locked lasers on the same laser bar. The autocorrelation data fits well with a hyperbolic secant pulse shape, and the pulsewidth is 1.4 ps. This is achieved at only 14 dBm of RF power. Unlike the mode-locked lasers with external cavities, which suffer from multiple-pulse generation within a single cycle, only a single pulse is observed in this monolithic CPM laser. Nearly 100% intensity modulation is demonstrated with a peak power of 10 mW. The average optical output power of 0.5 mW is measured from the collimated beam.

Chen et al. [46] have achieved ultrafast subpicosecond optical-pulse generation with passive CPM of monolithic MQW InGaAsP LDs. Transform-limited optical pulses with durations of 1.1, 0.83, 1.0, and 0.64 ps are achieved at repetition rates of 40, 80, 160, and 350 GHz, respectively, without using any external ac sources.

REFERENCES

[1] For example, Weiss, C. O., and R. Vilaseca, *Dynamics of Lasers*, Weinheim: VCH Verlagsgesellschaft mbH, 1991.

[2] Arecchi, F. T., R. Meucci, G. Puccioni, and J. Tredicce, "Experimental Evidence of Subharmonic Bifurcations, Multistability, and Turblence in a Q-Switched Gas Laser," *Phys. Rev. Lett.*, Vol. 49, No. 17, October 1982, pp. 1217–1220.

[3] Ivanov, D. V., Ya. I. Khanin, I. I. Matorin, and A. S. Pikovsky, "Chaos in a Solid-State Laser With Periodically Modulated Losses," *Phys. Lett.*, Vol. A89, No. 5, May 1982, pp. 229–230.

[4] Kawaguchi, H., "Optical Bistability and Chaos in a Semiconductor Laser With a Saturable Absorber," *Appl. Phys. Lett.*, Vol. 45, No. 12, December 1984, pp. 1264–1266.

[5] Lee, C.-H., T.-H. Yoon, and S.-Y. Shin, "Period Doubling and Chaos in a Directly Modulated Laser Diode," *Appl. Phys. Lett.*, Vol. 46, No. 1, January 1985, pp. 95–97.

[6] Kao, Y. H., H. T. Lin, and C. S. Wang, "Noise Enhancement of Period Doubling in Strongly Modulated Semiconductor Lasers," *Japanese J. Appl. Phys.*, Vol. 31, No. 7A, July 1992, pp. L846–L849.

[7] Agrawal, G. P., "Effect of Gain Nonlinearities on Period Doubling and Chaos in Directly Modulated Semiconductor Lasers," *Appl. Phys. Lett.*, Vol. 49, No. 16, October 1986, pp. 1013–1015.

[8] Hemery, E., L. Chusseau, and J.-M. Lourtioz, "Dynamic Behaviors of Semiconductor Lasers Under Strong Sinusoidal Current Modulation: Modeling and Experiments at 1.3 μm," *IEEE J. Quantum Electronics*, Vol. 26, No. 4, April 1990, pp. 633–641.

[9] Tang, M., and S. Wang, "Simulation Studies of the Dynamic Behavior of Semiconductor Lasers With Auger Recombination," *Appl. Phys. Lett.*, Vol. 50, No. 26, June 1987, pp. 1861–1863.

[10] Ngai, W. F., and H.-F. Liu, "Observation of Period Doubling, Period Tripling, and Period Quadrupling in a Directly Modulated 1.55 μm InGaAsP Distributed Feedback Laser Diode," *Appl. Phys. Lett.*, Vol. 62, No. 21, May 1993, pp. 2611–2613.

11] Kawaguchi, H., and K. Otsuka, "Generation of Single-Longitudinal-Mode Gigabit-Rate Optical Pulses From Semiconductor Lasers Through Harmonic-Frequency Sinusoidal Modulation," *Electron. Lett.*, Vol. 19, No. 17, August 1983, pp. 668–669.

12] Gallagher, D. F. G., I. H. White, J. E. Carroll, and R. G. Plumb, "Gigabit Pulse Position Bistability in Semiconductor Lasers," *J. Lightwave Technology*, Vol. LT-5, No. 10, October 1987, pp. 1391–1398.

13] Kawaguchi, H., unpublished.

14] Kawaguchi, H., "Optical Bistability in Active Devices," *J. Physique*, Colloque C2, Supplement au No. 6, Tome 49, June 1988, C2-63.

15] Winful, H. G., Y. C. Chen, and J. M. Liu, "Frequency Locking, Quasiperiodicity, and Chaos in Modulated Self-Pulsing Semiconductor Lasers," *Appl. Phys. Lett.*, Vol. 48, No. 10, March 1986, pp. 616–618.

16] Arnol'd, V. I., "Small Denominators. I, Mappings of the Circumference Onto Itself," *Am. Math. Soc. Transl. Ser. 2*, Vol. 46, 1965, pp. 213–284.

17] Allen, T., "On the Arithmetic of Phase Locking: Coupled Neurons as a Lattice on R^2," *Physica*, Vol. 6D, 1983, pp. 305–320.

18] Yariv, A., *Optical Electronics*, 4th edition, Philadelphia: Saunders College Publishing, 1991.

19] Ho, P.-T., "Picosecond Pulse Generation With Semiconductor Diode Lasers," in *Picosecond Optoelectronic Devices*, Lee, C. H., ed., Orlando: Academic Press, 1984.

20] Haus, H. A., "Theory of Modelocking of a Laser Diode in an External Resonator," *J. Appl. Phys.*, Vol. 51, No. 8, August 1980, pp. 4042–4049.

21] Haus, H. A., "Theory of Mode Locking With a Fast Saturable Absorber," *J. Appl. Phys.*, Vol. 46, No. 7, July 1975, pp. 3049–3058.

22] Haus, H. A., "Theory of Mode Locking With a Slow Saturable Absorber," *IEEE J. Quantum Electronics*, Vol. QE-11, No. 9, September 1975, pp. 736–746.

23] Shank, C. V., "Generation of Ultrashort Optical Pulses," in *Ultrashort Laser Pulses: Generation and Applications*, 2nd ed., W. Kaiser, ed., Berlin: Springer-Verlag, 1993.

24] Fork, R. L., B. I. Greene, and C. V. Shank, "Generation of Optical Pulses Shorter Than 0.1 psec by Colliding Pulse Mode Locking," *Appl. Phys. Lett.*, Vol. 38, No. 9, May 1981, pp. 671–672.

25] Garmire, E. M., and A. Yariv, "Laser Mode-Locking With Saturable Absorbers," *IEEE J. Quantum Electronics*, Vol. QE-3, No. 6, June 1967, pp. 222–226.

26] Lee, T.-P., and R. H. R. Roldan, "Repetitively Q-Switched Light Pulses From GaAs Injection Lasers With Tandem Double-Section Stripe Geometry," *IEEE J. Quantum Electronics*, Vol. QE-6, No. 6, June 1970, pp. 339–352.

27] Basov, N. G., V. V. Nikitin, and A. S. Semenov, "Dynamics of Semiconductor Injection Lasers," *Sov. Phys. Uspekhi*, Vol. 12, No. 2, September-October 1969, pp. 219–240.

28] Morozov, V. N., V. V. Nikitin, and A. A. Sheronov," Self-Synchronization of Modes in a GaAs Semiconductor Injection Laser," *JETP Lett.*, Vol. 7, No. 9, 1968, pp. 256–258.

29] Schrans, T., R. A. Salvatore, S. Sanders, and A. Yariv, "Subpicosecond (320 fs) Pulses From CW Passively Mode-Locked External Cavity Two-Section Multiquantum Well Lasers," *Electron. Lett.*, Vol. 28, No. 16, July 1992, pp. 1480–1482.

30] Schell, M., A. G. Weber, E. Schöll, and D. Bimberg, "Fundamental Limits of Sub-ps Pulse Generation by Active Mode Locking of Semiconductor Lasers: The Spectral Gain Width and the Facet Reflectivities," *IEEE J. Quantum Electronics*, Vol. 27, No. 6, June 1991, pp. 1661–1668.

31] Derickson, D. J., R. J. Helkey, A. Mar, J. R. Karin, J. E. Bowers, and R. L. Thornton, "Suppresion of Multiple Pulse Formation in External-Cavity Mode-Locked Semiconductor Lasers Using Intrawaveguide Saturable Absorbers," *IEEE Photonics Tech. Lett.*, Vol. 4, No. 4, April 1992, pp. 333–335.

32] Lau, K. Y., "Narrow-Band Modulation of Semiconductor Lasers at Millimeter Wave Frequencies (>100 GHz) by Mode Locking," *IEEE J. Quantum Electronics*, Vol. 26, No. 2, February 1990, pp. 250–261.

33] Vasil'ev, P. P., and A. B. Sergeev, "Generation of Bandwidth-Limited 2 ps Pulses With 100 GHz Repetition Rate From Multisegmented Injection Lasers," *Electron. Lett.*, Vol. 25, No. 16, August 1989, pp. 1049–1050.

[34] Vasil'ev, P. P., "Ultrashort Pulse Generation in Diode Lasers," *Optical and Quantum Electronics*, Vol 24, 1992, pp. 801–824.

[35] Belenov, E. M., and P. P. Vasil'ev, "Suppression of Phase Relaxation in Semiconductors; Coherent Emission of the Active Medium of a Picosecond Injection Laser," *JETP Lett.*, Vol. 48, No. 8, October 1988, pp. 456–459 .

[36] Sanders, S., L. Eng, J. Paslaski, and A. Yariv, "108 GHz Passive Mode Locking of a Multiple Quantum Well Semiconductor Laser With an Intracavity Absorber," *Appl. Phys. Lett.*, Vol. 56, No. 4, January 1990, pp. 310–311.

[37] Sanders, S., L. Eng, and A. Yariv, "Passive Mode-Locking of Monolithic InGaAs/AlGaAs Double Quantum Well Lasers at 42 GHz Repetition Rate," *Electron. Lett.*, Vol. 26, No. 14, July 1990, pp. 1087–1089.

[38] Haus, H. A., "Parameter Range for CW Passive Mode Locking," *IEEE J. Quantum Electronics*, Vol. QE-12, No. 3, March 1976, pp. 169–176.

[39] Paslaski, J., and K. Y. Lau, "Parameter Ranges for Ultrahigh Frequency Mode Locking of Semiconductor Lasers," *Appl. Phys. Lett.*, Vol. 59, No. 1, July 1991, pp. 7–9.

[40] Lau, K. Y., and J. Paslaski, "Condition for Short Pulse Generation in Ultrahigh Frequency Mode-Locking of Semiconductor Lasers," *IEEE Photonics Tech. Lett.*, Vol. 3, No. 11, November 1991, pp. 974–976.

[41] Wu, M. C., Y. K. Chen, T. Tanbun-Ek, R. A. Logan, M. A. Chin, and G. Raybon, "Transform-Limited 1.4 ps Optical Pulses From a Monolithic Colliding-Pulse Mode-Locked Quantum Well Laser," *Appl. Phys. Lett.*, Vol. 57, No. 8, August 1990, pp. 759–761.

[42] Bowers, J. E., P. A. Morton, A. Mar, and S. W. Corzine, "Actively Mode-Locked Semiconductor Lasers," *IEEE J. Quantum Electronics*, Vol. 25, No. 8, June 1989, pp. 1426–1439.

[43] Yang, W., and A. Gopinath, "Study of Passive Mode Locking of Semiconductor Lasers Using Time-Domain Modeling," *Appl. Phys. Lett.*, Vol. 63, No. 20, November 1993, pp. 2717–2719.

[44] Tucker, R. S., U. Koren, G. Raybon, C. A. Burrus, B. I. Miller, T. L. Koch, and G. Eisenstein, "40 GHz Active Mode-Locking in a 1.5 μm Monolithic Extended-Cavity Laser," *Electron, Lett.*, Vol. 25, No. 10, May 1989, pp. 621–622.

[45] Morton, P. A., J. E. Bowers, L. A. Koszi, M. Soler, J. Lopata, and D. P. Wilt, "Monolithic Hybrid Mode-Locked 1.3 μm Semiconductor Lasers," *Appl. Phys. Lett.*, Vol. 56, No. 2, January 1990, pp. 111–113.

[46] Chen, Y. K., M. C. Wu, T. Tanbun-Ek, R. A. Logan, and M. A. Chin, "Subpicosecond Monolithic Colliding-Pulse Mode-Locked Multiple Quantum Well Lasers," *Appl. Phys. Lett.*, Vol. 58, No. 12, March 1991, pp. 1253–1255.

Chapter 8
Wavelength Conversion

Wavelength converters transform information on one wavelength to another (see Figure 8.1). Conventional wavelength conversion techniques used second-harmonic generation [1] or frequency shifting [2] and were found to be difficult to apply to optically functional systems.

Wavelength conversion using an LD, however, has wide tunability and fast operation for multiwavelength photonic systems. The function of wavelength conversion is accomplished by the use of (1) DFB LDs, DBR LDs, and Y-lasers, where a part of the active device is replaced by a saturable absorber; (2) gain saturation in LDs; (3) gain saturation in LD amplifiers; and (4) FWM in LD amplifiers. The first method is used for intensity-modulation-to-intensity-modulation (IM/IM) conversion. IM/IM, IM/FM (intensity modulation to frequency modulation), and FM/FM are realized by using the other three methods.

8.1 TANDEM-TYPE BISTABLE LASER DIODES WITH SATURABLE ABSORBER

A tunable wavelength conversion device can be constructed by monolithic integration of a tunable wavelength LD and an optically triggered gate. Some device structures have been proposed as shown in Figure 8.2. The function was first demonstrated using a

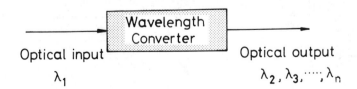

Figure 8.1 Operation principle of wavelength converter.

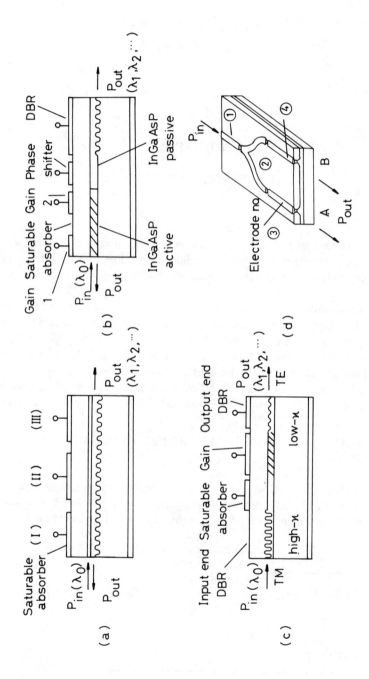

Figure 8.2 Structures of wavelength conversion devices: (a) three-section DFB LD; (b) DBR LD with phase shifter section; (c) DBR LD with asymmetric coupling coefficient structure; (d) Y-laser.

tandem-type bistable DFB LD with three divided sections by Kawaguchi et al. (Figure 8.2(a)) [3,4].

The basic operation principle of the device is shown in Figure 8.3. The device has two main functions: (1) the device emits coherent light only when light is injected, and (2) the output wavelength can be continuously tuned by changing the driving current. Therefore, this device is used for IM/IM conversion. The function of the first one is explained as follows. The driving current injected into the electrode located on one end (region (I) in the figure) is set at zero or at a low level depending on the resistance between the divided electrodes. The region acts as a saturable absorber whose absorption coefficient decreases with increasing input optical power. When optical input is injected into the saturable absorber, the absorption coefficient decreases, light intensity in the laser resonator increases, and positive feedback is established. Therefore, light output power increases abruptly at the laser threshold. On the other hand, when the light input intensity decreases, the output decreases abruptly through the reverse process. The current-light output curve shows a bistable or a differential gain characteristic, depending on the level of saturable absorption. Therefore, the device emits coherent light only when light is injected, provided that the bias current is set just below the laser turn-off threshold. The bistable operation is not essential in the tunable wavelength conversion device. However, if the amount of the saturable absorption is increased to get a high ON-OFF ratio, we

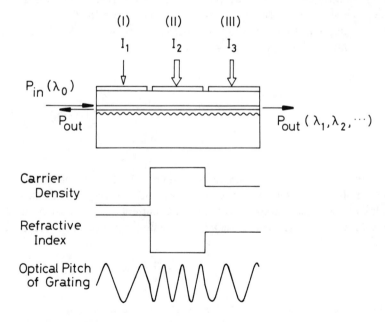

Figure 8.3 Basic operation principle of tunable optical-wavelength conversion device. Region (I) acts as a saturable absorber. The output wavelength is continuously tuned by the current in regions (II) and (III). (©1988 IEEE. From [4].)

will naturally see the bistable characteristics. The wavelength tunability can be realized by changing the driving-current distribution in a manner similar to that of an ordinary multielectrode DFB LD. Independent adjustment of the driving current to electrodes (II) and (III) results in a nonuniform distribution of carriers in the active layer. This changes the refractive index of the different sections of the active layer and increases the difference between the effective grating pitches of each region. The larger difference increases the threshold current of the laser. As a result, the output wavelength becomes shorter through the carrier-density-dependent refractive index as the current nonuniformity becomes more pronounced.

8.1.1 Coupled-Mode Theory for Two-Section DFB LDs [4]

The lasing threshold of the device can be analyzed using a standard eigenvalue equation for the propagation constants [5]. The device, which consists of two parts with different effective grating pitches depending on injection current, is analyzed. The model is schematically shown in Figure 8.4(a). Regions 1 and 2 in Figure 8.4(a) are equivalent to regions (II) and (III), respectively, in Figure 8.3. This model is applicable for explaining the experimental results described in Section 8.1.2, because the injected current and optical input into region (I) and the injected current into region (II) almost remained constant. Therefore, regions (I) and (II) can be considered one region for the output wavelength analysis. Coupling equations for counterpropagating two waves, $R(z)$ and $S(z)$, can be expressed as follows. For region p ($p = 1, 2$),

$$-\frac{d}{dz}R_p(z) + (\alpha_p - i\delta_p)R_p(z) = i\kappa e^{-i\Omega}S_p(z) \tag{8.1}$$

$$\frac{d}{dz}S_p(z) + (\alpha_p - i\delta_p)S_p(z) = i\kappa e^{i\Omega}R_p(z) \tag{8.2}$$

where α_p is the mode gain per unit length in region p, δ_p is the detuning of the propagation constant from the Bragg condition in region p, κ is the coupling coefficient, and Ω is the phase value of the grating at $z = 0$. When driving currents in the two regions differ, the propagation constants differ, and a frequency detuning from the Bragg frequency δ_p can be expressed as follows.

$$\delta_p = \beta_p - \pi/\Lambda_0 \tag{8.3}$$

where $\beta_p = n_p\omega/c$ is the propagation constant in the region p and Λ_0 is the pitch of the first-order grating. Here, κ is assumed to be the same in both regions, and gain is assumed to vary independently of the refractive index. The solutions of (8.1) and (8.2) are assumed in the form

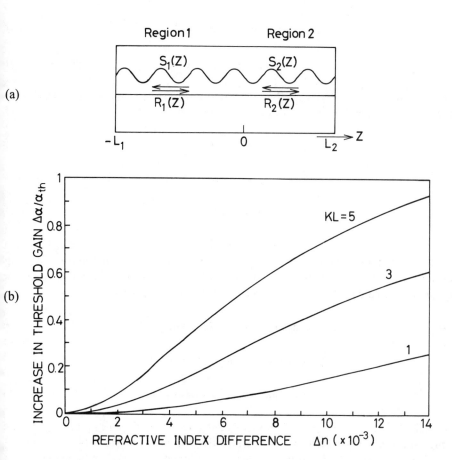

Figure 8.4 (a) Schematic diagram of a two-electrode DFB LD; (b) calculated relationship between increase in threshold gain versus refractive index nonuniformity. (©1988 IEEE. From [4].)

$$R_1(z) = r_{11} \, e^{\gamma_1 z} + r_{12} \, e^{-\gamma_1 z} \qquad (8.4)$$

$$S_1(z) = s_{11} \, e^{\gamma_1 z} + s_{12} \, e^{-\gamma_1 z} \qquad (8.5)$$

$$R_2(z) = r_{21} \, e^{\gamma_2 z} + r_{22} \, e^{-\gamma_2 z} \qquad (8.6)$$

$$S_2(z) = s_{21} \, e^{\gamma_2 z} + s_{22} \, e^{-\gamma_2 z} \qquad (8.7)$$

where r_{pq} and s_{pq} (p,q = 1,2) are constants. γ_1 and γ_2 satisfy the following dispersion relations.

$$\gamma_1^2 = (\alpha_1 - i\delta_1)^2 + \kappa^2 \tag{8.8}$$

$$\gamma_2^2 = (\alpha_2 - i\delta_2)^2 + \kappa^2 \tag{8.9}$$

Since the two regions are connected at $z = 0$ and the facet reflectivities are assumed to be zero, boundary conditions at both facets are given by

$$R_1(0) = R_2(0)$$

$$S_1(0) = S_2(0)$$

$$S_2(L_2) = 0$$

$$R_1(-L_1) = 0 \tag{8.10}$$

Substitution of (8.4) to (8.7) and (8.10) in (8.1) and (8.2) results in the following eigenvalue equation. The threshold gain α_{th} can be obtained by solving this equation, when the threshold gain for regions 1 and 2 is assumed to be the same.

$$\{ \gamma_1 \cosh(\gamma_1 L_1) - (\alpha_{th} - i\delta_1) \sinh(\gamma_1 L_1) \}$$

$$\times \{ \gamma_2 \cosh(\gamma_2 L_2) - (\alpha_{th} - i\delta_2) \sinh(\gamma_2 L_2) \}$$

$$+ 4\kappa^2 \sinh(\gamma_1 L_1) \sinh(\gamma_2 L_2) = 0 \tag{8.11}$$

In the calculation, $L_1 = L_2 = L$, $\beta_1 = (n_0 - (\Delta n/2))\omega/c$, and $\beta_2 = (n_0 + (\Delta n/2))\omega/c$ are assumed for simplicity, where Δn is the refractive index difference between regions 1 and 2. The calculated relationship between $\Delta\alpha_{th}/\alpha_{th0}$ and Δn, where $\Delta\alpha_{th}$ is the rate of increase in the threshold gain due to carrier nonuniformity (i.e., $\Delta\alpha_{th} = \alpha_{th} - \alpha_{th0}$) is shown in Figure 8.4(b). Here, α_{th0} is the threshold gain without carrier nonuniformity. $L = 150$ μm, $\lambda = 1.55$ μm, and $n_0 = 3.5$ in the calculation. Refractive index nonuniformity decreases optical feedback through grating and leads to an increase in the lasing threshold. The output wavelength of the wavelength conversion device depends on the threshold current through the carrier-dependent refractive index. The typical experimental value of the output wavelength shift is Δf (GHz) = +25 GHz/mA for the conventional DFB LD with $\lambda = 1.55$ μm and $I_{th} = 16$ mA. The refractive index change can be estimated from Δf. Therefore, when the current nonuniformity is $\Delta I = I_1 - I_2 = 1$ mA, Δn becomes 4.5 × 10^{-4}. On the other hand, $\Delta f = 4.5$ GHz/mA and $\Delta n = 0.80 \times 10^{-4}$ in an LD with $I_{th} = 75$ mA. That is, Δf and Δn are inversely proportional to I_{th}. From Figure 8.4(b), when $\kappa L = 3$, the lasing threshold gain increases 0.3% for an $I_{th} = 16$ mA device (about a 0.05-mA increase in I_{th} if gain linearly depends on current I) from that of the uniformly pumped LD. This leads to a 1.3-GHz increase in oscillation frequency. Thus, output wavelength can be tuned continuously by changing current distribution through the change in the threshold current.

8.1.2 Experimental Results on Three-Section DFB LDs

The structure of the wavelength conversion device is the same as that shown in Figure 3.4 in Chapter 3. The device consists of the wavelength-tunable DFB LD segment with two separate electrodes and the part of the saturable absorber into which light is injected [4]. The 1.55-μm BH GaInAsP/InP DFB LD with a first-order grating was fabricated by a metal-organic vapor-phase epitaxy (MOVPE)/LPE hybrid growth method. The p-type electrode was divided into three 100-μm sections. Injection currents for the divided sections are denoted by I_1, I_2, and I_3, respectively. Resistance between p-type electrodes R was in the 100Ω to 700Ω range, which was sufficiently large relative to the resistance between the p- and n-type electrodes. Thus, the divided regions can be excited independently through the electrodes.

A typical measured light output versus current ($I_2 + I_3$) curve of the device with $R = 100\Omega$ for the current ratio $I_1:I_2:I_3 = 0:2:1$ was shown in Figure 3.5(a). The bistable current region (the difference between the turn-on and turn-off threshold currents) decreased with increasing I_1 because saturable absorption decreased. The curve was also slightly dependent on the $I_2:I_3$ ratio, though it is not indicated in the figure. For example, the turn-on threshold registered a 2-mA increase and the bistable current region showed a 1.3-mA decrease when the $I_3/(I_2 + I_3)$ ratio changed from 0.2 to 0.67.

When the bias current ($I_b = I_1 + I_2 + I_3$) is set to just below the turn-off threshold, as shown in Figure 3.5(a), the optical input-output curve exhibits bistability (see Figure 3.5(b)). This is the result of the optical power injection into the saturable absorber. Bias current values are indicated by the closed circles (1) to (3) in Figure 3.5(b)(i). The bistable input power region increases with the changing of the bias current from zero to the turn-off threshold. When an input optical pulse that exceeds the turn-on threshold is injected, the device emits coherent light greater than the input power. The bistable characteristic is not essential for the wavelength conversion function. However, the device also acquires a pulse-shaping function as a result of this characteristic. A wavelength conversion device with a memory function can also be realized through this device.

The experimental setup for measuring optical-wavelength conversion function is shown in Figure 8.5(a). LD light was injected into the wavelength conversion device through an optical isolator. Polarization of the injected light was set to the same (TE) with the wavelength conversion device output or orthogonal (TM).

Lasing wavelength in relation to current ratio $I_3/(I_2 + I_3)$ for both the TE- and TM-polarized light inputs is shown in Figure 8.5(b). Resistance R of the device used in this experiment was 700Ω. I_1 is set at 10 mA to optimize the size of the bistable current region. Lasing wavelength was swept about 5Å for both TE and TM input without mode hopping when the ratio $I_2/(I_2 + I_3)$ changed from 0.4 to 0.65. Lasing wavelength did not

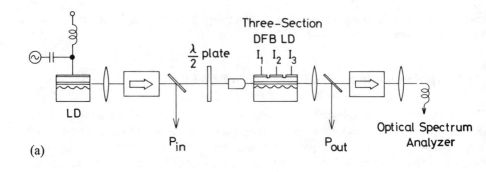

(a)

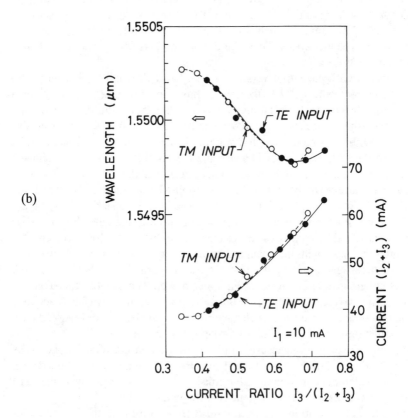

(b)

Figure 8.5(a,b) (a) Experimental setup for measuring the optical-wavelength conversion function. LD light is injected into a saturable absorber of the wavelength conversion device (three-section DFB LD) through an optical isolator. The lasing wavelength of the wavelength conversion device is changed by changing in I_2 and I_3. (b) Lasing wavelength against current ratio $I_3/(I_2 + I_3)$ for both TE- and TM-polarized light inputs. Bias current $(I_2 + I_3)$ is also shown. Input wavelength is 1.54742 μm.

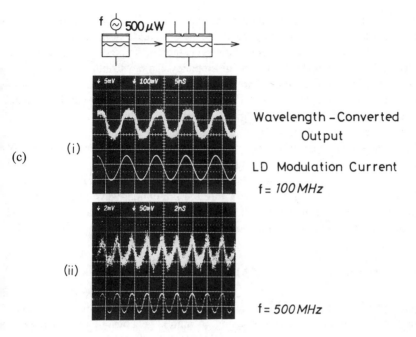

Figure 8.5(c) Driving-current waveforms of the LD for optical injection (lower trace) and light output waveforms from the optical-wavelength conversion device (upper trace): (i) 100 MHz; (ii) 500 MHz. (©1988 IEEE. From [4].)

depend on input polarization. Input power was 330 μW and input wavelength was 1.54742 μm for both polarizations. The laser operated in a single longitudinal TE mode and output power was about 1 mW in both cases. Coupling losses were assumed to be about 10 dB for input and 5 dB for output in the above power estimation. Input wavelength was set to be 23Å to 28Å less than the lasing wavelength in this case. Input wavelength was selected to be 3Å to 7Å less than lasing wavelength in another experiment [3]. No difference was found between them. No substantial increase is expected in optical set pulse power when input wavelength is in the gain wavelength region of the wavelength conversion device as described in Section 3.1. A laser oscillation wavelength of a DFB LD can be freely selected in the gain wavelength region by setting the grating pitch. Therefore, when the laser oscillation wavelength is set at the shorter wavelength side in the gain wavelength region, optical-wavelength conversion that shifts to the shorter wavelength might be possible.

The bias current ($I_2 + I_3$), which is just below the lasing threshold, is also shown against the current ratio $I_3/(I_2 + I_3)$ in Figure 8.5(b). I_1 is 10 mA. When the ratio $I_3/(I_2 + I_3)$ changed from 0.4 to 0.65, the current changed from about 40 to 60 mA. It can be seen that there is a strong correlation between the change in bias current and that in output wavelength. From the theoretical estimation previously described, because $\Delta I = 24$ mA (ΔI is the current nonuniformity), an approximate 15% increase in the threshold current and a 4.8Å wavelength shift are expected. However, there is some discrepancy between the estimated and experimental values. This discrepancy can be attributed to factors such as residual facet reflectivity, which are not considered in the calculation. The threshold current was at minimum at $I_3/(I_2 + I_3) \sim 0.35$, not at 0.5. This may be due to the influence of I_1. The current distribution becomes almost symmetric at 0.35.

Switching time was measured from the response to a TE-mode input pulse train generated from a conventional DFB LD by direct modulation. A device with $R = 700\Omega$ was used. LD driving-current waveforms (100, 500 MHz) for optical injection and light output waveforms from the optical-wavelength conversion device are shown in Figure 8.5(c). The input signal entering the detector through the optical-wavelength conversion device is negligible in this experiment. The device could be operated at up to 500 MHz under the condition that input optical power is about 500 μW.

8.1.3 Wavelength Conversion in DBR LDs

Performances of wavelength conversion devices can be improved by using DBR LDs. A tuning range of 3.5 nm was achieved by using an inhomogeneously excited LD with a DBR section and a phase shifter section, shown in Figure 8.2(b) [6]. High-speed operation at 1 Gbps has also been demonstrated with an extremely low input power (6 μW) in an LD with the same device structure. To achieve fast wavelength conversion, it is important to optimize the bias currents of device and the input power. Typical turn-off time of LDs with saturable absorbers is 2 or 3 ns, which is much slower than the few-hundred-picosecond turn-on time. Therefore, turn-off time limits the operation speeds. Turn-off starts when the injection light stops. At this time, the absorber is saturated and has very low absorption in the large internal power of the lasing light. In this situation, the laser and its internal light power turn off very slowly due to the absorber's very low absorption rate. Therefore, to obtain faster turn-off, it is necessary to increase the absorber's absorption or lower the power of the initial internal light. This will result in much faster turn-off response because feedback acceleration may occur between the decreasing internal light and increased absorption. Both input power and bias current should be lowered to attain a low initial internal power for fast turn-off. Figure 8.6 shows the measured turn-off time depending on input power and ΔI_b, where ΔI_b is defined by the current deviation from the turn-off threshold current. Under a fixed bias, lower input power caused faster turn-off. However, turn-off was unstable under low-power input light. This is the limiting point in the plots. The larger input drastically increased turn-off time and this is the effect of large saturation in the absorber's absorption. Under fixed input power, larger bias deviation caused faster turn-off, but placed a larger limitation on input power. The condition with ΔI_b at 0.2 mA is the optimum point, and the fastest turn-off time of 250-ps at 6 μW of input power was obtained.

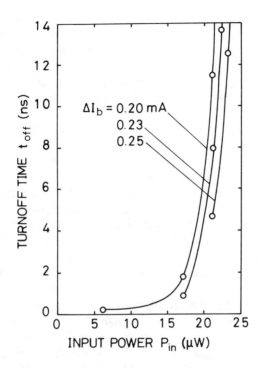

Figure 8.6 Turn-off-time dependence on input power. For lower input power than the terminal points of the traces, the operation is unstable. The fastest turn-off time is 250 ps with ΔI_b at 0.2 mA and P_{in} at 6 μW. ($^{\circ}$1992 IEEE. From [6].)

Paradisi and Montrosset analyzed the static performance of a wavelength converter obtained by a multisection DBR LD with a saturable absorber between two gain sections [7]. The performances are evaluated in terms of conversion efficiency, suppression ratio, and minimum input power needed for wavelength conversion.

A tunable wavelength conversion device with a multielectrode DBR structure (see Figure 8.2(b)) has optical logic functions of an inverter (NOT) and an exclusive OR (XOR) [8]. These optical functions are easily changed by varying bias current settings in gain sections. At just above the turn-on threshold, the device acts as an optical inverter. At just below the turn-off threshold, its acts as an optical XOR element. The mechanism of both NOT and XOR operations based on gain quenching accelerated through a saturable absorber is a major feature of the wavelength conversion device.

Usual optical-frequency conversion devices output the optical-frequency-converted light to both input and output ends of the devices. A device that is monolithically integrating a corrugation-grating–type filter with a fairly high coupling coefficient into the input end of the device emits low output power from the input end. The device structure is schematically shown in Figure 8.2(c) [9–12]. A scanning electron microscope (SEM) photograph of the device is shown in Figure 8.7(a). The input-end DBR mirror acts as a mode filter for TE- and TM-mode light. The Bragg wavelengths of a corrugation grating waveguide for TE- and TM-mode light differ because of the difference in equivalent refractive indexes of the waveguide for those modes. The high-reflectivity wavelength range becomes wider as the coupling coefficient of the DBR mirror becomes larger. The wavelength range where TE-polarized light cannot pass but TM-polarized light can pass through the DBR mirror is maximized by optimizing the coupling coefficient of the DBR mirror. Figure 8.7(b) shows the measured TE- and TM-polarized light transmission spectra of the higher-κ DBR mirror. The Bragg wavelength difference for TE- and TM-polarized light is about 4 nm and the high-reflectivity wavelength ranges for TE- and TM-polarized light are about 4 nm. The TE-polarized light cannot pass but TM-polarized light can pass through the DBR waveguide within the wavelength range of 1.551 to 1.555 μm. The coupling coefficient of the output-end DBR mirror is set lower than that of the input-end DBR mirror for single-mode operation of the optical-frequency conversion device. The converted light wavelength was scanned by injecting current into the output-end DBR mirror region.

Optical-frequency conversion is obtained using this device. The converted light wavelength was scanned from 1.556 to 1.552 μm by injecting current to the output-DBR mirror region. The wavelength scanning range was 4 nm and was coincident with the high-reflectivity wavelength range of the higher-κ DBR mirror. Figure 8.7(b) also shows the dependence of converted light power on injected light wavelength. The converted light wavelength was set to 1.5546 μm in this measurement. Open circles indicate the converted light power when the injected light had TE polarization, while solid circles show that for TM-polarized injected light. Optical-frequency conversion was achieved when the injected light had TM polarization and the injected light wavelength ranged

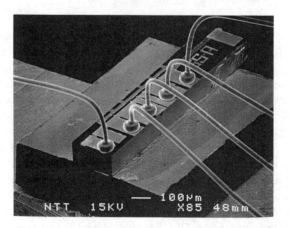

Figure 8.7(a) SEM photograph of a frequency conversion device with asymmetric κ-DBR structure (courtesy of NTT).

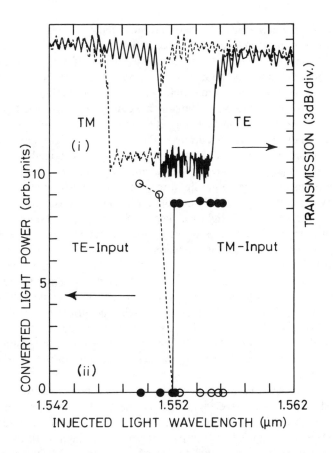

Figure 8.7(b) Transmission spectra and converted light power of optical-frequency conversion device with asymmetric κ-DBR structure: (i) transmission spectra of input-end DBR mirror; (ii) converted light power against injected light wavelength. Open circles: TE input; solid circles: TM input. (From [10].)

from 1.552 to 1.556 μm, where the reflectivity of the input-end DBR mirror was low for TM-polarized light. In this wavelength range, optical-frequency conversion was not achieved when the injected light had TE polarization. The optical-frequency conversion for TE-polarized light input was achieved in the injected light wavelength range below 1.5520 μm. The injected light wavelength and polarization dependence on the optical-frequency conversion characteristics reflects the characteristics of the higher-κ input-end DBR mirror. The ratio of converted light output from the input end of the device to the output end was -30 dB.

The frequency response of the unidirectional output optical-frequency conversion device was measured [13]. The light-output-versus-current characteristic of the device showed hysteresis characteristics. The bias current to the gain region was set just below the turn-off threshold for frequency conversion. The device was operated by sinusoidally intensity-modulated TM-polarized light with a wavelength of 1.554 μm, corresponding to the range where the input-end high-κ DBR region is highly reflective to TE-polarized light. In this case, the converted output light waveform from the device was almost the same as the input signal waveform. The wavelength of the converted light was set to 1.556 μm, even though the frequency-converted light wavelength could be varied from 1.556 to 1.552 μm by changing the current to the output-end DBR region. The IM index of the input intensity-modulated TM-polarized light was maintained at unity. The IM index is expressed as $(P_{max} - P_{min})/(P_{max} + P_{min})$, where P_{max} and P_{min} denote maximum and minimum intensities of the IM output light, respectively. The IM index of the converted light was unity when the input light's modulation frequency was lower than 300 MHz. The intensity of the input signal light was kept at 5 mW$_{p-p}$. The coupling coefficient of the signal light to the saturable absorber region was estimated to be 0.1, so the net input light intensity coupled to the saturable absorber region was estimated to be 0.5 mW$_{p-p}$. The converted light intensity was measured to be 0.5 mW$_{p-p}$, and the power conversion efficiency, which denotes the ratio of the converted signal intensity to the input signal intensity, was estimated to be unity. Above this frequency, both the IM index and the power conversion efficiency decreased gradually with increasing modulation frequency. The 3-dB bandwidth, which denotes the modulation frequency at which the IM index became half of that at zero modulation frequency, was measured to be about 800 MHz. This value was thought to be limited by the carrier lifetime in the saturable absorber region (about 1 ns).

Light injection can be used to manipulate the optical output of multisegment devices as described above. If short-wavelength light is injected to one facet, the free carriers generated due to absorption push the device from below to above the threshold. In the case of the Y-laser (see Figure 8.2(d)), two different optical input ports can be used, and the response of the device corresponds to optical logic OR or AND connections of the two inputs [14]. An optical input data stream at $\lambda_{in} = 1.31$ μm is injected to one input, and the same data stream appears at the output, but transformed to $\lambda_{out} = 1.55$ μm. It should be noted that the output wavelength λ_{out} of the device is electrically presettable within the tuning range of the Y-laser. Furthermore, since the incoming light is absorbed, wavelength, spectrum, and polarization are not critical. This means that wavelength conversion from about 850 nm or about 1.3 to 1.55 μm is feasible, and the output is single-mode, even if the input is multimode. Applications of the described wavelength conversion could be, for example, where networks operating at a short wavelength (optical interconnection, low-cost subscriber networks) are to be connected with 1.55-μm transmission systems, or where transformation of 1.3-μm Fabry-Perot multimode to 1.55-μm single-mode signal is needed.

8.1.4 Hybrid Devices

If we use hybrid configuration, the wavelength tuning range becomes much wider. A tunable wavelength converter was constructed using an MQW BLD and a liquid crystal abry-Perot interferometer (LC-FPI), which was inserted into the external cavity [15]. An MQW BLD consists of a gain region and a saturable absorption region, where the bsorption coefficient nonlinearly decreases, depending on the injected number of photons, nd the amount of the absorption is controlled by the applied voltage. This causes differen- ial gain or bistable current-light output characteristics, depending on voltage applied to he saturable absorption region of the MQW BLD. One of the facets of an MQW BLD s AR coated to configure an external cavity and an LC-FPI is inserted to control the asing wavelength. The wavelength converter is set at the threshold characteristics and ias current is injected into the gain region just below the threshold. When a signal pulse s injected, it will be amplified in the gain region and absorbed in the saturable absorption egion, and it reduces the loss inside the cavity and causes lasing. The wavelength converter ases at the wavelength selected by the LC-FPI only when a signal light pulse is injected. nput light pulses with a wavelength of 1.495 to 1.515 μm are converted to light pulses /ith a wavelength of 1.485 to 1.525 μm.

.2 GAIN SATURATION IN LASER DIODES

Vavelength conversion can be performed by the depletion of carriers in LDs, where the ignal bandwidth is determined by the relaxation frequency of the LD. The operation rinciple is schematically shown in Figure 8.8. The output data becomes the inversion of 1e input data.

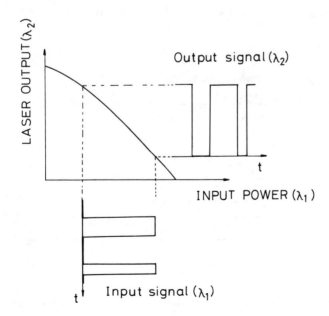

igure 8.8 Schematic of the principle of LD output modulation that is used in wavelength converters.

By using this method, penalty-free wavelength conversion of 2.5-Gbps signals over more than 18 nm and with an internal amplification of 13 dB has been achieved [16]. The device used as a wavelength converter is a three-section DBR LD (see Figure 8.9). The operation principle is conceptually simple: an amplitude-modulated signal with wavelength λ_1 is injected into the gain section of the DBR LD in which the carriers are depleted due to the input signal. The change in carrier number will generate both IM and FM of the output signal from the DBR LD, which is oscillating at λ_2. Obviously, both IM/IM and IM/FM conversion can be realized. The efficiency of these two modes of operation depends on the bias conditions of the DBR LD. The output wavelength λ_2 can be changed by the current to the Bragg (and phase) section of the DBR LD. The possible range of the input wavelength is given by the gain bandwidth of the gain section; therefore the conversion principle allows wavelength conversion from both shorter to longer and longer to shorter wavelengths.

Wavelength conversion of 5- to 10-Gbps data channels has also been demonstrated by the use of a three-section DFB LD [17]. A penalty-free wavelength shift of 5 nm was obtained for a data rate up to 8 Gbps. The coupled optical input power is -1 dBm and converted output power is 15 dBm at the facet.

Using the lasing frequency shift due to carrier depletion by externally injected light, a 1.534-μm FM light was converted to a 1.550-μm FM light [18]. The fundamental configuration for the wavelength conversion is illustrated in Figure 8.10. A Mach-Zehnder (M-Z) optical filter, an optical-power adjusting element, and a DFB LD are connected in series. The DFB LD is assumed to oscillate in the TE mode at a wavelength of λ_2. In this figure, FM(λ_i) and IM(λ_i) denote FM and IM light with a wavelength of λ_i.

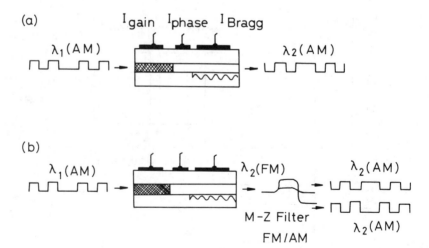

Figure 8.9 Principle of operation: (a) AM to AM; (b) AM to FM (and FM to AM by Mach-Zehnder interferometer). (©1993 IEEE. From [16].)

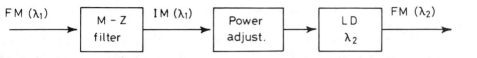

Figure 8.10 Fundamental configuration for wavelength conversion using a Mach-Zehnder optical filter, an optical-power adjusting element, and a DFB LD. FM(λ_i) and IM(λ_i) denote FM and IM light with a wavelength of λ_i.(From [18]. Reproduced by permission of IEE.)

FM(λ_1), which will be converted, is input to the M-Z filter. The M-Z filter has the configuration shown in Figure 8.11(a), where two 3-dB couplers are connected by two optical paths with different lengths (i.e., an asymmetric M-Z interferometer). Transmittance of this configuration as a function of optical frequency is shown in Figure 8.11(b). The frequency interval is determined by the optical-path difference. We assume that the frequency-shift keying (FSK) light is input to the M-Z filter. When the transmittance of the M-Z filter is adjusted to match the mark (f_m) and space (f_s) frequencies, as shown in Figure 8.11(b), the FM light is converted to IM light. This is the FM/IM conversion function of the M-Z filter. Thus, IM(λ_1), which has a modulation signal identical to the input FSK signal, can be obtained at the M-Z filter output. Note that this process has a waveform-shaping function, since slight deviations of mark or space frequencies become

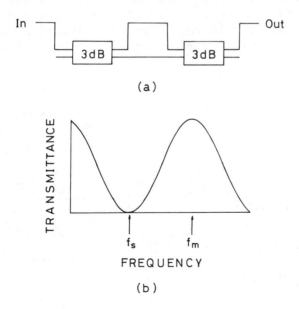

Figure 8.11 (a) Configuration and (b) transmittance of an M-Z optical filter. (From [18]. Reproduced by permission of IEE.)

flat at the top or bottom of the sine-shape transmittance of the M-Z filter when they ar
converted into IM signals.

Next, IM(λ_1) is input to the DFB LD through an optical-power adjusting elemen
(e.g., an optical amplifier or an attenuator). Here, IM(λ_1) is assumed to be TM-polarize
light. When external light is injected into a single-mode oscillating DFB LD, the carrie
density changes, the refractive index changes, and the lasing frequency shifts. Through thi
optical-injection-induced frequency shift, the DFB LD is frequency modulated according t
the IM(λ_1) that has an identical signal to the FM(λ_1). Thus, an FM(λ_1) replica can b
obtained at the λ_2 wavelength by adjusting the intensity of the light injected into the DF
LD. This is how the wavelength conversion from λ_1 to λ_2 for FM light is accomplished

TM-polarized light is injected into the DFB LD, not TE-polarized light. When TE
polarized light at a wavelength close to λ_2 is injected, it interacts with the original oscillatin
light and causes injection locking or multimode oscillation. For application to dens
wavelength-division multiplexing (WDM) or frequency-division multiplexing (FDM) sys
tems, conversion between narrow-spacing wavelengths is important and these phenomen
(injection locking, multimode oscillation) must be avoided. This is why TM-polarize
light is injected in this wavelength conversion.

The configuration of the wavelength conversion experiment that was carried out i
shown in Figure 8.12. Light emitted from a DFB LD, LD1 ($\lambda_1 = 1.534$ μm), was inpu
to a silica-based waveguide M-Z filter with a frequency period of 4 GHz. LD1 wa
frequency modulated by direct-current modulation with a 120-Mbps pulse pattern of (1,0)
where the deviation frequency was set at 2 GHz. The output from the M-Z filter wa
injected into the DFB LD (LD2). An LD amplifier was inserted between the M-Z filte
and LD2 in order to inject an appropriate light intensity to induce a frequency shift of

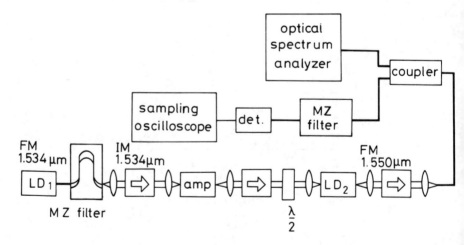

Figure 8.12 Experimental setup demonstrating wavelength conversion. (From [18]. Reproduced by permissio
of IEE.)

;Hz in LD2. Here, the injected light was changed to TM-polarized light by a half-wave plate. The output light from LD2 was demodulated by an M-Z filter with a frequency period of 10 GHz.

The output spectrum in the wide wavelength region is shown in Figure 8.13(a). Although the input light transmitting through LD2 is barely observed at 1.534 μm, we can say that the single-wavelength light ($\lambda_2 = 1.550$ μm) appears at the output. The frequency spectrum at 1.550 μm is shown in Figure 8.13(b). It is shown in Figure 8.13(b) that LD2 was (1,0) pattern frequency modulated with a deviation frequency of 2 GHz. From the experimental results, peak power required for the injected light to induce a frequency shift of 2 GHz is estimated to be 0.5 mW for the LD2 chip.

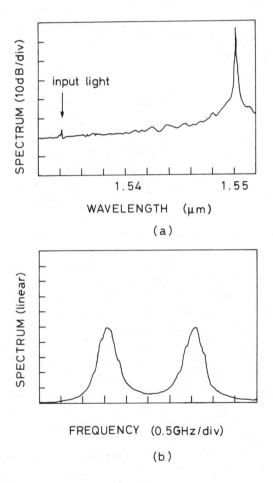

Figure 8.13 Output spectra of LD2 obtained from wavelength conversion experiment: (a) wide wavelength region; (b) frequency spectrum at 1.55 μm. (From [18]. Reproduced by permission of IEE.)

Wavelength conversion with a self-wavelength-selecting function is also possible [19]. The converter uses the FM/IM conversion function of an M-Z filter and the light-injection-induced frequency shift in a DFB LD. FM light is converted to another wavelength that can be tuned by a wavelength selection signal contained in the original input light.

The Y-laser (see Figure 8.2(d)) can be operated as a tunable frequency converter through gain saturation [14,20,21]. If external light of around 100-μW input power (TE) from a DFB source is launched, the Y-laser switches from its electrically preselected interferometric mode λ_Y to the optical input wavelength λ_0. By this method we achieve up to 6-THz tunable frequency conversion of 2.5-Gbps data stream. Using an asymmetric-κ DBR laser described in Section 8.1.3, unidirectional output optical-frequency conversion has been achieved for 10-Gbps nonreturn-to-zero (NRZ) signals through the gain saturation [13]. Wide tunable wavelength conversion of 40 nm has been obtained by using a super-structured grating LD at a 10-Gbps data rate [22].

8.3 GAIN SATURATION IN LASER DIODE AMPLIFIERS

Optical-wavelength conversion is also possible using an LD amplifier. The operation principle is simple and relies on gain saturation in the LD amplifier. An injected CW channel is modulated according to the bit pattern of the input signal channel due to gain saturation in the LD amplifier. The complementary data signal is encoded on the wavelength-shifted beam because the gain of the amplifier is high when the input signal is low and low when the input signal is high. This device does not actually transfer data from one wavelength to another, but rather copies data onto a signal at a different wavelength.

Conversion of data from 1.3 to 1.5 μm has been demonstrated by Barnsley and Fiddyment using a two-section Fabry-Perot LD amplifier [23]. Conversion of the data between the two wavelengths is achieved by altering the gain of a Fabry-Perot nonlinear amplifier at 1.55 μm by the injection of 1.3-μm light into a saturable absorber region within the amplifier structure. Durhuus et al. have also demonstrated, by using a two-section Fabry-Perot LD amplifier, an optically controlled gate/frequency converter that operates at 2.5 Gbps with a coupled optical control signal of -11.7 dBm [24]. They used a 1.553-μm control light input and a 1.548-μm signal light input. Joergensen et al. [25] used LD amplifiers for efficient wavelength conversion up to 4 Gbps. Rise time of 120 ps, fall time of 70 ps, and an extinction ratio up to 11 dB at a coupled input power of -12 dBm were measured. A system experiment demonstrated wavelength conversion over 17 nm at 4 Gbps.

A wavelength shifter can be constructed using the gain saturation nonlinearity of a traveling-wave LD amplifier [26]. In this device, an input IM signal at bit rates up to 2.5 Gbps, which is capable of saturating the gain of an LD amplifier by as much as 10 dB, is injected into the amplifier along with a CW beam at a shifted wavelength. The optical amplifier is a traveling-wave device, so the wavelength shifter operates over the broad, approximately 50-nm gain bandwidth of the amplifier without wavelength restrictions imposed by Fabry-Perot resonances of the chip.

The cascadability of wavelength shifters using gain saturation in LD amplifiers has been demonstrated [27]. Data were transmitted through two such wavelength shifters with negligible penalty at 1 Gbps and 2 dB at 2 Gbps. Fan-out using two probe beams was also obtained. The wavelength shifter can copy data from a pump signal simultaneously to several probe signals at different wavelengths. In the multicasting experiment, the two probe wavelengths were separated by 0.4 nm and each carried the same complementary data of the pump signal. The results were obtained with an input pump power of −2 dBm and input probe power of −13 dBm. Multicasting to a large number of probe signals will require decreasing input power per probe signal, so that the sum of probe powers remains below the saturation power of the LD amplifier.

8.4 FOUR-WAVE MIXING IN LASER DIODE AMPLIFIERS

Phase conjugation through DFWM has been successfully used in such various fields as holography, adaptive optics, and laser resonators [28]. In DFWM experiments, the conjugated beam is separated from pump waves by noncollinear geometry. Better conversion efficiencies could be achieved by using a collinear interaction. Beam separation can be obtained in this case by having pump and probe beams with slightly different frequencies, ω and $\omega - \delta\omega$, respectively. In this NDFWM, the conjugated beam will appear at frequency $\omega + \delta\omega$ thus providing a frequency-selective beam geometry.

FWM in LDs and LD amplifiers has an extremely high conversion efficiency when the pump-probe detuning is kept below a few gigahertz. This is because the carrier density is easily modulated at the pump-probe detuning frequency and induces the dynamic gain and refractive index gratings inside the active layer. The large linewidth enhancement factor of the active layer can drastically improve the Bragg scattering efficiency of the index grating generated in this process. Since the pioneering experiment done by Nakajima and Frey [29], considerable work on NDFWM has been reported so far using a collinear interaction, because usual LDs and LD amplifiers have a single-mode waveguide. Lucente et al. [30] observed NDFWM in a broad-area LD, which permits the use of different injection geometries to control the excited modes of the device.

The other mechanism responsible for FWM is the nonlinear gain, which is well known as the gain suppression due to an intense light field. In the case of FWM, the variation of the photon density at the pump-probe detuning frequency directly creates the dynamic gain and index gratings via the nonlinear gain effect without accompanying the actual carrier density modulation [31]. The scattering efficiency caused by this process is much smaller than that by the carrier density modulation; however, the relaxation time governing this effect is believed to be less than 1 ps, and hence FWM induced by the nonlinear gain may become dominant only when the pump-probe detuning is far above the cutoff frequency of the carrier density modulation. However, the physical mechanisms causing the gain nonlinearity are still not fully understood. Two likely mechanisms are carrier heating and spectrum hole burning, and will be discussed later in detail.

8.4.1 Four-Wave Mixing Theory [32]

We consider FWM in TWAs, where the pump, probe, and signal lights propagate in the same direction [32]. Let the pump, probe, and signal frequencies be $f_p, f_q,$ and f_s, respectively. The pump-probe detuning f_d is $f_p - f_q$, and the frequency of the signal light f_s is given as $f_p + f_d$. A schematic drawing of a frequency spectrum obtained in FWM is shown in Figure 8.14.

The mode amplitudes, E_p of the pump light, E_q of the probe light, and E_s of the signal light, obey the following equations, which include contributions from both the carrier density modulation and the nonlinear gain, when $|E_q|$ and $|E_s|$ are much smaller than $|E_p|$:

$$\frac{dE_p}{dz} = \frac{1}{2}g\{1 - i\alpha - (1 - i\beta)\kappa|E_p|^2\}E_p \tag{8.12}$$

$$\frac{dE_q}{dz} = \frac{1}{2}g(1 - i\alpha)E_q - \xi_1|E_p|^2E_q - \zeta_1E_p^2E_s^* \tag{8.13}$$

$$\frac{dE_s}{dz} = \frac{1}{2}g(1 - i\alpha)E_s - \xi_2|E_p|^2E_s - \zeta_2E_p^2E_q^* \tag{8.14}$$

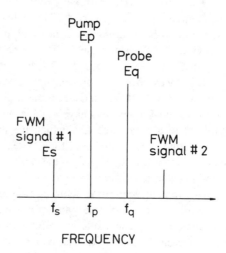

Figure 8.14 Frequency spectrum obtained in FWM.

he parameter α denotes the linewidth enhancement factor, and g the saturated gain,
iven as

$$g = \frac{g_0}{1 + |E_p|^2/I_s} \tag{8.15}$$

where g_0 denotes the unsaturated gain, and I_s the saturation intensity determined by the
carrier lifetime τ_s. The nonlinear gain coefficient κ is related to the more conventional
notation ϵ by

$$\kappa = \frac{2\epsilon_0 n^2}{h f_p} \epsilon \tag{8.16}$$

where n is the refractive index. The change in the real refractive index may be associated
with the nonlinear gain. In (8.12), the ratio of the real refractive index change to the
imaginary refractive index change is represented by β.

In (8.12) to (8.14), we assume that the phase-matching condition is satisfied:

$$2n_p = n_q + n_s \tag{8.17}$$

where n_p, n_q, and n_s denote refractive indexes at the pump frequency, the probe frequency,
and the signal frequency, respectively. In the experiment shown in Figure 8.15, the
maximum detuning $|f_d|$ is 400 GHz and (8.17) is approximately held.

The nonlinear parameters $\xi_{1,2}$ and $\zeta_{1,2}$ generating the FWM signal are expressed as

$$\xi_{1,2} = \frac{g}{2} \left\{ \frac{(1 - i\alpha)/I_s}{1 + |E_p|^2 \pm i2\pi f_d \tau_s} + (1 - i\beta)\, \kappa \left(1 + \frac{1}{1 \pm i2\pi f_d T_c} \right) \right\} \tag{8.18}$$

$$\zeta_{1,2} = \frac{g}{2} \left\{ \frac{(1 - i\alpha)/I_s}{1 + |E_p|^2 \pm i2\pi f_d \tau_s} + (1 - i\beta)\frac{\kappa}{1 \pm i2\pi f_d T_c} \right\} \tag{8.19}$$

where T_c is the relaxation time governing the nonlinear gain effect. Note that the first
term in (8.18) and (8.19) originates from the carrier density modulation, whereas the
second term is from the nonlinear gain effect. The signal power at the amplifier output
can be obtained as a function of the pump-probe detuning f_d using (8.12) to (8.14).

Agrawal attributed the origin of the nonlinear gain to spectral hole burning in his
theory of highly nondegenerate four-wave mixing (HNDFWM) [31]. Equations (8.12) to
8.19) coincide with those derived by Agrawal [31] if we assume that T_c in (8.18) and
8.19) correspond to the intraband scattering time of electrons in the conduction band
and ignore the transverse scattering time T_2 of the polarization in Agrawal's formation,
considering that T_2 is shorter than T_c. The nonlinear gain coefficient κ is written in terms
of more fundamental parameters as

$$\kappa = \left(\frac{h^2}{\theta^2 T_2 (T_c + T_v)} \right)^{-1}$$

(8.20)

where θ denotes the dipole moment and T_v is the intraband scattering time for electron in the valence band. The parameter β is determined from the slope of the gain spectrum [33] and is usually much smaller than unity in contrast with α if the change in the gain spectrum is due to the spectral hole burning, which is nearly symmetric.

Instead of the numerical calculation of the FWM spectrum using (8.12) to (8.14) we derive more simplified formulas to obtain a physical understanding of experimental results. We assume that the gain is not saturated (i.e., $1 + |E_p|^2/I_s \simeq 1$) and that the signal power is much smaller than the pump and probe power. In such a case, the electric field of the pump and probe lights approximately obey the following equation obtained from (8.12) and (8.13).

$$\frac{dE_{p,q}}{dz} = \frac{1}{2} g_0 (1 - i\alpha) E_{p,q}$$

(8.21)

By substituting solutions of (8.21) into (8.14) and solving the differential equation, we have the signal amplitude E_s written as

$$E_s = E_{s1} + E_{s2}$$

(8.22)

$$E_{s1} = -\frac{1}{2} \frac{1 - i\alpha}{1 - i2\pi f_d \tau_s} \frac{E_p^2}{I_s} E_q^*$$

(8.23)

$$E_{s2} = -\frac{1}{2} \frac{1 - i\beta}{1 - i2\pi f_d T_c} \kappa E_p^2 E_q^*$$

(8.24)

where all the amplitudes are referred at the amplifier output. Equations (8.23) and (8.24) represent the contribution from the carrier density modulation and the nonlinear gain respectively.

When we assume that the gain grating is dominant ($|\beta| \gg 1$) in the nonlinear gain process, that the carrier density modulation mainly generates the index grating ($|\alpha| \gg 1$) and that the pump-probe detuning frequency is between two cutoff frequencies (i.e., $1/(2\pi T_c) \gg |f_d| \gg 1/(2\pi \tau_s)$), (8.23) and (8.24) are further reduced to the following simple forms.

$$E_{s1} \sim -\frac{1}{2} \frac{\alpha}{2\pi f_d \tau_s} \frac{E_p^2}{I_s} E_q^*$$

(8.25)

$$E_{s2} \sim -\frac{1}{2} \kappa E_p^2 E_q^*$$

(8.26)

quation (8.25) shows that the FWM signal power due to the carrier density modulation
reduced at the rate of 20 dB/decade for both signs of f_d, but the phase of the electric
eld is reversed by the change in the sign of f_d. On the other hand, the FWM signal power
ue to the nonlinear gain is independent of f_d, as shown by (8.26). Note that E_{s1} and E_{s2}
re additive when $f_d > 0$, whereas they are subtractive when $f_d < 0$ because $\alpha > 0$. Therefore,
ne FWM signal power measured when $f_d > 0$ decreases at a rate slower than 20 dB/
ecade, but the signal power measured when $f_d < 0$ is reduced more steeply than 20 dB/
ecade.

4.2 Experiment on Four-Wave Mixing

he experimental setup used by Kikuchi et al. for the measurement of NDFWM is shown
a Figure 8.15(a). A 1.5-μm InGaAsP TWA used in the experiment has an angled-facet
tructure, and the effective facet reflectivity is less than 0.01%. The gain peak wavelength
; around 1.50 μm. The wavelength of the pump DFB LD is 1.542 μm. At this wavelength,
ne gain of the amplifier is about 10 dB. The linewidth enhancement factor b is measured
) be as large as 11 because of the large positive detuning of the pump light from the
ain peak. The frequency of the pump light is kept constant. On the other hand, the
requency of the probe laser is swept by changing the laser temperature. The total range
f the pump-probe detuning f_d, which is covered by three probed DFB LDs, reaches ±400
;Hz in this experimental setup.

The FWM signal generated at f_s is detected by a heterodyne receiver to improve
he receiver sensitivity and the frequency resolution. The signal power at the IF stage is
neasured by a spectrum analyzer following the heterodyne receiver. One of three DFB
.Ds is used as a local oscillator, depending on the signal frequency. The frequency of
he local laser is also precisely set near the signal frequency by changing the laser
emperature.

The squares in Figure 8.15(b) and circles in Figure 8.15(c), respectively, show the
'WM signal power measured as a function of the positive detuning ($f_d > 0$) and negative
letuning ($f_d < 0$). The FWM signal power reduces at the rate of 20 dB/decade, shown
)y solid lines when $|f_d| < 10$ GHz. The important point to note in Figure 8.15(a) and (b)
s the asymmetry appearing in the FWM spectrum when $|f_d| > 10$ GHz. When the detuning
s positive ($f_d > 0$), the FWM signal power decreases slower than 20 dB/decade, and
ipproaches a constant value as the pump-probe detuning frequency increases above 200
;Hz. However, the signal power for the negative detuning ($f_d < 0$) decreases more steeply
han 20 dB/decade, and a dip is observed at $f_d = 250$ GHz.

As the origin of the gain nonlinearity effect in LDs, two mechanisms are possible
is described before: carrier heating and spectral hole burning. Kikuchi et al. showed that
he asymmetry appearing in the FWM spectrum is explained only by the gain grating
:aused by the nonlinear gain due to the spectral hole burning [32]. The estimated relaxation
ime governing the nonlinear gain effect is less than 0.3 ps. This fact also supports the

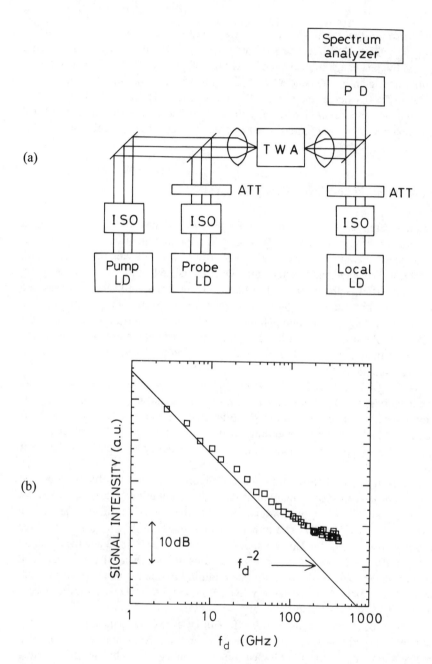

(a)

(b)

Figure 8.15(a,b) (a) Experimental setup for the measurement of nondegenerate FWM. The FWM signal i detected by a heterodyne receiver. (b) FWM signal power measured as a function of the positive detuning.

(c)

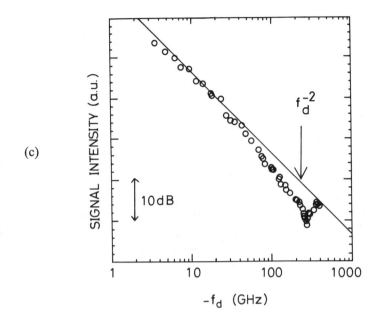

Figure 8.15(c) FWM signal power measured as a function of the negative detuning. (©1992 IEEE. From [32].)

ole of the spectral hole burning in the nonlinear gain effect rather than the dynamic arrier heating.

Tiemeijer performed NDFWM experiments at high input powers and large frequency separations [34]. The magnitude of the new generated frequencies and their asymmetry, and, furthermore, the asymmetric gain of the signal waves themselves, cannot be explained by carrier density modulations alone. A new mechanism of NDFWM with a characteristic time of 650 fs was needed to explain the experimental data. This mechanism is concluded to be carrier heating exhibiting the same characteristic time as that obtained previously [35] with a pump-probe technique, while spectral hole burning is much faster and is unable to explain the observed asymmetries. The results show that carrier heating is an important source of nonlinear gain in LD active media. NDFWM at even higher frequency separations and power levels will probably show the predicted high NDFWM due to spectral hole burning.

Vahala et al. measured NDFWM in a compressively strained QW TWA operating at 1.5 μm [36]. The relative signal power as a function of detuning frequency is plotted in Figure 8.16(a,b). A corner frequency of about 200 GHz, corresponding to a subpicosecond time constant of 650 fs, is visible in the plot. Also shown in the plots is a theoretical fit based on a model employing a single saturation power and characteristic lifetime for both the interband and intraband processes. In addition to the relaxation time constant, the data fit yields a nonlinear gain coefficient of 4.3×10^{-23} m^3 in good agreement with

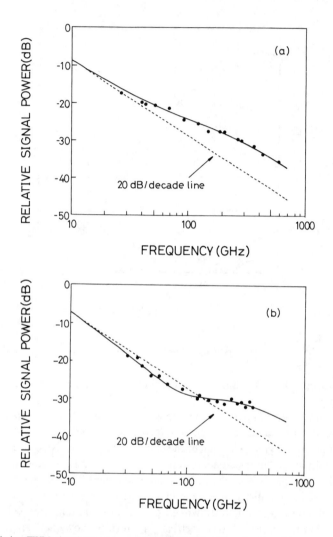

Figure 8.16 Relative FWM signal power as a function of (a) positive detuning frequency and (b) negative detuning frequency. (From [36]. Reproduced by permission of OSA.)

results of independent measurements of this parameter in compressively strained QW lasers and a value of −2.2 for the ratio of real and imaginary parts of refractive index change induced by gain nonlinearity. This value and the relaxation time of 650 fs suggest that dynamic carrier heating is the principal contribution to the gain nonlinearity at the wavelength studied.

Zhou et al. [37] have also measured ultrafast dynamics in a 1.5-μm tensile-strained QW optical amplifier by HNDFWM as a function of detuning frequency from −1.7 to

.5 THz, which provides a 94-fs equivalent temporal resolution. Frequency response data indicate the presence of two ultrafast physical processes with characteristic relaxation times of 650 fs and <100 fs. The longer time constant is believed to be associated with the dynamic carrier heating effect.

.4.3 Highly Nondegenerate Four-Wave Mixing

Murata et al. observed HNDFWM with more than 1-THz detuning in 1.5-μm MQW DFB LDs [38] and in 1.5-μm Fabry-Perot LDs [39]. They reported an optical-frequency conversion experiment on 1-Gbps IM signals in a 1-THz conversion range using the HNDFWM process in a 1.5-μm InGaAsP Fabry-Perot LD [39]. The converted signals were enhanced by laser-cavity resonance. The laser they used for FWM was a compressively strained MQW LD. A 1.5-μm DFB LD was used as a master laser. The FWM LD, which was biased above the threshold, was injection locked by the master laser and operated in a single longitudinal mode. The lasing mode acted as a pump light in the HNDFWM process. The probe light source consisted of a 1.5-μm DFB LD and a LiNbO$_3$ optical intensity modulator. The master light and the IM probe light were injected into the FWM laser from the same facet. The converted signal was output from the other facet and separated by two Fabry-Perot-type optical-frequency filters.

In the experiment, the pump frequency was adjusted to the FWM LD gain peak. Fabry-Perot cavity resonance modes were observed to occur on either side of the pump frequency. When the probe frequency was tuned to one of the resonance modes, a converted signal appeared at the corresponding resonant mode on the other side. The probe and converted-signal outputs were enhanced by the cavity resonance, a phenomenon similar to that occurring in a resonant-type LD amplifier.

As long as the probe frequency is adjusted so that the converted-signal frequency corresponds to one of the resonant-frequency peaks, the converted-signal output will remain essentially constant in a detuning range of approximately 1 THz for constant pump and probe outputs. This is possible because HNDFWM is based on an ultrafast optical nonlinear process with subpicosecond response time, and it may be applied to a terahertz-range frequency converter. They have estimated third-order nonlinear susceptibility $|\chi^{(3)}|$ for HNDFWM to be 0.6×10^{-15} (m/V)^2. Since the maximum modulation bandwidth of converted signals is determined by the resonance bandwidth at the cavity-resonance mode where the converted signal is enhanced, this resonance bandwidth is an important factor in device applications. Measured resonance bandwidth was about 5 GHz, which indicates the possibility of gigabit-per-second optical-signal conversion. This bandwidth may be increased by shortening the FWM LD cavity length (i.e., by reduction of the cavity Q-factor).

Bit error rates for probe and converted-signal lights, modulated with a 1-Gbps NRZ $2^{15} - 1$) pattern, was measured. The IM probe light was injected at the longer wavelength side than the pump wavelength of an FWM LD. Separation of the converted signal was approximately 1 THz and the average converted-signal output power was −11 dBm a the FWM LD output facet. The average probe input level measured in terms of the FWM LD photocurrent was 4 μA. Power penalty due to the frequency conversion was less tha 1 dB.

Koch and Telle demonstrated that LDs can act as efficient and convenient optica mixing elements, being sufficiently fast to bridge terahertz frequency gaps in the optica region [40]. The mixing product was investigated by heterodyne detection with the ai of an additional LD acting as local oscillator. The coherent nature of the process has bee made evident by phase locking the beat note between local oscillator and mixing produc to an RF reference signal. The phase-locked loop (PLL) consists of an RF preamplifie a double balanced mixer that generates the phase error signal, and a junction field-effec transistor (JFET) for controlling the pump LD injection current. The obtained frequenc spectrum shows the characteristic δ-spike of a PLL at the RF oscillator frequency an servo-loop-induced noise sidebands. The δ-spike contains more than 80% of the tota mixing product power. The steep roll-off of the mixing efficiency for fundamental interval $\delta\nu \geq 1.5$ THz can be explained by phase matching and dispersion effects in the mixe LD cavity.

Schnabel et al. performed an FWM experiment in a traveling-wave 1.5-μm strained layer MQW LD amplifier with one of the injected light waves being a train of $\leq$ 30-p pulses [41]. The frequency conversion range was 3.8 THz. This experiment clearly show the application of the LD amplifier as a functional device operating in a time regime tha is faster than the free-carrier lifetime.

A DFB LD (power P_1 at the output of the LD amplifier) and a mode-locked LI (P_2) were coupled into the strained MQW LD amplifier. The mode-locked laser produce a train of pulses with a repetition rate of 3.1 GHz. The frequency spacing Δf betwee both lasers was about 1.9 THz. The gain peak of the saturated LD amplifier is near th wavelength of 1.58 μm. Figure 8.17(a) shows the optical spectrum at the output of th LD amplifier measured with an optical spectrum analyzer and normalized to the outpu power P_1 (+12 dBm) of the DFB LD. It shows the powers P_1 and P_2 of the amplifie injected light waves and the power P_3 of one of the two sidebands generated by FWM The other sideband (designated P_4) is not seen in Figure 8.17(a) at the given frequency spacing $\Delta f = 1.9$ THz of the two injected light waves.

To perform measurements in the time domain on the FWM-generated signal (P_3) optical filters and an additional optical amplifier were used at the output of the LI amplifier to select the FWM component P_3 before it was viewed on an oscilloscope Figure 8.17(b) depicts measurements in the time domain. The upper trace in the figur shows the pulses of the mode-locked laser in front of the LD amplifier detected with th same photodetector. The pulsewidth in this trace corresponds to the bandwidth of th photoreceiver. The actual pulsewidth was measured to be less than 30 ps with a faste but less sensitive receiver. The lower trace in Figure 8.17(b) depicts the pulses (P_3 generated by FWM. A comparison of the pulsewidths in the upper and lower trace reveal no pulse broadening of the FWM-generated pulses. The measured pulsewidth seems no

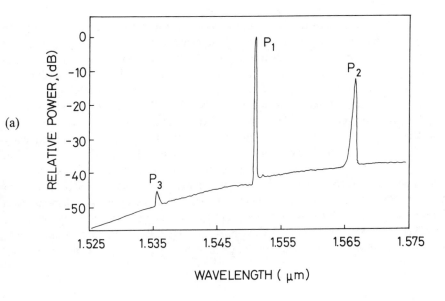

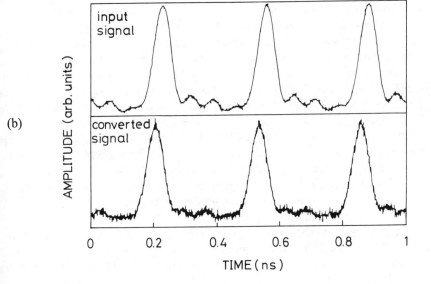

Figure 8.17 (a) Optical spectrum at output of LD amplifier in pulse conversion experiment. P_3 is the 3.8-THz converted-pulse spectrum of P_2; P_1 is the CW DFB LD. (b) Measurements of pulse train in time domain. Upper trace: pulse train in front of LD amplifier; lower trace: frequency-converted pulse train. (From [41]. Reproduced by permission of IEE.)

to be affected by the free-carrier lifetime. We also note that no FWM-generated pulse was observed when the polarizations of the two injected light waves where perpendicular to each other. This proves that the measured time domain signal is caused by the FWM-generated sideband (P_3). In an additional experiment, they replaced both lasers with tunable external cavity lasers and performed FWM experiments with CW light waves. The frequency spacing of both lasers was varied between 450 GHz and 2.4 THz and the ratio P_1/P_3 was measured against Δf. From this experiment, the time constant $\tau = 200 + 20$ fs was obtained as the characteristic time for the nonlinear gain effect. Consequently they conclude, based on the measurements in the frequency domain, that the limitation on the frequency spacing in the FWM experiments is given by the 200-fs time constant.

Schnabel et al. also demonstrated [42] wavelength conversion of an 18-Gbps data signal over 15 nm (corresponding to about 1.9 THz). Iannone et al. [43] demonstrated the 5.0-nm wavelength conversion of a mode-locked optical pulse train by cavity-enhanced HNDFWM in a 1.53-μm injection-locked bulk InGaAsP V-groove laser. Input pulses having 17-ps duration are broadened to 61 ps due to filtering imposed by the mixing laser.

8.4.4 Highly Nondegenerate Four-Wave Mixing With Two Pump Waves

In a conventional NDFWM experiment, a signal wave S_{in} and a pump wave P_1 both traveling with the same polarization in the same direction through an LD amplifier generate two new light waves, FW1 and FW2, as indicated in Figure 8.18(a) [44]. If S_{in} is modulated and P_1 unmodulated, it is obtained from calculations and experiment that FW2 is a frequency-converted replica of S_{in}, whereas FW2 degenerates to a δ-function (for phase-shift keying (PSK) modulation). This frequency conversion scheme has the disadvantage that the frequency spacing $2\Delta f_1$ between the input signal S_{in} and the frequency-converted signal FW2 is limited to less than 5 GHz by the efficiency of NDFWM in an LD amplifier. Here Δf_1 is the frequency spacing between P_1 and S_{in}. To enlarge this frequency spacing, they used a second pump wave P_2 (see Figure 8.18). The resulting spectrum of the input and generated waves, which is shown schematically in Figure 8.18(a), is only valid if the frequency spacing Δf_2 between P_1 and P_2 is larger than 20 GHz. The power of the generated light waves S_{out1} and S_{out2} is proportional to the product of the powers of P_2, P_1, and S_{in}, and are located with a frequency spacing Δf_1 at each side of P_2. S_{out1} and S_{out2} are each a frequency-converted replica of the input signal S_{in}. In contrast to the strong dependence of the NDFWM efficiency on Δf_1, S_{out1} and S_{out2} are only weakly dependent on Δf_2. Therefore, the frequency conversion range Δf_2 can be as large as 4 THz (the 3-dB bandwidth of the LD amplifier).

All-optical frequency conversion with the use of 140-Mbps differential phase-shift keying (DPSK) signal transmission has been demonstrated using the experimental setup and is schematically depicted in Figure 8.18(b). Three tunable external-cavity LDs (1.3 μm, spectral width less than 100 kHz, equipped with isolators) provide the light waves

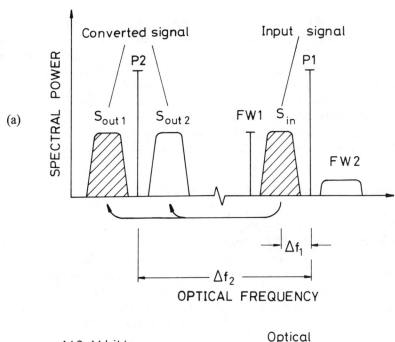

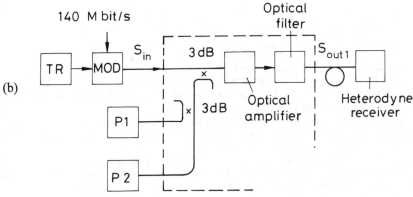

Figure 8.18 (a) Schematic of the light spectrum at the output of the optical amplifier as frequency converter. (b) Block diagram of the experimental setup for frequency conversion. (From [44]. Reproduced by permission of IEE.)

P_1, P_2, and S_{in}. The frequency spacings are $\Delta f_1 = 1$ GHz between S_{in} and P_1 and $\Delta f_2 = 1,500$ GHz between P_1 and P_2. The light wave S_{in} is modulated by a LiNbO$_3$ phase modulator with a 140-Mbps DPSK signal. The three light waves are fed via fiber couplers into an LD amplifier with input light powers 14.5, 28, and 1.2 μW for P_1, P_2, and S_{in}, respectively. The LD amplifier exhibits an internal TE single-pass gain of 21 dB with a gain ripple of less than 2 dB. The LD amplifier output waves are indicated schematically in Figure 8.18(a). With the data given above, the converted signal power S_{out1} at the output of the LD amplifier is 3 μW. An optical filter (grating) following the LD amplifier transmits the light components S_{out1}, S_{out2}, and P_2 with a loss of 2 dB and suppresses the light components P_1, S_{in}, FW1, and FW2 by at least 50 dB. The local laser of the heterodyne receiver is tuned to select the converted signal S_{out1} (intermediate frequency 1.5 GHz, seven stage Chebyshev filter). Transmission experiments with a BER of 10^{-9} were performed. If we consider the frequency converter as defined by all components inside the broken line in Figure 8.25, the fiber-to-fiber conversion efficiency of S_{in} to S_{out1} is better than -10 dB.

8.4.5 Enhancement of NDFWM via Electrical Modulation

By modulating an LD amplifier at a frequency determined by the detuning between the pump and probe in a collinear FWM geometry, enhancements in gain and reflectivity of 2 to 3 orders of magnitude are possible for remarkably small modulation depths ($m = 10^{-2}$) [45]. The experiments were performed on a DFB LD amplifier operating at 1.53 μm [46]. With the electrical modulation in phase with the optical beat modulation of the carrier density, the population pulsations, and, therefore, frequency mixing, were enhanced. Correspondingly, if they are out of phase, the interaction is suppressed.

REFERENCES

[1] Franken, P. A., A. E. Hill, C. W. Peters, and G. Weinreich, "Generation of Optical Harmonics," *Phys. Rev. Lett.*, Vol. 7, No. 4, August 1961, pp. 118–119.

[2] Izutsu, M., S. Shikama, and T. Sueta, "Integrated Optical SSB Modulator/Frequency Shifter," *IEEE J. Quantum Electronics*, Vol. QE-17, No. 11, November 1981, pp. 2225–2227.

[3] Kawaguchi, H., K. Oe, H. Yasaka, K. Magari, M. Fukuda, and Y. Itaya, "Tunable Optical-Wavelength Conversion Using a Multielectrod Distributed-Feedback Laser Diode With a Saturable Absorber," *Electron. Lett.*, Vol. 23, No. 20, September 1987, pp. 1088–1090.

[4] Kawaguchi, H., K. Magari, H. Yasaka, M. Fukuda, and K. Oe, "Tunable Optical-Wavelength Conversion Using an Optically Triggerable Multielectrode Distributed Feedback Laser Diode," *IEEE J. Quantum Electronics*, Vol. QE-24, No. 11, November 1988, pp. 2153–2159.

[5] For example, see Streifer, W., R.D. Burnham, and D.R. Scifres, "Effect of External Reflectors on Longitudinal Modes of Distributed Feedback Lasers," *IEEE J. Quantum Electronics*, Vol. QE-11, No. 4, April 1975, pp. 154–161.

[6] Kondo, K., M. Kuno, S. Yamakoshi, and K. Wakao, "A Tunable Wavelength-Conversion Laser," *IEEE J. Quantum Electronics*, Vol. QE-28, No. 5, May 1992, pp. 1343–1348.

[7] Paradisi, A., and I. Montrosset, "Analysis of Wavelength Conversion Using a Multisection DBR Laser With a Saturable Absorber," *IEEE J. Quantum Electronics*, Vol. 29, No. 5, May 1993, pp. 1285–1294.

[8] Nobuhara, H., K. Kondo, S. Yamakoshi, and K. Wakao, "Optical Logic Functions Using a Tunable Wavelength Conversion Laser Diode," *IEEE J. Quantum Electronics*, Vol. 28, No. 7, July 1992, pp. 1722–1726.

[9] Yasaka, H., K. Takahata, K. Kasaya, M. Okamoto, Y. Kondo, M. Naganuma, and M. Ikeda, "One-Directional Output Optical Frequency Conversion Device With Asymmetric κ-DBR Structure," *ECOC'92*.

[10] Takahata, K., K. Kasaya, and H. Yasaka, "Wavelength Dependence of Optical Frequency Conversion Device With Asymmetric κ-DBR Structure," *Electron. Lett.*, Vol. 28, No. 22, October 1992, pp. 2078–2079.

[11] Takahata et al., to be published in *IEEE J. Quantum Electronics*.

[12] Kasaya, K., K. Takahata, and H. Yasaka, "Optical Frequency Conversion Device With Asymmetric κ-DBR Structure," *IEEE Photonics Tech. Lett.*, Vol. 5, No. 3, March 1993, pp. 321–324.

[13] Yasaka, H., K. Takahata, K. Kasaya, and K. Oe, "Frequency Response of a Unidirectional Output Optical Frequency Conversion Device," to be published in *IEEE Photonics Tech. Lett.*

[14] Hildebrand, O., M. Schilling, W. Idler, D. Baums, G. Laube, and K. Wünstel, "The Integrated Interferometric Injection Laser (Y-Laser): One Device Concept for Various System Applications," *ECOC'91*, pp. 39–46.

[15] Tsuda, H., K. Hirabayashi, H. Iwamura, and T. Kurokawa, "Tunable Wavelength Conversion Using a Liquid Crystal Filter and a Bistable Laser Diode," *Appl. Phys. Lett.*, Vol. 61 No. 17, October 1992, pp. 2006–2008.

[16] Durhuus, T., R. J. S. Pedersen, B. Mikkelsen, K. E. Stubkjaer, M. Öberg, and S. Nilsson, "Optical Wavelength Conversion Over 18 nm at 2.5 Gb/s by DBR-Laser," *IEEE Photonics Tech. Lett.*, Vol. 5, No. 1, January 1993, pp. 86–88.

[17] Mikkelsen, B., R. J. S. Pedersen, T. Durhuus, C. Braagaard, C. Joergensen, and K. E. Stubkjaer, "Wavelength Conversion of High Speed Data Signals," *Electron. Lett.*, Vol. 29, No. 19, September 1993, pp. 1716–1718.

[18] Inoue, K., "Wavelength Conversion for Frequency-Modulated Light Using Optical Modulation to Oscillation Frequency of a DFB Laser Diode," *J. Lightwave Technology*, Vol. 8, No. 6, June 1990, pp. 906–911.

[19] Inoue, K., "Wavelength Conversion With Self Wavelength Selection Using Mach-Zehnder Filter and DFB-LD," *Electron. Lett.*, Vol. 25, No. 25, December 1989, pp. 1707–1708.

[20] Schilling, M., W. Idler, D. Baums, G. Laube, K. Wünstel, and O. Hildebrand, "Multifunction Photonic Switching Operation of 1500 nm Y-Coupled Cavity Laser (YCCL) With 28 nm Tuning Capability," *IEEE Photonics Tech. Lett.*, Vol. 3, No. 12, December 1991, pp. 1054–1057.

[21] Schilling, M., W. Idler, D. Baums, K. Dütting, G. Laube, K. Wünstel, O. Hildebrand, "6 THz Range Tunable 2.5 Gb/s Frequency Conversion by a Multiquantum Well Y Laser," *IEEE J. Quantum Electronics*, Vol. 29, No. 6, June 1993, pp. 1835–1843.

[22] Yasaka, H., H. Ishii, K. Takahata, and K. Oe, "Wavelength Conversion Using Super-Structured Grating (SSG) Laser Diode," *Extended Abstracts, The 54th Autumn Meeting, 1993, The Japan Society of Applied Physics*, No. 3, 28a-G-9, p. 1014 (in Japanese).

[23] Barnsley, P.E., and P.J. Fiddyment, "Wavelength Conversion From 1.3 to 1.5 μm Using Split Contact Optical Amplifiers," *IEEE Photonics Tech. Lett.*, Vol. 3, No. 3, March 1991, pp. 256–258.

[24] Durhuus, T., B. Fernier, P. Garabedian, F. Leblond, J. L. Lafragette, B. Mikkelsen, C.G. Joergensen, and K. E. Stubkjaer, "High-Speed All-Optical Gating Using a Two-Section Semiconductor Optical Amplifier Structure," *Conf. on Lasers & Electro-Optics*, Anaheim, California, 1992, CTh S4, pp. 552–554.

[25] Joergensen, C., T. Durhuus, C. Braagaard, B. Mikkelsen, and K. E. Stubjaer, "4 Gb/s Optical Wavelength Conversion Using Semiconductor Optical Amplifiers," *IEEE Photonics Tech. Lett.*, Vol. 5, No. 6, June 1993, pp. 657–660.

[26] Glance, B., J. M. Wiesenfeld, U. Koren, A. H. Gnauck, H. M. Presby, and A. Jourdan, "High Performance Optical Wavelength Shifter," *Electron. Lett.*, Vol. 28, No. 18, August 1992, pp. 1714–1715.

[27] Wiesenfeld, J. M., and B. Glance, "Cascadability and Fanout of Semiconductor Optical Amplifier Wavelength Shifter," *IEEE Photonics Tech. Lett.*, Vol. 4, No. 10, October 1992, pp. 1168–1171.

[28] Fisher, R. A., ed., *Optical Phase Conjugation*, New York: Academic 1983.

[29] Nakajima, H., and R. Frey, "Intracavity Nearly Degenerate Four-Wave Mixing in a (GaAl)As Semiconductor Laser," *Appl. Phys. Lett.*, Vol. 47, No. 8, October 1985, pp. 769–771.

[30] Lucente, M., G. M. Carter, and J. G. Fujimoto, "Nonlinear Mixing and Phase Conjugation in Broad Area Diode Laser," *IQEC'88*.

[31] Agrawal, G. P., "Population Pulsations and Nondegenerate Four-Wave Mixing in Semiconductor Lasers and Amplifiers," *J. Opt. Soc. Am. B*, Vol. 5, No. 1, January 1988, pp. 147–159.

[32] Kikuchi, K., M. Kakui, C. E. Zah, and T. P. Lee, "Observation of Highly Nondegenerate Four-Wave

Mixing in 1.5 μm Traveling-Wave Semiconductor Optical Amplifiers and Estimation of Nonlinear Gai Coefficient," *IEEE J. Quantum Electronics*, Vol. QE-28, No. 1, January 1992, pp. 151–156.

[33] Agrawal, G. P., "Gain Nonlinearity in Semiconductor Lasers: Theory and Application to Distribute Feedback Lasers," *IEEE J. Quantum Electronics*, Vol. QE-23, No. 6, June 1987, pp. 860–868.

[34] Tiemeijer, L. F., "Effects of Nonlinear Gain on Four-Wave Mixing and Asymmetric Gain Saturation i a Semiconductor Laser Amplifier," *Appl. Phys. Lett.*, Vol. 59, No. 5, July 1991, pp. 499–501.

[35] Hall, K. L., J. Mark, E. P. Ippen, and G. Eisenstein, "Femtosecond Gain Dynamics in InGaAsP Optic: Amplifiers," *Appl. Phys. Lett.*, Vol. 56, No. 18, April 1990, pp. 1740–1742.

[36] Vahala, K. J., J. Zhou, N. Park, J. W. Dawson, M. A. Newkirk, U. Koren, and B. I. Miller, "Measureme: of Gain Nonlinearity in a Strained Multiple Quantum Well Optical Amplifier by Highly Nondegenera: Four-Wave Mixing," *CLEO'93*, JThA6, pp. 448–449.

[37] Zhou, J., N. Park, J. W. Dawson, K. J. Vahala, M. A. Newkirk, and B. I. Miller, "Terahertz Four-Wav Mixing Spectroscopy for Study of Ultrafast Dynamics in a Semiconductor Optical Amplifier," *Appl. Phy Lett.*, Vol. 63, No. 9, August 1993, pp. 1179–1181.

[38] Murata, S., A. Tomita, J. Shimizu, M. Kitamura, and A. Suzuki, "Observation of Highly Nondegenera: Four-Wave Mixing (>1 THz) in an InGaAsP Multiple Quantum Well Laser," *Appl. Phys. Lett.*, Vol. 5: No. 14, April 1991, pp. 1458–1460.

[39] Murata, S., Tomita, A., Shimizu, J., and A. Suzuki, "THz Optical-Frequency Conversion of 1 Gb/s-Signal Using Highly Nondegenerate Four-Wave Mixing in an InGaAsP Semiconductor Laser," *IEEE Photonic Tech. Lett.*, Vol. 3, No. 11, November 1991, pp. 1021–1023.

[40] Koch, Ch., and H. R. Telle, "Bridging THz-Frequency Gaps in the Near IR by Coherent Four-Wav Mixing in GaAlAs Laser Diodes," *Optics Communications*, Vol. 91, 1992, pp. 371–376.

[41] Schnabel, R., W. Pieper, R. Ludwig, and H. G. Weber, "Multiterahertz Frequency Conversion of Picosecond Pulse Train Using Nonlinear Gain Dynamics in a 1.5 μm MQW Semiconductor Laser Ampl fier," *Electron. Lett.*, Vol. 29, No. 9, April 1993, pp. 821–822.

[42] Schnabel, R., W. Pieper, A. Ehrhardt, M. Eiselt, and H. G. Weber, "Ultrafast Signal Processing Usin: Semiconductor Laser Amplifier," *LD amp '93*, PD4.

[43] Iannone, P. P., P. R. Prucnal, G. Raybon, U. Koren, and C. A. Burrus, "Broadband Wavelength Shifte for Picosecond Optical Pulses at 1.5 μm," *Electron. Lett.*, Vol. 29, No. 17, August 1993, pp. 1518–152(

[44] Grosskopf, G., R. Ludwig, and H. G. Weber, "140 Mbit/s DPSK Transmission Using an All-Optic: Frequency Converter With a 4000 GHz Conversion Range," *Electron. Lett.*, Vol. 24, No. 17, Augu: 1988, pp. 1106–1107.

[45] McCall, M. W., "Enhanced Multiwave Mixing Interactions in Semiconductor Optical Amplifiers via Pum: Modulation," *IEEE J. Quantum Electronics*, Vol. 28, No. 1, January 1992, pp. 9–15.

[46] George, D. S., and M. W. McCall, "Enhanced Nearly Degenerate Four-Wave Mixing in a Distribute Feedback Semiconductor Laser Amplifier via Coherent Electrical Modulation," *CLEO'93*, CTuN14, p: 160–161.

Chapter 9

Wavelength Selection and Wavelength-Selective Photodetection

9.1 WAVELENGTH SELECTION

We can use a narrowband transmission filter to reject the unwanted channels as shown in Figure 9.1. If the filter is tunable, the center wavelength (frequency) λ_0 can be shifted by changing, for example, the voltage or the current applied to the filter, and we can select one channel that we want to receive. We can classify tunable filters into three categories: passive, active, and tunable LD amplifiers, as shown in Table 9.1 [1].

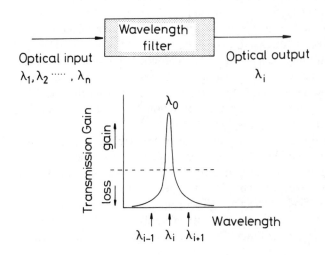

Figure 9.1 Operation principle of wavelength selection.

Table 9.1
Tunable Filter Types and Their Characteristics

	Type	Resolution	Range	No. Channels	Speed
Passive	Etalon ($F \sim 200$)			~30*	ms*
	Fiber Fabry-Perot ($F \sim 200$)			~30*	ms*
	Two-stage ($F \sim 3000$)			~1,000*	ms*
	Waveguide Mach-Zehnder	0.38Å (5 GHz)	45Å (600 GHz)	128*	ms
Active	Electro-optic TE/TM	6Å	160Å	~10	ns*
	Acousto-optic TE/TM	10Å	400nm	~100	~10 μs
Laser diode amplifiers	DFB amplifier	1–2Å	4–5Å	2–3	1 ns
	2-section DFB amplifier	0.85Å	6Å	8	ns*
	Phase-shift-controlled DFB amplifier	0.32Å (4 GHz)	9.5Å (120 GHz)	18	ns*

*Estimated.

   The passive category is composed of those wavelength-selective components that are basically passive and can be made tunable by varying some mechanical element of the filter, such as mirror position or etalon angle. This includes Fabry-Perot etalons, tunable fiber Fabry-Perot filters, and tunable M-Z filters. For the Fabry-Perot filters, the number of resolvable wavelengths is related to the value of the finesse F of the filter. Normal Fabry-Perot filters and etalons have finesses of up to 200 and, for example, in FSK systems, this implies a number of channels of $F/6$, or about 30. This can be increased by raising the finesse or by operating two filters in tandem with suitable isolation, for which the finesse may reach 3,000 and the number of channels 1,000. The advantages of such filters are the very fine frequency resolution that can be achieved. The disadvantages are, primarily, their tuning speed and losses.

   The M-Z integrated-optic interferometer tunable filter, listed in Table 9.1, is a waveguide device with $\log_2(N)$ stages, each of which passes every other incoming wavelength in a divide-by-two fashion until only the desired wavelength remains. Here, N is the number of wavelengths. This filter was demonstrated with 100 wavelengths separated by 10 GHz in optical frequency, and with thermal control of the exact tuning [2]. The measured tuning times were on the order of milliseconds. The number of simultaneously resolvable wavelengths is limited by the number of stages required and the loss incurred in each stage.

   In the active category, there are two filters based on wavelength-selective polarization transformation by either electro-optic or acousto-optic means. In both cases, the orthogonal polarizations of the waveguide are coupled together at a specific tunable wavelength. In the electro-optic case, the wavelength selected is tuned by changing the dc voltage on the electrodes; in the acousto-optic case, the wavelength is tuned by changing the frequency of the acoustic-drive frequency.

A filter bandwidth (FWHM) of about 1 nm has been achieved by both filters. However, the acousto-optic tunable filter has a much broader tuning range (the entire 1.3 to 1.56 μm range) than the electro-optic type, typically 16 nm. This is because the effective grating in the acousto-optic case is dynamic and its pitch changes with the acoustic-drive frequency, whereas for the electro-optic case, the grating is fixed by the metallization pattern.

Both types of filter are controlled electronically and can be tuned to a reasonably high speed. The acousto-optic filter is limited to tuning times of a few microseconds by the acoustic propagation velocity, whereas the electro-optic filter can be tuned typically in the nanosecond range. Fairly large voltages are required for the electro-optic case, which may determine its actual achievable tuning speed. The acousto-optic drive power for an integrated acousto-optic case has been demonstrated to be on the order of 100 to 200 mW. Another significant difference is that the acousto-optic filter can be driven with driving signals of many frequencies simultaneously, thereby selecting more than one wavelength simultaneously. Such a filter opens the possibility of switching and rearranging wavelengths just as if they were bits in a register.

The third category of filter in Table 9.1 is that of LD amplifiers as tunable filters. Operation of a resonant laser structure, such as a DFB or DBR laser, below the threshold results in narrowband amplification. These types of filters offer the following important advantages: narrow electronically controlled bandwidth, the possibility of electronic tuning of the central frequency, net gain (as opposed to loss in passive filters), small size, and integrability. The principles and performance of this type of filter are described in this chapter in detail.

9.2 TUNABLE NARROWBAND LASER DIODE AMPLIFIER FILTERS

A DFB LD can act as a tunable filter when biased just below the laser threshold. This filter enables us to tune on nanosecond time scales. A conventional one-section DFB LD, as shown in Figure 9.2(a), was used for a narrowband optical filter by Kawaguchi et al. [3]. A $\lambda/4$-shifted DFB LD was used to obtain a narrow bandwidth (Figure 9.2(b)) [4,5]. The frequency of peak amplification can be changed by the LD amplifier bias current through the carrier density dependence of the active layer refractive index. However, in these devices, changing the driving current for spectrum selections has the adverse effect of changing the insertion gain and gain bandwidth.

Using a two-section DFB LD amplifier, a narrow spectrum selection with wide frequency tunability was achieved while maintaining a constant gain and a constant gain bandwidth (Figure 9.2(c)) [6,7]. A 1.5-μm InGaAsP/InP DFB LD amplifier whose p-

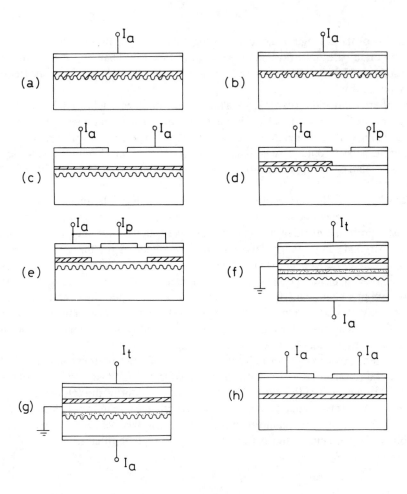

Figure 9.2 Tunable optical-wavelength filters: (a) one-section DFB LD amplifier; (b) one-section λ/4-shifted DFB LD amplifier; (c) two-section DFB LD amplifier; (d) phase-controlled DBR LD amplifier; (e) phase-shift-controlled DFB LD amplifier; (f) tunable twin-guide DFB LD amplifier; (g) λ/4-shifted twin-guide DFB LD amplifier; (h) Fabry-Perot LD amplifier.

type electrode was divided into two parts was used in the experiment. The maximum tunability in the gain center frequency was 33.3 GHz without mode jumping. This amplifier could maintain a constant maximum insertion gain of 10.0 ± 0.6 dB and a constant FWHM gain bandwidth of 3.6 ± 0.4 GHz. Optical-frequency selection with an extinction ratio better than −20 dB can be obtained for two optical inputs separated by 15.4 GHz and having the same input power.

To achieve a wider wavelength tuning range, several kinds of structures have been studied: a phase-controlled DBR LD filter (Figure 9.2(d)) [8–10], a phase-shift-controlled

DFB LD filter (Figure 9.2(e)) [11], a tunable twin-guide DFB (TTG-DFB) LD filter, (Figure 9.2(f)) [12], and a $\lambda/4$-shifted twin-guide DFB LD filter (Figure 9.2(g)) [13]. In the phase-shift-controlled DBR LD filter, a 120-GHz continuous wavelength tuning range with a 24.5-dB constant gain and a 4-GHz constant bandwidth (FWHM) has been achieved. In the TTG-DFB structure, the tuning layer is formed above a conventional DFB laser structure. The currents to the tuning and active layers can be independently injected into each region. Tuning is done by changing equivalent refractive index by current injection in the tuning layer. The transmission spectrum shifts over its stop band without changing shape. A 1.35-nm tunability with TTG-DFB structures has been demonstrated [12]. The limitation of the tuning was due to the insufficient suppression of the second-peak transmission gain. To achieve wide tunability, suppression of other wavelengths except the transmission wavelength is important. When the second-peak transmission gain can be suppressed, tuning can be extended over a second peak wavelength. In the $\lambda/4$-shifted twin-guide DFB active filter with an independent current-injected tuning waveguide, a wide wavelength tunability of 4.2 nm, no change in gain spectrum shape, and a large anticipated suppression ratio have been achieved [13]. A 1.4-μm-composition tuning waveguide layer with independent current injection and a $\lambda/4$-shifted corrugation with a large coupling coefficient (70 to 80 cm^{-1}) were adopted.

The transmission wavelength for the DFB LD-type filters is close to the Bragg wavelength, which is determined by the grating pitch and the effective refractive index of the waveguide. The effective refractive index of the waveguide depends on the thickness and the width of the waveguide. Thus, the transmission wavelength changes according to the dimension of the waveguide. For example, the transmission wavelength distributes with a standard deviation of several tens of angstroms when a wafer, from which tunable wavelength filters are taken, is prepared by conventional LPE. Fabry-Perot resonators have many resonance modes (transmission wavelengths). LDs have the gain spectrum whose FWHM is over 100Å. Thus, it is easy for the transmission wavelength to fit the preassigned wavelength when the Fabry-Perot resonators are used as wavelength-tunable optical filters (Figure 9.2(h)). The idea to use a Fabry-Perot amplifier as a wavelength-tunable optical filter (a channel selector) have been proposed in [14], where the transmission wavelength is tuned by simultaneous adjustment of injection current and temperature to achieve constant-gain and constant-bandwidth tuning. A tuning range as wide as 188 GHz (15Å) with 23-dB constant gain and 5-GHz constant bandwidth has been achieved by Numai in a 1.5-μm optical filter [15,16]. A 25-channel wavelength selection with less than -10 dB of crosstalk is expected with this filter. Pan and Dagenais experimentally demonstrated the use of a Fabry-Perot LD amplifier as a wavelength filter [17]. Its performance at a data rate of 140 Mbps was studied in a two-channel experiment. A minimum channel spacing of 12 GHz with usable optical gain of 10 dB was obtained.

9.3 COUPLED-WAVE THEORY FOR DFB LASER DIODE AMPLIFIERS

A DFB LD amplifier is the same device as a DFB LD except for its bias level, and is biased at just below the lasing threshold. The amplifier amplifies only one (or two)

narrowband by the frequency selectivity of the grating. The existence of two waves $R(z)$ and $S(z)$, which propagate in z and $-z$ directions, respectively, are assumed, as shown in Figure 9.3 [7]. $R(z)$ and $S(z)$ are coupled to each other by the grating in the active layer. An electrical field $E(z)$ is written using $R(z)$ and $S(z)$ as

$$E(z) = R(z)\, e^{-i\beta_0 z} + S(z)\, e^{i\beta_0 z} \tag{9.1}$$

where β_0 is the Bragg frequency of the grating [18,19].

As the electrical field $E(z)$ satisfies Maxwell's equations, a pair of coupled-mode equations are obtained:

$$-\frac{d}{dz}R(z) + (\alpha - i\delta)R(z) = i\kappa e^{-i\Omega}S(z) \tag{9.2}$$

$$\frac{d}{dz}S(z) + (\alpha - i\delta)S(z) = i\kappa e^{i\Omega}R(z) \tag{9.3}$$

$$2\alpha = \Gamma g_0 - \alpha_0 \tag{9.4}$$

where α is the modal gain per unit length, δ is the detuning of the propagation constant from the Bragg condition, κ is the coupling coefficient, L is the cavity length of the amplifier, Γ is the confinement factor, g_0 is the local gain per unit length, α_0 is the loss coefficient per unit length, and Ω is the phase value of the grating at $z = 0$.

The right sides of (9.2) and (9.3) represent the optical feedback from $S(z)$ to $R(z)$ and from $R(z)$ to $S(z)$, respectively. κ expresses the strengths of the optical feedback effect by the grating. The relationship between g_0 and α is given in (9.4). Using δ, the frequency detuning $\Delta\lambda$ from the Bragg frequency is given by

$$\Delta\lambda = -\frac{\delta}{2\pi}\lambda\Lambda \tag{9.5}$$

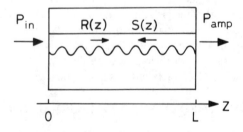

Figure 9.3 Analytical model of DFB LD amplifier. (©1988 IEEE. From [7].)

where λ is the wavelength of the injected optical signal and Λ is the grating pitch.

The solutions of (9.2) and (9.3) are assumed in the form

$$R(z) = r_1 e^{\gamma z} + r_2 e^{-\gamma z} \tag{9.6}$$

$$S(z) = s_1 e^{\gamma z} + s_2 e^{-\gamma z} \tag{9.7}$$

where γ_1, γ_2, s_1, and s_2 are constants. γ is a complex propagation constant and satisfies the dispersion relation

$$\gamma^2 = (\alpha - i\delta)^2 + \kappa^2 \tag{9.8}$$

The parameters γ_1, γ_2, s_1, and s_2 can be expressed using $R(0)$, $R(L)$, $S(0)$, and $S(L)$, which are the forward and backward waves at the interfaces of $z = 0$ and $z = L$. Using (9.2) to (9.8), it is shown that four waves at the end of the cavity, $R(0)$, $R(L)$, $S(0)$, and $S(L)$, have the following relations.

$$\begin{pmatrix} R(L) \\ S(L) \end{pmatrix} = \begin{pmatrix} F_{11} & F_{12} \\ F_{21} & F_{22} \end{pmatrix} \begin{pmatrix} R(0) \\ S(0) \end{pmatrix} \tag{9.9}$$

where

$$F_{11} = \cosh(\gamma L) + \frac{(\alpha - i\delta)L}{\gamma L} \sinh(\gamma L)$$

$$F_{12} = -i\frac{\kappa L}{\gamma L} \sinh(\gamma L) \tag{9.10}$$

$$F_{21} = i\frac{\kappa L}{\gamma L} \sinh(\gamma L)$$

$$F_{22} = \cosh(\gamma L) - \frac{(\alpha - i\delta)L}{\gamma L} \sinh(\gamma L)$$

An optical signal injected into the DFB LD amplifiers is amplified by the optical gain. In the calculations, both facet reflectivities of the amplifier are assumed to be zero. Then P_{in} and an amplified transmission output power P_{amp} are connected with $|R(0)|^2$ and $|R(L)|^2$ by boundary conditions, respectively. A signal gain G, which is defined by the ratio of P_{amp} to P_{in}, is given by using (9.9) and (9.10):

$$G = \frac{P_{amp}}{P_{in}} = \frac{|R(L)|^2}{|R(0)|^2} = \left| \cosh(\gamma L) - \frac{(\alpha - i\delta)L}{\gamma L} \sinh(\gamma L) \right|^{-2} \tag{9.11}$$

The Bragg frequency shift $\Delta\lambda_{\text{Bragg}}$ through the refractive index change is due to the injected carrier, and is expressed using b as

$$\Delta\lambda_{\text{Bragg}} = 2\Delta n\Lambda = -2b\frac{c}{\omega}\Delta\alpha\Lambda \tag{9.12}$$

where b is the ratio of real-to-imaginary refractive index change.

On the other hand, as the input power increases in strength, the saturation of the mode gain becomes pronounced, and the refractive index changes. This is because the increase of the input power causes the increase of the stimulated emission and the decrease of the injected carrier in the active layer. When the carrier decreases, this causes both the gain saturation and the increase in the refractive index. Assuming the internal optical intensity to be uniform along the cavity, the internal optical intensity P_{av} is introduced into this analysis. P_{av} is the averaged value of $R(z)$ and $S(z)$ in the cavity:

$$P_{\text{av}} = \frac{1}{L}\int_0^L (|R(z)|^2 + |S(z)|^2)\mathrm{d}z \tag{9.13}$$

α and δ under high optical input is given by

$$2\alpha = \Gamma g_m - \alpha_0 \tag{9.14}$$

$$\delta L = \delta_0 L + \frac{g_0 Lb}{2}\frac{P_{\text{av}}/P_s}{1 + P_{\text{av}}/P_s} \tag{9.15}$$

$$g_m = \frac{g_0}{1 + P_{\text{av}}/P_s} \tag{9.16}$$

The first term of the right-hand side in (9.14) refers to the gain saturation due to optical input power and the second term of the right side in (9.15) refers to the detuning caused by the gain saturation due to the optical input power. P_s is the saturation power defined by

$$P_s = \frac{h\nu}{\Gamma\tau(\mathrm{d}g_m/\mathrm{d}N)} \tag{9.17}$$

where $h\nu$ is the photon energy corresponding to the optical input frequency. τ is the carrier lifetime. Assuming $\tau = 2$ ns, N is the carrier density in the active layer, and $\mathrm{d}g_m/\mathrm{d}N = 2.7 \times 10^{-16}$ cm^2, P_s is calculated to be about 9.4×10^5 W/cm^2 from (9.16). Assuming a cross section of 0.49 μm^2 for the active layer of the amplifier, the estimated saturation power is about 5 mW. κ is 80 cm^{-1} in this calculation.

9.4 CALCULATED RESULTS ON DFB LASER DIODE AMPLIFIERS

9.4.1 Low Input Power Condition

First, the amplification characteristics are calculated under the low input power condition where the gain saturation does not occur. The calculated results for the signal gain G versus the frequency detuning $\Delta\lambda$ from the Bragg condition are shown in Figure 9.4(a) using (9.5), (9.8), and (9.11). The parameters used in the calculations are $\kappa L = 2.9$, $\lambda = 1.3$ μm, $\Gamma = 0.3$, $\alpha_0 L = 1.2$, $b = 2.5$, $\Lambda = 3{,}962$Å, and $P_s = 5$ mW. The gain spectrum of the amplifier has two identical peaks that are symmetric to the Bragg frequency. The injected light is hardly amplified in the vicinity of the Bragg frequency, and the region is called the *stop band*. In the device used in the experiment, as described later, one mode was strongly amplified because of the residual reflectivities (about 5%) at the facets. The DFB mode on the other side of the stop band was as weak as the next DFB mode and was about one-fifth of the main DFB mode at $0.98I_{th}$. Gain profiles in the lower frequency (longer wavelength) side from the Bragg frequency are calculated using (9.11) as shown in Figure 9.4(a). As the gain g_0 increases, the maximum signal gain G_{max} increases and the FWHM gain bandwidth becomes narrow. As g_0 increases, the gain maximum frequency shifts toward the higher frequency side according to (9.12). In this figure, the zero point of the frequency detuning was set to the Bragg frequency in the threshold condition. The gain spectrum can be characterized by two parameters, G_{max} and the FWHM bandwidth. G_{max} is uniquely related to the FWHM bandwidth for a fixed κL. The relationship between G_{max} and FWHM bandwidth with changing L and constant κ are calculated using (9.11) and are shown in Figure 9.4(b). For longer L, the maximum gain to maintain the constant FWHM bandwidth becomes smaller. In this model, the signal gain approaches infinity because the spontaneous emission effect was ignored. In real devices, as the gain approaches the threshold gain, the mode gain saturates because the spontaneous emission power becomes high. Therefore, this model cannot be applied in the case where the driving current is in the immediate vicinity of the threshold.

9.4.2 High Input Power Condition

When high input power is injected into the amplifier, gain saturation occurs. This is because the stimulated emission causes a decrease in the carrier density. The input power dependence of the gain spectra calculated using (9.5), (9.8), (9.11), and (9.13) to (9.15) is shown in Figure 9.5. The cavity lengths are $L = 250$ μm and $L = 1{,}250$ μm. In the two cases, the unsaturated G_{max} is set at 25 dB and is shown at $P_{in} = -93$ dBm in the figure. The FWHM bandwidth for both $L = 250$ μm and $L = 1{,}250$ μm are 8.7 GHz and 0.1 GHz, respectively. As L becomes long, the gain profile becomes sharp. As expected, the gain maximum frequency shift Δf_g from the unsaturated gain maximum frequency becomes large as the input power becomes high. At input intensity higher than $P_{in} = -43$

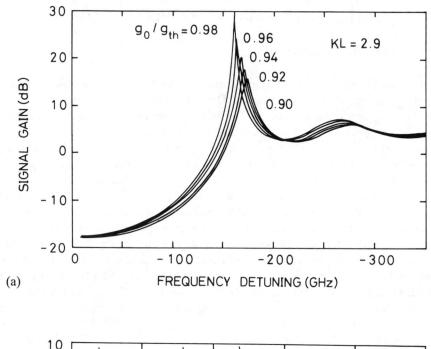

(a)

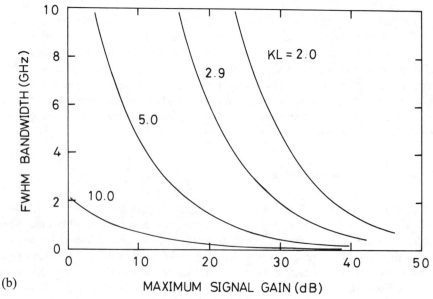

(b)

Figure 9.4 (a) Calculated gain profiles of DFB LD amplifier. $\kappa L = 2.9$, $L = 300$ μm. (b) Relationship between FWHM bandwidth and maximum signal gain G_{max}. $\kappa = 80$ cm^{-1}. ($^{\circledcirc}$1988 IEEE. From [7].)

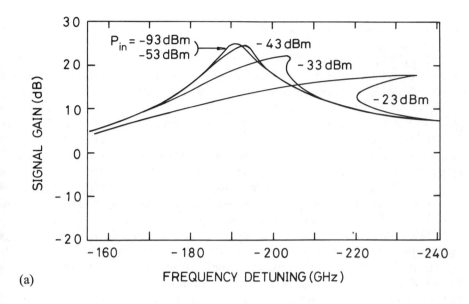

(a)

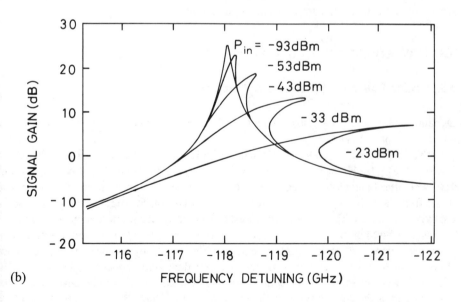

(b)

Figure 9.5 Optical input power P_{in} dependence of gain profiles with an unsaturated maximum signal gain of 25 dB: (a) $\kappa = 80$ cm^{-1} and $L = 250$ μm ($\kappa L = 2$); (b) $\kappa = 80$ cm^{-1} and $L = 1,250$ μm ($\kappa L = 10$). ($^{©}$1988 IEEE. From [7].)

dBm in both cases, two signal gain values exist on the lower frequency side and the signal shows bistability.

The input power P_{in} dependence of the various characteristics of the amplifier, such as G_{max}, Δf_g, and the FWHM bandwidth, can be described as follows. As the input power becomes high, G_{max} decreases sharply. As L becomes longer (κL becomes larger) and G_{max} larger, the saturation input power P_{3dB} becomes lower. Here, P_{3dB} is defined by the input power that causes the signal gain to decrease 3 dB from the unsaturated value. The higher P_{in}, the shorter cavity length (smaller κL), and the larger G_{max} cause the larger gain maximum frequency shift. The FWHM bandwidth is narrow for the amplifier with the long cavity length. But the FWHM bandwidth of the amplifier with the short cavity and the small G_{max} does not broaden up to the relatively high input power. The amplifier with the long cavity is effective for narrowband amplification with a low input power. But the amplifier with the short cavity has a wide input power tolerance.

The amplifier with large G_{max} results in pronounced gain saturation for the high optical input. Therefore, if the DFB LD amplifier is used as an optical filter with subgigahertz bandwidth, it is desirable to use the amplifier with the long cavity and with small unsaturated signal gain because of the small effect of the gain saturation. The DFB LD amplifier with the short cavity can be used as an optical filter with several-gigahertz FWHM bandwidth. Since the amplifier with the short cavity does not cause the gain saturation effect up to a relatively high input power, a large G_{max} can be obtained by setting the driving current to just near the threshold.

9.5 RATE EQUATION ANALYSIS [20]

9.5.1 Basic Equations for Resonant Amplifiers

Dynamic characteristics of narrowband tunable resonant optical amplifiers as active filter are studied theoretically.

The theoretical model of a resonant amplifier with external input is illustrated in Figure 9.6(a) [20]. Higher cavity gain levels lead to higher cavity Q and correspondingly narrower filter bandwidth. As an active element, the width and position of the resonance in the frequency domain can be controlled by the bias current and the refractive index of the cavity. When the bias is below the threshold but above the transparency point, *incoherent photons* build up their energy inside the cavity in response to the internal spontaneous emission source, and their dynamic behavior is governed by an intensity rate equation. When a coherent external source is injected into the cavity, *coherent photons* build up with their field governed by an amplitude rate equation. Both incoherent and coherent photons obtain their gains through the same electronic pool. Because both the gain and the index of refraction are functions of carrier density, changing the carrier density can change the width and the center frequency of the active filter. When the external source is weak compared with the noise source, the output can only follow the cavity characteri

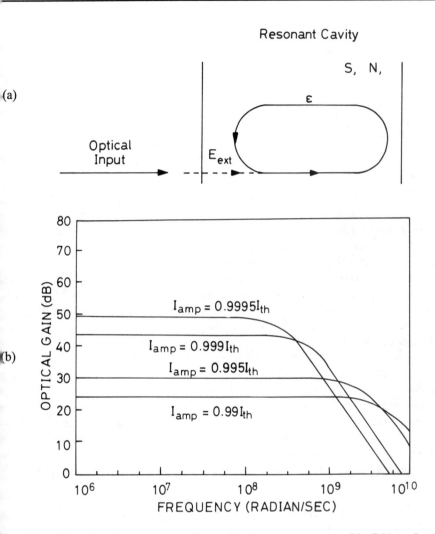

Figure 9.6 (a) Illustration of the resonant amplifier model; (b) frequency response of the field as a function of amplifier bias. (©1988 IEEE. From [20].)

cs and the filter function is normally performed. When the external source is strong, the generated coherent photon density becomes comparable or larger than the incoherent photon density and generates a noticeable amount of carrier density variation in the electronic pool. The filter characteristics will thus be changed by the external source. In this sense, we use the terms *weak input* and *strong input* here.

The above statements can be modeled by following Lang's formal treatment of the injection-locking scheme with some modification of the derivation and physical interpretation [21]:

$$\frac{d\epsilon}{dt} = \frac{1}{2}\left[AV_g\Gamma(N - N_t) - \frac{1}{\tau_p}\right]\epsilon - i(\Omega - \omega)\epsilon + \kappa E_{ext} \qquad (9.18)$$

$$\frac{dS}{dt} = \left[AV_g\Gamma(N - N_t) - \frac{1}{\tau_p}\right]S + \frac{\beta N}{\tau_s} \qquad (9.19)$$

$$\frac{dN}{dt} = -AV_g(N - N_t)(|\epsilon|^2 + S) - \frac{N}{\tau_s} + \frac{J}{ed} \qquad (9.20)$$

In the above equations, ϵ is the coherent field, which is generated in response to the external source E_{ext}. Both ϵ and E_{ext} are normalized such that their absolute value square equals the photon density. The parameter S is the incoherent photon density generated in response to the spontaneous emission in (9.19). J is the current density, e is the unit charge, d is the active layer thickness, N and N_t are the carrier density and transparency carrier density, respectively, τ_p is the photon lifetime, and V_g is the group velocity of the photon. Here A is the gain coefficient with a unit of area, τ_s is the carrier lifetime, Γ is the photon confinement factor, β is the rate of the spontaneous emission into the cavity modes, κ is the coupling constant (equal to $c/2nL$ where L is the cavity length), n is the refractive index, and c is the light speed. The coupling constant κ can be easily derived by finding the relation of the fields inside and outside a Fabry-Perot cavity and their relationship with the photon buildup or decay time of the cavity. However, with our definition of κ, the exact meaning of E_{ext} is the amount of the external field that just passes the interface into the cavity, and ϵ is the final field built up in the cavity. The advantage of such a definition is that we can separate experimentally coupling losses into and out of the cavity from our gain calculation. Here ω is the instantaneous frequency of the external source input E_{ext}, which may be time-varying for FSK input signals. The resonant frequency term Ω in (9.18) is a function of N and ω, and can be expressed as

$$\Omega(N,\omega) = \Omega_0 + \frac{\partial\Omega}{\partial N}\delta N - \left(\frac{n_g}{n} - 1\right)(\omega - \Omega_0) \qquad (9.21)$$

where $N = N_0 + \delta N$. Ω_0 is the center frequency of resonance when N equals N_0 and when there is no input, and n_g is the group index given by $n_g = n + \Omega_0(\partial n/\partial\Omega)$. The second term of (9.21) is a result of the carrier dependence of the index. The third term is due to the material dispersion that produces the frequency pulling effect in a laser system. Because the α factor can be expressed as $\alpha = -2(\partial\Omega/\partial N)/(\partial G/\partial N)$, where $G = AV_g\Gamma(N - N_t)$ is the gain, we have $\partial\Omega/\partial N = -(\Gamma\alpha AV_g/2)$. The equations are very similar to those used in the injection-locking scheme [21]. However, in this case, the bias is below the threshold and the carrier density is not clamped (N_0 is a variable). Also, S represents total incoherent photons instead of side modes. Because the bias is below the threshold, none of the modes can be excited without external input.

5.2 AC Small-Signal Analysis

The ac small-signal frequency response can be derived analytically. It is a function of as current, input intensity, and input optical frequency. We can write the complex field in (9.18) as

$$\epsilon = E(t) \, e^{i\phi(t)} \tag{9.22}$$

where $E(t)$ and $\phi(t)$ are real and ϕ is the phase angle between ϵ and E_{ext}. Putting (9.22) to (9.18) and separating the real and imaginary parts results in

$$\frac{dE}{dt} = \frac{1}{2} \left[AV_g\Gamma(N_0 - N_t) - \frac{1}{\tau_p} + AV_g\Gamma\delta N \right] E + \kappa E_{ext} \cos \phi \tag{9.23}$$

and

$$E\frac{d\phi}{dt} = E\left[-\frac{1}{2}\alpha AV_g\Gamma\delta N - \frac{n_g}{n}(\omega - \Omega_0) \right] - \kappa E_{ext} \sin\phi \tag{9.24}$$

We have four coupled equations with five variables: E, E_{ext}, N, S, and ϕ.

The transfer functions like $|E'/E'_{ext}|^2$ can be derived analytically by the standard procedure to linearize the rate equations. Here, E' and E'_{ext} are the small-signal amplitudes, which are assumed to be much smaller than their steady-state values of E and E_{ext}. Because this is very lengthy, we show only the calculated results. Figure 9.6(b) shows the frequency response of the small-signal gain $|E'/E'_{ext}|^2$ at different biases. Each curve is calculated at zero frequency detuning ($\Delta\nu = (1/2\pi)(\omega - \Omega_0) = 0$) and about -72-dBm dc optical input power ($E_{ext0} = 1 \times 10^6 \sqrt{photon/m^3}$). The field response has a wider 3-dB frequency bandwidth when the bias is lower. This can simply be explained with the wider filter bandwidth at low bias.

5.3 Large-Signal Analysis

To observe the time evolution of the optical and electrical (carrier) output, we can numerically integrate the four coupled equations (9.23), (9.24), (9.19), and (9.20) and use different optical signals like amplitude-shift keying (ASK) or FSK as the input. The numerical integration is accomplished using a fourth-order Runge-Kutta method. We can simulate the case of pure ASK by sending a modulated bit pattern as the external input, $E_{ext}(t)$, in (9.22) and (9.23), keeping the input frequency constant. For a pure FSK signal, we fix the amplitude of E_{ext} and feed the bit pattern to the frequency input, $\omega(t)$, in (9.23) by specifying the two tones of the oscillation frequency to represent the 0s and 1s of the bit pattern. A chirped signal can be simulated by using a combination of these two conditions. We proceed in what follows from simpler cases like a weak ASK input, to more complex ones like a strong FSK input. Whether the signal is weak or strong has interesting effects on the nonlinearity of the filter. The choice of ASK or FSK input clearly affects the choice of the center frequency, and for FSK we also must specify the width of the frequency sweeping. The generated output can be presented either as a distorted 0-1 string or as a corresponding eye pattern. We use these tools to evaluate the performance of the device under different operating conditions.

We conclude the ASK performance by using Figure 9.10 as a summary. The bias-dependent eye opening for each different input is shown in the figure. The input signal is assumed to have a pseudorandom pattern at 500 Mbps, in which the 0s and 1s are roughly equally weighted. Here we define the term *eye opening* as the maximum vertical distance inside the eye, and for easy comparison we show the absolute value of the opening in each curve in units of photon density. For a given input, the eye opening is an increasing function with the bias. When the high resonance effect appears, the increasing starts to saturate. After this maximum point, increasing the bias will only decrease the opening. For the curve of high-level input, each point is optimized in the frequency detuning to avoid oscillation and generate the largest eye opening. The eye opening degradation at high resonance is less serious for higher power input due to the effective filter width broadening.

To conclude the performance of the FSK input, we again calculate the eye opening as a function of bias and input power (Figure 9.7). Each point is optimized with the frequency detuning and the frequency sweeping width. Compared with the ASK case, the FSK signal seems able to generate higher eye opening by using Δf as a new degree of freedom. Here, $\Delta f = (\omega_H - \omega_L)/2\pi$ is the frequency difference of the two tones.

9.6 EXPERIMENTAL RESULTS ON SIGNAL SELECTION

9.6.1 Characteristics of DFB Laser Diode Amplifiers

The one-section InGaAsP/InP DFB LD amplifier with BH was fabricated by LPE growth [3]. The amplifier had a second-order grating with a pitch of 3,962Å. The cavity length is 300 μm and the coupling coefficient κL is 2.9. Both facets were formed by cleavage and coated for antireflection. The measured insertion gain profiles are shown in Figure 9.8. The input power in the amplifier was maintained at −43 dBm. As the driving current I_{amp}/I_{th}, increases, the insertion gain increases almost linearly, the FWHM narrows, and the gain peak shifts linearly toward the higher frequency side in the range of $0.92I_{th}$ to $0.98I_{th}$. The gain peak shifted at a rate of 4.5 GHz/mA in this device. The maximum insertion gain was 14.5 dB at $I_{amp}/I_{th} = 0.98$. The narrowest FWHM bandwidth of 3.4 GHz was obtained at the same bias level.

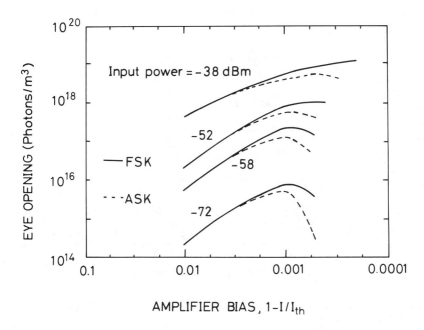

Figure 9.7 Eye opening as a function of bias and input power level for both ASK and FSK input. ([©]1991 IEEE. From [20].)

Changing the driving current of the conventional one-section DFB LD amplifier to shift the gain maximum frequency causes the optical gain and the FWHM bandwidth to change. The two-section DFB LD amplifier could be used to tune the gain maximum frequency by adjusting the carrier density in each electrode to maintain a constant mode gain.

An InGaAsP/InP two-section DFB LD amplifier with BH was fabricated by the MOVPE/LPE hybrid growth method. The thickness and width of the active layer were 0.13 and 1.0 μm, respectively. The gain maximum wavelength (frequency) of the amplifier was 1.547 μm (193.7 THz). The LD amplifier's p-type electrode was divided into two sections. The resistance between the divided electrodes was 146Ω. It had a first-order grating with 2,400Å pitch. Both facets were formed by cleavage, and the facet of the input side was coated for antireflection. The cavity length was 300 μm.

The insertion gain profiles for the various operation currents are shown in Figure 9.9 [3]. The input power was maintained at a −41-dBm constant. I_1 and I_2 are the front and rear side currents. Gain profiles with a constant gain and a constant FWHM bandwidth by the multielectrode DFB LD amplifier were obtained. The gain maximum frequency tunability was 33.3 GHz without mode jumping. By changing the current I_1 and I_2 (then $I_t = I_1 + I_2$ also changes), this amplifier can maintain a constant maximum insertion gain of 10.0 ± 0.6 dB and a constant FWHM bandwidth of 3.6 ± 0.4 GHz. As I_1/I_t diverts

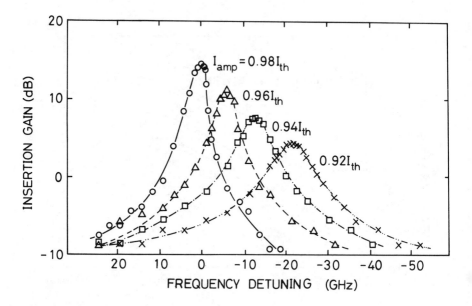

Figure 9.8 Gain profiles for four driving-current levels in the one-section DFB LD amplifier. (©1988 IEEE From [7].)

from 0.5, the gain peak frequency shift Δf_g increases toward the higher frequency side from the uniform condition ($I_1/I_t = 0.5$, a condition that corresponds to the driving condition in conventional DFB LD amplifiers), and I_t increases. Moreover, I_t and Δf_g have minimum values at the uniform condition. Thus, Δf_g and I_t have the same tendency for I_1/I_t. The reason for this is thought to be as follows: as the carrier density difference between the front- and rear-side regions becomes large, the refractive index difference in the two regions increases. The nonuniformity of the refractive index reduces the optical feedback through the grating between the front and rear sides. Therefore, in order to maintain a constant insertion gain, I_t must be increased. Since the increase of I_t causes the carrier density to increase and the refractive index to decrease, Δf_g increases toward the higher frequency side. In other words, the uniform condition has the minimum I_t and the lowest gain maximum frequency, because the optical feedback works most efficiently in that case. Thus, the tunability of the gain maximum frequency can be obtained by varying I_t while maintaining a constant gain and a constant gain bandwidth.

9.6.2 Filtering by DFB Laser Diode Amplifiers

Optical-frequency selectivity was experimentally measured on a two-section DFB LD amplifier. By changing I_1 and I_2, the gain maximum frequency could be tuned to the one optical input frequency of the two inputs. The experimental setup for measuring the

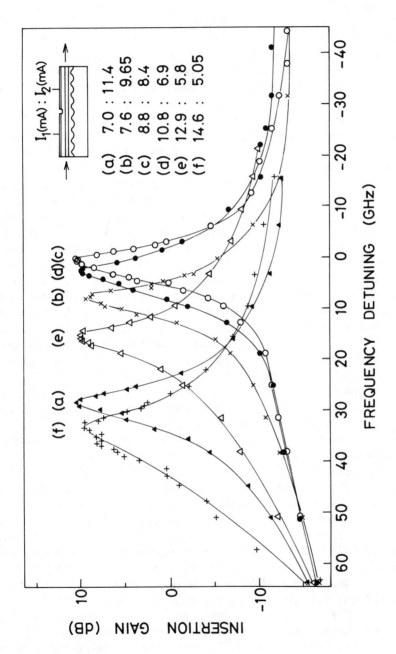

Figure 9.9 Gain profiles for various driving currents in the two-section DFB LD amplifier. By adjusting I_1 and I_2, gain maximum frequency can be tuned with constant insertion gain and constant FWHM bandwidth. (©1988 IEEE. From [7].)

optical-frequency selectivity of the amplifier is shown in Figure 9.10(a). Two single frequency laser beams were injected into the two-section DFB LD amplifier through a optical isolator. Polarization of LD1 and LD2 were matched to that of the LD amplifie. The lasing frequencies of LD1 and LD2 were λ_1 and λ_2, respectively. λ_2 was set to frequency 9.8 GHz higher than λ_1 by adjusting both the heat-sink temperature and th bias current of LD1 and LD2. The injection currents of LD1 and LD2 were sinusoidall modulated with $f_1 = 10.6$ MHz and $f_2 = 12.8$ MHz, respectively. The input power leve were almost the same for both frequencies and were −39 dBm for λ_1 and −38 dBm fc λ_2. The amplified optical-frequency spectra were measured using a Fabry-Perot interferom eter with a 23.7-GHz free spectral range and a finesse of 28. The modulated amplificatio signal spectra were measured using an RF spectrum analyzer.

The results of one optical-signal selection from two optical inputs are shown i Figure 9.10: (a) shows the optical-frequency spectra, and (b) shows the RF signal spectr. The two-section DFB LD amplifier was biased at (i) $I_1 = 6.4$ mA, $I_2 = 9.6$ mA, and (ii $I_1 = 5.8$ mA, $I_2 = 10.9$ mA. The gain maximum optical frequency corresponded to λ_1 i condition (i). Therefore, only the λ_1 signal was amplified. An extinction ratio of bette than −20 dB was measured from the RF signal spectrum, and an extinction ratio of −2 dB was also measured from the optical spectrum without modulation. A 5.6-dB insertio gain was also obtained. When the bias current was set to condition (ii), the gain maximur optical frequency corresponded to λ_2. Only the λ_2 signal was amplified. An extinctio ratio of better than −20 dB was measured from the RF spectrum. When an RF modulatio was not applied, an extinction ratio of better than −23 dB was measured from the optica spectrum. An insertion gain of 3.6 dB was obtained in this case. Since cases (i) and (ii have the same internal gain (i.e., they have the same gain bandwidths), the difference i their insertion gains will be caused by the coupling-loss difference between λ_1 and λ signals in optical-ray axis alignment. When more closely spaced input signals are used one optical selection can also be obtained but the extinction ratio deteriorates.

Since two optical signals with 15.4-GHz separation were selected with an extinctio ratio of better than −20 dB, three optical signals can be selected by the two-section DFI LD amplifier because the tuning range of the gain maximum frequency was 33.3 GHz.

Usually, the filtering bandwidth is measured by feeding into an amplifier filter wavelength-tunable continuous light signal having a very narrow linewidth to obtai sufficient resolution in the measurement. Instead of this conventional technique, Nakajim and De Faria [22] used 50-ps optical pulses exhibiting a wide spectral width to measur the filtering bandwidth, filtering profile, and time response of a three-contact straine MQW amplifier filter. When the spectral width of the signal light exceeds the filterin bandwidth, the spectral filtering induces an exponential decay tail in the output pulses In the strained MQW amplifier filter, the time response is characterized by two time constants (τ_1 and τ_2). However, the rise time of the output pulses remained almost th same. These behaviors may decrease the transfer bit rate when the spectral width of th signal light exceeds the filtering bandwidth. The maximum transfer bit rate can be obtaine from the relationship $\tau_2 \times \Delta f = 1/\pi$. Here, τ_2 is the long time constant of about 160 p

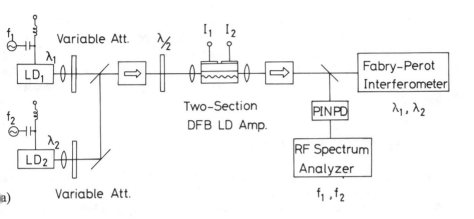

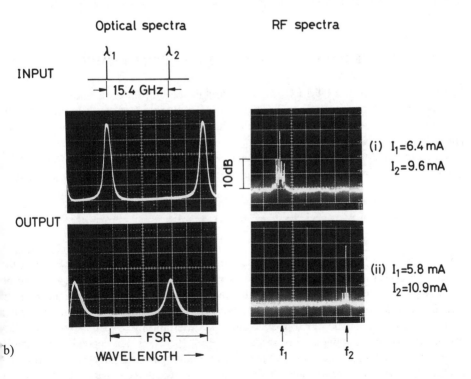

Figure 9.10 (a) Experimental setup for measuring optical selection from two inputs. (b) Two-channel selectivities obtained by the two-section DFB LD amplifier: (i) measured optical-frequency spectra; (ii) measured RF spectra. Gain maximum frequency corresponds to λ_1 for bias level (i) and λ_2 for bias level (ii), respectively. Only λ_1 and f_1 are amplified for (i), and only λ_2 and f_2 are amplified for (ii). ($^{©}$1988 IEEE. From [7].)

and Δf is the FWHM spectral bandwidth of the output spectra. However, this limitation will be overcome by choosing a sufficiently wider filtering bandwidth than that obtained from the above relationship (or spectral width of the signal). It has been suggested that the filtering bandwidth broadening is related to the spectral power density of the input light rather than the optical power. Accordingly, the filtering bandwidth may depend on the linewidth of the input light.

9.6.3 Fast Tuning of Center Wavelength

Fast wavelength tuning has been demonstrated using a DFB LD as a tunable optical amplifier [23]. Tuning over 0.3 nm (38 GHz) with 1-ns switching times and dynamic tunability $\Delta\lambda/\Delta I_a = -0.19$ nm/mA have been demonstrated by current injection. Selection between two wavelength-multiplexed channels indicates that transient-response requirements for a packet switch operating at 1.2 Gbps can be realized with a BER less than 10^{-9}.

9.7 NOISE PROPERTIES AND DEVICE STRUCTURE OPTIMIZATION

The noise generated by a DFB LD amplifier exhibits the simple dependence on the signal and spontaneous emission power [24,25]. When the output of an optical amplifier falls on a photodetector, the photocurrent contains fluctuations due to interference among signal and spontaneous emission fields and due to shot noise. An analysis [24] of these fluctuations based on the photon statistics of the resonant, single-mode optical amplifier shows that, ignoring the small shot-noise terms, the photocurrent signal-to-noise ratio is given by

$$\frac{<i_s>^2}{\text{var}(i)} \approx \frac{1}{2}\left[\frac{(<P_s>/<P_{ASE}>)^2}{1 + (<P_s>/<P_{ASE}>)}\right] \times \left[\frac{(\Delta\nu/B)^2}{(\Delta\nu/B) + \exp(-\Delta\nu/B) - 1}\right] \quad (9.25)$$

Here, $<P_s>$ is the mean signal power measured at the amplifier output, $<P_{ASE}>$ is the mean ASE power at the output, $\Delta\nu$ is the amplifier's optical bandwidth (assumed to be much larger than the source laser's linewidth), and B is the detector's electrical bandwidth. This expression is plotted in Figure 9.11, showing the dependence of the photocurrent signal-to-noise ratio on the signal-to-spontaneous-power ratio $<P_s>/<P_{ASE}>$, with the bandwidth ratio $\Delta\nu/B$ as parameter. For values of $<P_s>/<P_{ASE}> > 1$, the dependence is nearly linear, since the mean squared signal photocurrent is proportional to $<P_s>^2$, whereas the photocurrent variance, dominated in this regime by signal-spontaneous beat noise, is proportional to $<P_s>$.

When such an amplifier is used in an NRZ IM digital system, $<i_s>^2/\text{var}(i)$ determines the error probability P_e, if Gaussian statistics are assumed, with $P_e = 10^{-9}$, requiring that $<i_s>^2/\text{var}(i) \sim 150$ at the input to the digital system's threshold detector. Thus, according

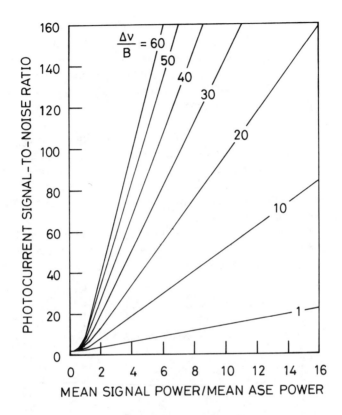

Figure 9.11 Theoretical dependence of the photocurrent signal-to-noise ratio, at the output of a resonant active filter, on the signal-to-spontaneous-power ratio $<P_s>/<P_{ASE}>$ for various values of the bandwidth ratio $\Delta\nu/B$. $\Delta\nu$ is the optical bandwidth of the active filter; B is the RF bandwidth of the detection system. (©1991 IEEE. From [25].)

to (9.25), low error rates can be obtained either by operating at large values of $<P_s>/$ $<P_{ASE}>$, which correspond to low relative signal-spontaneous beat-noise levels, or at large values of $\Delta\nu/B$, in which case most of the beat-noise spectral components generated in the photocurrent lie outside the receiver's RF bandwidth and therefore do not degrade the system's performance.

The dependence of the photocurrent signal-to-noise ratio on the signal-to-spontaneous-power ratio was experimentally verified. An InGaAsP DFB LD was used to illuminate a similar wavelength-matched DFB device, biased at 17.7 mA (98.3% of threshold) and thus operated as an active optical filter. The source laser operated at a linewidth of 40 MHz. The amplified output light was detected, electrically low-pass filtered, and passed to instruments used for measuring the quantities appearing in (9.25). The amplifier's FWHM optical bandwidth $\Delta\nu$ at this bias level was 1,330 MHz, considerably larger than the source linewidth, as required for the validity of (9.25). The receiver's RF bandwidth B was fixed at 60 MHz by a low-pass electrical filter. Therefore, $\Delta\nu/B = 22.2$. The power ratio $<P_s>/<P_{ASE}>$ was varied approximately from 1 to 4 by adjusting the incident optical power between -47 and -41 dBm. At sufficiently large input-power levels, saturation effects in such resonant LD amplifiers result in instabilities that can introduce additional noise. To avoid the unstable regime, the incident power was set at a value smaller than -40 dBm.

Sahlén considered four-section DBR filters [26], with two Bragg sections (AR-coated facets), surrounding an active section (with gain) and a passive phase control section. There are two main reasons to examine DBR filters. The first is that DBR LDs have a wider tuning range than multisection DFB LDs. The second is that the saturation problems can be partially overcome by using a tapered waveguide structure in order to reduce the irradiance in the active section.

Figure 9.12 shows the upper limits set by saturation and the lower limit set by beat noise to achieve BER = 10^{-9} for a nonoptimized device. An ideal receiver (limited by the filter beat noise) is assumed. Assuming $n_{sp} = 4$ and the requirement of the side-mode

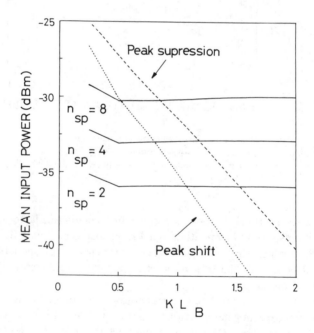

Figure 9.12(a) Calculated maximum mean input power allowed before saturation effects become too large (i.e., before $P > P_{3dB}$ and $\delta\nu > \Delta\nu/2$), and minimum mean input power required to reach BER = 10^{-9} due to beat noise in the DBR filter, as functions of κL_B ($\partial g/\partial N = 2 \times 10^{16}$ cm^2, $\partial n/\partial N = -1 \times 10^{-20}$ cm^{-3}, $G = 20$ dB): dynamic range for nonoptimized device: $A_{mode} = 7 \times 10^{-9}$ cm^2, $L_a = 100$ μm, $L_\phi = 100$ μm, $\kappa = 100$ cm^{-1}, $\tau_{rec} = 1.5$ ns, $\eta = -5$ dB.

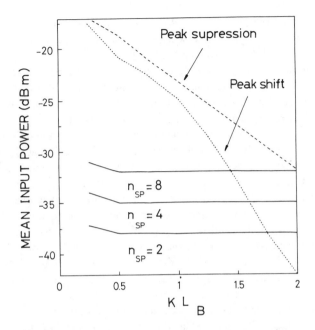

Figure 9.12(b) Calculated maximum mean input power allowed before saturation effects become too large (i.e., before $P > P_{3dB}$ and $\delta\nu > \Delta\nu/2$), and minimum mean input power required to reach BER $= 10^{-9}$ due to beat noise in the DBR filter, as functions of κL_B ($\partial g/\partial N = 2 \times 10^{16}$ cm^2, $\partial n/\partial N = -1 \times 10^{-20}$ cm^{-3}, $G = 20$ dB): dynamic range for optimized transmission device: $A_{mode} = 5 \times 10^{-8}$ cm^2, $L_a = 100$ μm, $L_\phi = 200$ μm, $\kappa = 12$ cm^{-1}, $\tau_{rec} = 1$ ns, $\eta = -3$ dB. ($^{©}$1992 IEEE. From [26].)

rejection ratio (SMRR) > 10 dB ($\kappa L_B \approx 1$ for that design) gives zero dynamic range. Here, n_{sp} is the spontaneous emission factor:

$$n_{sp} = \frac{1}{1 - \exp((\hbar\omega - \Delta E_F)/kT)} \times \frac{1}{1 - \alpha_{scatt}/g\Gamma} \qquad (9.26)$$

where ΔE_F is the quasi-Fermi level separation.

To improve performances of the filters, first, the geometry should be modified. The shortest sections that can be fabricated with good electric isolation are about 50 to 100 μm. Another problem with sections that are too short is that the resulting high carrier densities cause strong Auger recombination and heterobarrier leakage. Given this constraint, the coupling coefficient κ and the coupling strength κL_B (L_B is length of a Bragg section) have to be determined. The smaller κ is, the better, as long as the linewidth $\Delta\nu$ is sufficiently large and the cavity factor Q is low. One finds that coupling strengths $\kappa L_B \approx 0.4$ to 0.8 are optimal. Small κ means large L_B, which requires low-loss waveguides, even when tuning current is injected. We assume that $\kappa = 12$ cm^{-1} is feasible. The conclusion that short active and phase sections and weak gratings should be employed is consistent with the results of Kazovsky et al. [9].

Second, the mode area must be increased. A thin low-bandgap layer and wider, higher bandgap confinement layer makes this possible. Single-QW structures allow further increase in the mode area. Tapering also increases A_{mode}. The crucial problem is to make the taper adiabatic. Sahlén has analyzed this using the beam propagation method. The overall transmission to the single-mode output guide is 98% on the following condition. The active section, with a 10-μm width, is excited with its local fundamental mode. The width of the input waveguide is 1.5 μm and the length of the taper is 200 μm. The waveguide effective index is 3.22 and the substrate refractive index is 3.15. However, this is not sufficient per se. If a large fraction of the power locally excites higher order modes in the taper (but later reenters the fundamental mode), this will cause modal dispersion. Such dispersion would not be a problem in a traveling-wave semiconductor amplifier, since the intermodal delay time is less than 1 ps. On the other hand, for a resonant filter with a few-gigahertz linewidth, it would be disastrous (higher order modes are phase-mismatched and act as cavity loss). He has examined this problem by dividing the taper into a larger number of subsections. We excited each subsection with its local fundamental mode and propagated the light through the short subsection. At the end of the subsection, the irradiance was correlated with the (new) local fundamental mode. This was repeated for the whole structure. After passing through each subsection, we discarded the energy in higher order modes. The taper described above preserves 93% of the power in the local fundamental mode. Thus, tapering is feasible. On the other hand, with the same design except for steeper tapering (100 μm), although the total transmission is 93%, a much larger fraction of the light locally excites the third-order mode in the taper. With tapering, we cannot make structures with active puls phase sections much shorter than, say 300 μm together. This also implies that we must make devices with sufficiently small κ in order to avoid crosstalk.

An alternative form of tapering is a curvature-matched Bragg filter which ensures single-mode operation. It works only for reflection filters, but provides the best possibilities of overcoming the saturation problem. In conclusion, with a combination of tapering and a thinner active layer, the mode area can be increased by a factor of 5 to 20, with reasonable assumptions. Tapering cannot be employed in DFB filters. The drawback with tapering is that currents are required and hence more power is dissipated.

A third stage in the optimization procedure is to reduce τ_{rec}, which is automatically effected through higher carrier concentration (thin active layer). Fourth, the Bragg wavelength should be optimized with respect to the quasi-Fermi levels. One should operate at a point on the gain curve where n_{sp}, $\partial g/\partial N$, and $\partial n/\partial N$ are minimized. For a given Bragg wavelength and material composition, these factors depend on wavelength detuning and carrier density. Generally speaking, one should operate with a high value of $\Delta E_F - \hbar\omega_{Bragg}$. There is a substantial spread in published data on the differential gain and differential index coefficients. The values used here are typical of 1.55-μm LDs.

In Figure 9.12(b), these measures are combined for a $G = 20$ dB transmission device. For $\kappa \times L_B = 0.5$ to 0.6, the SMRR is larger than 13 dB and $\Delta\nu = 2$ to 3 GHz. A large increase in dynamic range (10 to 15 dB, depending on n_{sp}) is obtained.

9.8 WAVELENGTH-SELECTIVE PHOTODETECTION

9.8.1 Photodetection in Laser Diode Amplifiers

LD amplifiers can also function as photodetecting devices. Photodetection occurs as a result of the stimulated-emission-induced carrier recombination in the active layer. This changes the quasi-Fermi level separation and thus the external voltage across the forward-biased diode. This differs from the usual reverse-biased detector, where the incoming signal carriers and the current are of opposite sign [27].

A two-section LD amplifier can act as a device integrated with an optical preamplifier and a detector. Fortenberry et al. [28] reported that up to 16-dB improvement in detected output voltage is obtained for an LD amplifier with an internal gain of 30 dB. Experimental results shows a 4-dB improvement.

Light input into an LD amplifier causes stimulated emission resulting in a change in the carrier density. The relation between the change in photon density $\Delta S(z)$ at a distance z along the cavity from the input facet and the change in carrier density $\Delta N(z)$ can be obtained from the standard rate equation for carrier density as

$$\Delta N(z) = -\Delta S(z) t_s g v_g \qquad (9.27)$$

where $\Delta N(z)$ is the change in carrier density at any point z along the length of the LD amplifier, $\Delta S(z)$ is the change in photon density at the same point, t_s is the carrier lifetime, g is the material gain per unit length, and v_g is the group velocity within the amplifier. The change in junction voltage $\Delta V(z)$ at a point z on the junction is given by

$$\Delta V(z) = \eta\left(\frac{kT}{q}\right) \ln\left[\frac{N + \Delta N(z)}{N}\right] \qquad (9.28)$$

where η is the junction ideality factor, k is the Boltzmann constant, T is the temperature, q is the charge of an electron, and N is the steady-state carrier density. For small $\Delta N(z)$, the change in voltage is directly proportional to the change in carrier density and is thus directly proportional to the change in photon density.

If an optical amplifier has a single electrical contact along the entire length of the device and is operated below saturation, the optical intensity increases exponentially as it propagates along the length of the LD amplifier. If the amplifier length is L and the net gain coefficient is $g' = g - \alpha_{sc}$, where α_{sc} is the attenuation coefficient. The internal optical gain becomes $\exp(g'L)$. Given an input photon density ΔS_{in}, the output photon density is given by $\Delta S_{out} = \Delta S_{in}\exp(g'L)$. The contact voltage generated by a change in photon density is proportional to the average of ΔV below the contact. By integrating along the length of the device, it is easy to show that this average is

$$\Delta V_{1-section} = \Delta S_{in}K\left[\frac{\exp(g'L) - 1}{g'L}\right] \tag{9.29}$$

where K is a constant.

The detector response can be improved by dividing the electrical contact into two sections. The first section of the LD amplifier acts as an optical preamplifier and the second acts as a detector. This is illustrated in Figure 9.13, which shows a two-section amplifier with contacts of length L_1 and L_2. Note that, unlike previous work on integrated optical preamplifiers where the output of the preamplifier is absorbed by a photodiode, the detector section in the present device acts as an amplifier and allows the optical signal to pass through the two-section amplifier.

The voltage on the second contact is the gain of the first section $\exp(g'L_1)$, multiplied by the average ΔV under the second contact. Following the argument above, this average is

$$\Delta V_{2-section} = \Delta S_{in}K \exp(g'L_1)\left[\frac{\exp(g'L_2) - 1}{g'L_2}\right] \tag{9.30}$$

Comparing the output voltages of a one-section LD amplifier to a two-section LD amplifier, the improvement in responsivity obtained from the two-section LD amplifier, for $\exp(g'L)$ » 1, is

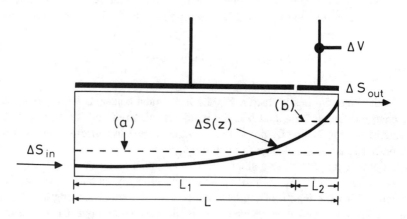

Figure 9.13 Diagram of two-section LD amplifier showing change in photon density along length. (From [28].)

$$\text{improvement (dB)} = 20 \log_{10}\left(\frac{\Delta V_{2-\text{section}}}{\Delta V_{1-\text{section}}}\right)$$

$$= 20 \log_{10}\left\{\frac{(L_1 + L_2)[1 - \exp(-g'L_2)]}{L_2}\right\} \qquad (9.31)$$

Equation (9.31) implies that the length of the second contact L_2 should be minimized for maximum improvement. However, the impedance of the second contact is an important consideration. To obtain a large detection bandwidth, the load impedance connected to the second contact should be relatively low. A 50Ω microwave amplifier input impedance would be suitable in many applications. Note that the contact resistance of an LD amplifier is inversely proportional to its length. A longer contact is desirable to reduce the series contact resistance, whereas a shorter contact is needed to improve the detector responsivity as indicated above. Thus, the optimum contact length will be a compromise between these two competing demands.

The optimum length for the second contact can be determined by modeling the LD amplifier detector as a simple circuit. The model consists of a low-impedance forward-biased junction connected in series with a resistor, representing the contact resistance, whose value is inversely proportional to the length of the second contact as described above. The load impedance of the electrical amplifier is connected in series. Note that increasing the length of the second contact decreases the contact resistance and increases the detected voltage obtained across the load impedance.

From (9.31), the improvement in detector responsivity is roughly proportional to the ratio $(L_1 + L_2)/L_2$. In addition, the contact resistance decreases as the second contact is lengthened. Thus, increasing the total length of the device $(L_1 + L_2)$ allows the second section to be sufficiently long to give a relatively low contact resistance while keeping the ratio $(L_1 + L_2)/L_2$ large. The best calculated improvement in output voltage is about 16 dB.

A three-section LD amplifier can be used to monitor amplifier gain by measuring the ac voltages at the facet contacts [29].

9.8.2 Wavelength-Selective Photodetection

A tunable narrowband optical filter can also be operated as a narrowband photodetector and, if desired, as an FSK discriminator [30–33]. Recent work has demonstrated FSK transmission up to 250 Mbps using an identical MQW DBR LD as both a transmitter and a receiver [30]. At the receiving end, the MQW DBR LD acts as a tunable filter, FSK discriminator, and photodetector. More recently, a high-bit-rate FSK transmission using two-electrode DFB LDs as both transmitter and receiver has been reported [31].

An optical FM receiver based on a one-section Fabry-Perot LD, with one AR-coated facet has been reported [34]. The device, biased far above the threshold, acts as an FSK discriminator/photodetector. An FSK transmission system, up to 1.5 Gbps with −30-dBm sensitivities for 10^{-9} BER, has been demonstrated using this laser. The speed is limited by the transmitter FM response, not by the receiver response.

REFERENCES

[1] Brackett, C. A., "Dense Wavelength Division Multiplexing Networks: Principles and Applications," *IEEE J. Selected Areas in Communications*, Vol. 8, No. 6, August 1990, pp. 948–964.

[2] Toba, H., K. Oda, K. Nakanishi, N. Shibata, K. Nosu, N. Takato, and M. Fukuda, "100-Channel Optical FDM Transmission/Distribution at 622 Mb/s Over 50 km," *Opt. Fiber Commun. Conf., OFC'90*, San Francisco, CA, 1990, postdeadline paper PD1.

[3] Kawaguchi, H., K. Magari, K. Oe, Y. Noguchi, Y. Nakano, and G. Motosugi, "Optical Frequency-Selective Amplification in a Distributed Feedback Type Semiconductor Laser Amplifier," *Appl. Phys. Lett.*, Vol 50, No. 2, January 1987, pp. 66–67.

[4] Numai, T., S. Murata, and I. Mito, "Tunable Wavelength Filters Using λ/4-Shifted Waveguide Grating Resonators," *Appl. Phys. Lett.*, Vol. 53, No. 2, July 1988, pp. 83–85.

[5] Numai, T., M. Fujiwara, N. Shimosaka, K. Kaede, S. Suzuki, and I. Mito, "1.5 μm λ/4-Shifted DFB LD Filter and 100 Mbit/s Two-Channel Wavelength Signal Switching," *Electron. Lett.*, Vol. 24, No. 4, February 1988, pp. 236–237.

[6] Magari, K., H. Kawaguchi, K. Oe, Y. Nakano, and M. Fukuda, "Optical Selection With a Constant Gain and a Gain Bandwidth by a Multielectrode Distributed Feedback Laser Amplifier," *Appl. Phys. Lett.*, Vol 51, No. 24, December 1987, pp. 70–72.

[7] Magari, K., H. Kawaguchi, K. Oe, and M. Fukuda, "Optical Narrow-Band Filters Using Optical Amplification With Distributed Feedback," *IEEE J. Quantum Electronics*, Vol. QE-24, No. 11, November 1988, pp. 2178–2190.

[8] Numai, T., S. Murata, and I. Mito, "1.5 μm Tunable Wavelength Filter With Wide Tuning Range and High Constant Gain Using a Phase-Controlled Distributed Feedback Laser Diode," *Appl. Phys. Lett.*, Vol. 53, No. 13, September 1988, pp. 1168–1169.

[9] Kazovsky, L. G., M. Stern, S. G. Menocal Jr., and C.-E. Zah, "DBR Active Optical Filters: Transfer Function and Noise Characteristics," *IEEE J. Lightwave Technology*, Vol. 8, No. 10, October 1990, pp. 1441–1451.

[10] Tessler, N., R. Nagar, G. Eisenstein, J. Salzman, U. Koren, G. Raybon, and C. A. Burrus, Jr., "Distributed Bragg Reflector Active Optical Filters," *IEEE J. Quantum Electronics*, Vol. 27, No. 8, August 1991, pp. 2016–2024.

[11] Numai, T., S. Murata, and I. Mito, "1.5 μm Tunable Wavelength Filter Using a Phase-Shift-Controlled Distributed Feedback Laser Diode With a Wide Tuning Range and a High Constant Gain," *Appl. Phys. Lett.*, Vol. 54, No. 19, May 1989, pp. 1859–1860.

[12] Schimpe, R., B. Bauer, and S. Illek, "Tunable Wavelength Filter Using Tunable Twin-Guide-(TTG-) Laser," *Conf. Dig. of 12th IEEE Int. Semiconductor Laser Conf.*, IEEE, 1990, L-17.

[13] Tanaka, K., T. Inoue, M. Matsuda, T. Yamamoto, H. Kobayashi, K. Wakao, and T. Mikawa, "Wide-Wavelength-Tunable Active Filter With a λ/4-Shifted Distributed Feedback Structure With Independently Current-Injected Tuning Waveguide," *PS'91*.

[14] Kazovsky, L., and J. Werner, "Multichannel Optical Communictions Using Tunable Fabry-Perot Amplifiers," *Applied Optics*, Vol. 28, No. 3, February 1989, pp. 553–555.

[15] Numai, T., "1.5 μm Optical Filter Using a Two-Section Fabry-Perot Laser Diode With Wide Tuning Range and High Constant Gain," *IEEE Photonics Tech. Lett.*, Vol. 2, No. 6, June 1990, pp. 401–403.

[16] Numai, T., "1.5 μm Two-Section Fabry-Perot Wavelength Tunable Optical Filter," *IEEE J. Lightwave Technology*, Vol. 10, No. 11, November 1992, pp. 1590–1596.

[17] Pan, Z., and M. Dagenais, "Bistable Diode Laser Amplifier as a Narrow Bandwidth High-Gain Filter for

Use in Wavelength Division Demultiplexing," *IEEE Photonics Tech. Lett.*, Vol. 4, No. 9, September 1992, pp. 1054–1057.

[18] Kogelnik, H., and C. V. Shank, "Coupled-Wave Theory of Distributed Feedback Lasers," *J. Appl. Phys.*, Vol. 43, No. 5, May 1992, pp. 2327–2335.

[19] Streifer, W., R. Burnham, and D. Scifres, "Effect of External Reflections on Longitudinal Modes of Distributed Feedback Lasers," *IEEE J. Quantum Electronics*, Vol. QE-11, No. 4, April 1975, pp. 154–161.

[20] Choa, F.-S., and T. L. Koch, "Static and Dynamical Characteristics of Narrow-Band Tunable Resonant Amplifiers as Active Filters and Receivers," *J. Lightwave Technology*, Vol. 9, No. 1, January 1991, pp. 73–83.

[21] Lang, R., "Injection Locking Properties of a Semiconductor Laser," *IEEE J. Quantum Electronics*, Vol. QE-18, No. 6, June 1982, pp. 976–983.

[22] Nakajima, H., and I. F. De Faria, Jr., "Transient Response of MS-DFB Amplifier-Filters," *IEEE Photonics Tech. Lett.*, Vol. 5, No. 9, September 1993, pp. 1032–1034.

[23] Kobrinski, H., M. P. Vecchi, E. L. Goldstein, and R. M. Bulley, "Wavelength Selection With Nanosecond Switching Times Using Distributed-Feedback Laser Amplifiers," *Electron. Lett.*, Vol. 24, No. 15, July 1988, pp. 969–971.

[24] Goldstein, E. L., and M. C. Teich, "Noise in Resonant Optical Amplifiers of General Resonator Configuration," *IEEE J. Quantum Electronics*, Vol. QE-25, No. 11, November 1989, pp. 2289–2296.

[25] Goldstein, E. L., and M. C. Teich, "Noise Measurements on Distributed-Feedback Optical Amplifiers Used as Tunable Active Filters," *IEEE Photonics Tech. Lett.*, Vol. 3, No. 1, January 1991, pp. 45–46.

[26] Sahlén, O., "Active DBR Filters for 2.5 Gb/s Operation: Linewidth, Crosstalk, Noise, and Saturation Properties," *IEEE J. Lightwave Technology*, Vol. 10, No. 11, November 1992, pp. 1631–1643.

[27] Nakajima, H., "High-Speed and High-Gain Optical Amplifying Photodetection in a Semiconductor Laser Amplifier," *Appl. Phys. Lett.*, Vol. 54, No. 11, March 1989, pp. 984–986.

[28] Fortenberry, R. M., A. J. Lowery, and R. S. Tucker, "Up to 16 dB Improvement in Detected Voltage Using Two-Section Semiconductor Optical Amplifier Detector," *Electron. Lett.*, Vol. 28, No. 5, February 1992, pp. 474–476.

[29] Newkirk, M. A., U. Koren, B. I. Miller, M. Chien, M. G. Young, T. L. Koch, G. Raybon, C. A. Burrus, B. Tell, and K. F. Brown-Goebeler, "Three-Section Semiconductor-Laser Amplifier for Monitoring of Optical Gain," *OFC/IOOC'93 Technical Dig.*, Th K6, pp. 217–218.

[30] Koch, T. L., F. S. Choa, F. Heismann, and U. Koren, "Tunable Multiple-Quantum-Well Distributed-Bragg-Reflector Lasers as Tunable Narrowband Receivers," *Electron. Lett.*, Vol. 25, No. 14, July 1989, pp. 890–892.

[31] Chawki, M. J., R. Auffret, and L. Berthou, "1.5 Gbit/s FSK Transmission System Using Two Electrode DFB Laser as a Tunable FSK Discriminator/Photodetector," *Electron. Lett.*, Vol. 26, No. 15, July 1990, pp. 1146–1148.

[32] Hui, R., N. Caponio, P. Gambini, M. Puleo, and E. Vezzoni, "DFB Active Filter/Detector for Multichannel FSK Optical Transmission," *Electron. Lett.*, Vol. 27, No. 22, 1991, October pp. 2016–2018.

[33] Chawki, M. J., R. Auffret, E. L. Coquil, P. Pottier, L. Berthou, H. Paciullo, and J. L. Bihan, "Two-Electrode DFB Laser Filter Used as a Wide Tunable Narrow-Band FM Receiver: Tuning analysis, Characteristics and Experimental FSK-WDM System," *J. Lightwave Technology*, Vol. 10, No. 10, October 1992, pp. 1388–1397.

[34] Chawki, M. J., L. L. Guiner, D. Dumay, and J. C. Keromnes, "1.5 Gbit/s Transmission System Using FP Laser as FSK Discriminator/Photodetector," *Electron. Lett.*, Vol. 28, No. 17, August 1992, pp. 1573–1574.

Chapter 10
Applications for Photonic Switching

In this chapter, we study applications of the functional LDs described in previous chapters for photonics switching. For system architectures, we might refer to a book written by Midwinter and Guo [1]. Computing is another important potential application area of the functional LDs [2], and will be briefly described in Chapter 11.

10.1 APPLICATIONS OF ULTRASHORT OPTICAL PULSES GENERATED FROM LASER DIODES

Short-pulse generation from LDs is currently a very active research area, as described in Section 7.2. This work is motivated by a need for compact, reliable, long-lived sources of short optical pulses for electro-optic sampling systems, as sources for soliton transmission systems, as tunable sources for measurements in physical science, and for telecommunication applications, including high-bit-rate externally modulated transmission systems and time-division-multiplexed transmission systems [3,4]. Here, we see some examples of applications of short optical pulses generated by mode-locking of LDs.

10.1.1 High-Speed Data Transmission Systems

Optical Time-Division Multiplexing Systems

Optical time-division multiplexing (TDM) is of great interest in high-speed data systems as a method to better use the potential capacity of optical fibers, without requiring the development of higher speed electronics. By combining lower-bit-rate optical data streams for transmission and demultiplexing before optoelectronic conversion, transmission capacity can be increased despite the electronic speed bottleneck. One of the main techniques, optical TDM, uses high-speed electro-optic switches to achieve multiplexing and demultiplexing of synchronized data streams.

The data of an experimental system using the multiple-laser architecture was reported by Tucker et al. [5]. A block diagram of the four-channel system is shown in Figure 10.1(a). The baseband data rate is 4 Gbps. A common 4-GHz clock drives the four transmitters via a series of microwave delay lines that are adjusted to provide correct timing of the optical pulses. A schematic of the optical transmitters is shown in Figure 10.1(b). The clock drives both the word generator and the laser, thereby ensuring that the electrical data is synchronized to the optical pulses. The lasers used in this experiment are of the fiber extended-cavity type. All lasers are within 8 nm of the wavelength of zero first-order chromatic dispersion in the fiber. The total spectral width of four multiplexed lasers is about 10 nm. The four-channel 16-Gbps system in Figure 10.1(a) serves as a good example of how optical multiplexing and demultiplexing can remove the need for very wideband electronics. This system transmits data over 8 km of fiber at 16 Gbps, but uses a maximum electronic bandwidth of only 2.6 GHz (in the drive electronics for the modulator and in the receiver). Since the demultiplexer (DMUX) switches are driven by sinusoidal excitation, there is negligible electrical bandwidth required in the DMUX.

Soliton Transmission Systems

The existence of solitons in optical fibers was suggested in 1973 [6], and solitons were observed experimentally in 1980 [7]. Remarkable progress during the decade of the 1980s

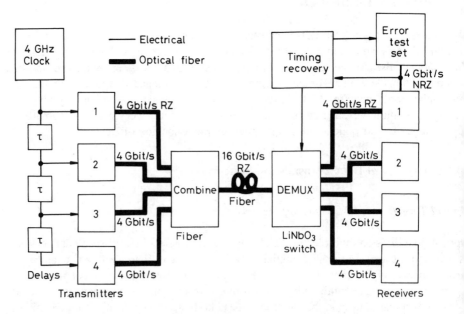

Figure 10.1(a) Block schematic of four-channel optical TDM system. The electrical time delays τ are 62.5 ps, corresponding to 1-bit period at 16 Gbps.

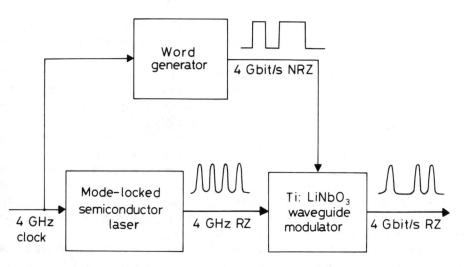

Figure 10.1(b) 4-Gbps optical transmitter using mode-locked LD. (©1988 IEEE. From [5].)

has transformed fiber solitons from a mathematical curiosity into a practical entity useful for optical communications [8].

With a mode-locked LD as a signal source, the transmission of 60-ps solitons at 2.4 Gbps over paths as great as 12,000 km in a recirculating loop of dispersion-shifted fiber and erbium-doped fiber amplifiers (EDFA) is achieved [9]. The recirculating loop (Figure 10.2) used the same 25-km lengths of dispersion-shifted fiber. Each EDFA was pumped directly, with pump light always propagating counter to the signals. In this way, the required pump power to each amplifier was reduced to less than 20 mW, the amplifier

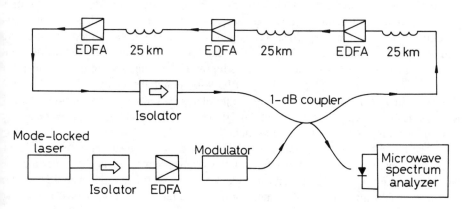

Figure 10.2 Experimental setup for the recirculating-loop configuration used to demonstrate soliton transmission over 12,000 km at 2.4 Gbps. (From [9]. Reproduced by permission of IEE.)

gains were more easily matched to the corresponding segment loss, and the potential for cross-phase modulation of the signals from amplitude fluctuations in the pump light was eliminated.

The mode-locked InGaAsP LD was embedded in an external cavity of 62.5-mm optical length, whose round-trip time (416.66 ps) corresponded to just one period of the 2.4-GHz pulse repetition rate; we thus avoided the periodic pulse-to-pulse amplitude fluctuations typical of a multiple-length cavity. The cavity end mirrors were formed by an uncoated facet of the diode and a 1,200-groove/mm grating, respectively; the grating was coupled to the other, AR-coated facet of the diode by a collimating lens (a x60 microscope objective, in later versions replaced by a sapphire ball lens of 2.8-mm effective focal length). The LD was driven simultaneously by about 0.6W of RF power at 2.4 GHz and about 30 mA of direct current, with the ratio of the two electrical drives carefully adjusted to yield minimum chirp and the best symmetry of the mode-locked pulses; the resultant time-bandwidth product ($\tau \Delta f$) was about 0.42.

Earlier versions of the laser used an internal, low-finesse (about 10) etalon of FWHM around 10 GHz, but when the ball lens was used, it became possible to fill the grating with a wide enough beam (about 3 mm in diameter) that the grating alone had the required selectivity. Nevertheless, the same etalon, applied externally, is useful in cleaning up the fringes of the laser's optical spectrum and in making it almost perfectly symmetrical. Finally, an erbium-doped fiber buffer amplifier boosted the laser's time-average output power (about 0.3 mW after coupling into a fiber) to a level large enough to establish solitons in the loop (peak pulse power at the input to each 25-km span around 1.5 mW), in spite of large (>11 dB) input coupling losses.

10.1.2 Clock Distribution

The distribution of the master clock in high-speed digital switching machines is a long-standing problem. In many applications, such as future-generation telecommunications switching machines or high-speed supercomputers, as many as 2,000 printed circuit boards require a common master clock. Usually, the fundamental hardware limit to the attainable system switching speed is not the switching limits of the transistors in the machine, but the clock skew limit imposed by the distribution technique employed [10]. Photonics offers new insight into the solutions of these problems [10]. Dynamic clock jitter as small as 50 ps has been reported for fan-outs of 10. Most of the jitter in the pulses generated by gain switching is due to LD turn-on delay.

The incorporation of a mode-locked LD in the distribution scenario virtually eliminates this source of jitter. The diagram of the arrangement used in the experiment is shown in Figure 10.3 [11]. The master clock is the hybrid mode-locked LD system operating at 830 nm. The laser system consists of three main components: (1) a low-power hybrid mode-locked oscillator emitting output pulses of 5 ps in duration, with an average power of 1 mW, (2) a traveling-wave LD amplifier that boosts the average power

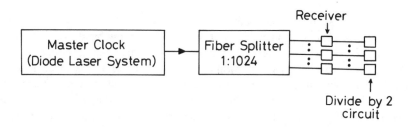

Figure 10.3 Optically distributed clock network. (©1991 IEEE. From [11].)

to 30 mW, and (3) a dual grating pulse compressor that reduces the pulse duration to 460
fs with 10 mW of average power. The hybrid nature of the mode-locking process takes
advantage of the pulse-shortening mechanism due to the saturable absorber and the temporal stability provided by the RF-synthesized source. The optical pulse train produced by
this system was distributed to 1,024 ports using an optical fiber splitter with 50-μm cores
as shown in Figure 10.3. At the output of the fiber splitter, the optical pulse train was
detected and amplified using a 700-MHz optoelectronic receiver with a 1-V/mW sensitivity
at 850 nm. The output of the receiver was then used to drive a divide-by-two emitter
coupled logic circuit to generate the square wave clocking waveform.

The pulse-to-pulse timing jitter of the mode-locked pulse train was first measured
in order to determine the lower bound of the timing jitter. This was accomplished by using
standard frequency domain techniques [12]. The timing jitter measurement is obtained by
detecting the mode-locked optical pulse train with a high-speed photodiode and analyzing
the power spectrum with an RF spectrum analyzer. The temporal jitter of the pulse train
manifests itself as a frequency-squared increase of the power contained in the phase noise
sidebands of the power spectrum [12]. Once this characteristic has been identified in the
measured power spectrum, the relative timing jitter can be obtained by using

$$\frac{\Delta T}{T} = \frac{1}{2\pi n} \left[\frac{P_{sb}}{P_c} \frac{\delta \nu}{B_r} \right]^{1/2} \tag{10.1}$$

In this equation, n is the harmonic number, P_{sb} and P_c are the powers contained in the
phase noise sideband and the carrier, respectively, $\delta \nu$ is the −3-dB bandwidth of the noise
band, B_r is the resolution bandwidth of the RF spectrum analyzer, T is the pulse period,
and ΔT is the timing jitter. From a measurement of the RF power spectrum, one extracts
a relative timing jitter of about 400 fs, with a measurement duration of 100 sec. This
jitter value is extremely low as compared to other standard laser systems, such as mode-
locked argon and YAG systems, but is typical of mode-locked LD systems. The extremely
low timing jitter and high power contained in the optical pulse train (over 70W peak) are
the key features that enable this optical source to be used as a master clock in an optically
distributed clocking network.

On the other hand, such frequency domain techniques are not suitable for measuring th
timing jitter of the optoelectronic receiver because of their low bandwidth. Delfyett et a
[11] used the correlator, which consists of a microwave mixer and delay lines, to measur
the timing jitter. Measurements of this kind on any two output ports yielded a total pulse
to-pulse jitter of approximately 12 ps over measurement periods of approximately 1 hou
This 12-ps result represents total excursion and includes all sources of jitter, correlate
and uncorrelated, rapidly varying and slowly varying, including optical reflections, possi
bilities of modal noise, and temporal and thermal variations of the electronics.

One can estimate the maximum fan-out capability of the present clock network b
examining the power budget. The maximum number of output ports N_{max} to be drive
can be given by

$$N_{max} = (N_{ph})(hc/\lambda)(B/P_{min})L^{-1} \qquad (10.2$$

Here, N_{ph} is the number of photons per bit, L is the loss factor, P_{min} is the minimur
detection power, B is the bit rate, and h, c, and λ are Planck's constant, the speed of ligh
and the operating wavelength, respectively. In the limiting case of no optical loss in th
coupling and splitting of the master optical clock signal ($L = 1$), and with a minimur
detectable power of 1 μW, one could expect the maximum fan-out capability to be abor
7,000, with a clock rate of 300 MHz. Higher clocking rates will result in a proportional
smaller split ratio due to the finite number of photons that one can extract from the lase
amplifier used in the experiment; that is, doubling the repetition rate reduces the spl
factor by two since the number of photons per pulse have been reduced by a factor of :
Additional reductions in the fan-out capabilities also rise at higher repetition rates owin
to the larger power required by the receivers.

10.2 ALL-OPTICAL-FIBER COMMUNICATION SYSTEMS

Present optical-fiber communication systems are composed of regenerative repeater
which are inserted into the line to perform timing recovery, data retiming, and regeneratic
of signals attenuated by long-distance fiber transmission (so-called *3R function*). In
conventional repeater, the optical signal is first converted to an electrical signal by
photodiode, then the electrical signal is processed 3R function by semiconductor electr
circuits. The electrical signal is converted to an optical signal with an LD. If the optic.
signal can be directly treated with low noise, the repeater can be smaller and less expensiv
than existing repeaters.

10.2.1 Clock Extraction

System synchronization is a serious problem in constructing all-optical signal-processir
systems, such as regenerative repeaters, time-division switching systems, and DMUX

n order to realize the system synchronization, an all-optical timing extraction circuit is
equired, which recovers timing information from an incoming optical data stream and
roduces an optical clock without an intermediate electric stage. Several techniques for
ptical timing extraction circuits have been reported:

1. Injection locking using a variety of self-pulsating optical oscillators [13–22];
2. Injection locking of mode-locked lasers [23,24];
3. Two-section semiconductor nonlinear optical amplifiers [25,26];
4. Optical phase-locked loop using an LD amplifier [27];
5. Optical tank circuits using a Fabry-Perot resonator [28].

Iere, the first technique (injection locking of self-pulsating optical oscillators) is described
i more detail. As described in Section 3.2, the inhomogeneously excited LD shows self-
ulsation when it contains a rapidly saturable absorber in its cavity. The locking of
epetition frequency to that of the injected signal can be obtained by the injection of
ptical pulses into the saturable absorber. A self-pulsating three-section DFB LD can be
jection locked to a pulse train launched from a master laser, whose repetition rate is
early equal to f_{FR}/N, where f_{FR} is the repetition rate of self-pulsating three-section DFB
D under free-running condition and N is an integer. For N larger than 2, we call
ie phenomenon *subharmonic injection locking*. Subharmonic injection locking is very
nportant from the system application point of view, as will be described later. It is worth
oting that the injection locking described in this section is different from the usual optical
jection locking described in Section 4.2. In usual injection locking, the oscillating
equency of the slave laser is locked to that of the master laser. On the other hand, the
ilse repetition rate of the output is locked to that of the input in this case.

The injection-locking characteristics were observed in both the time and spectral
gions and are shown in Figure 10.4(a). Figure 10.4(a)(i) is an oscilloscope trace and
ie microwave spectrum of the optical pulse train at a repetition rate 196 MHz from the
elf-pulsating three-section DFB LD under free running. When a pulse train, which is
icoded as ... 010000001000000010 ... at a data clock rate of 197 MHz as shown in
ie lower trace of the left-hand side of Figure 10.4(a)(ii), is injected into the self-pulsating
ree-section DFB LD, the laser is subharmonic injection locked ($N = 8$) to the incoming
ilse train. As a result, the repetition rate changed to that of an injected pulse train of
)7 MHz and a pulse train synchronized to the injected signal is output (Figure 10.4(a)(ii)
pper trace). This was easily confirmed by observing the microwave spectrum shown in
gure 10.4(a)(ii). If the clock frequency of the input signal is separated far from the free-
nning pulsation frequency, the three-section DFB LD cannot be locked into the input
gnal (see Figure 10.4(a)(iii)).

Precise measurements of lock-in ranges and relative phase changes for various
jection conditions were made with a network analyzer [13,14]. High optical injection
vels yield a large lock-in frequency range. In order to extract an optical timing clock
om a random input optical signal, subharmonic as well as harmonic injection locking
required. Figure 10.4(b) shows the injection-locking characteristics for four RZ signal

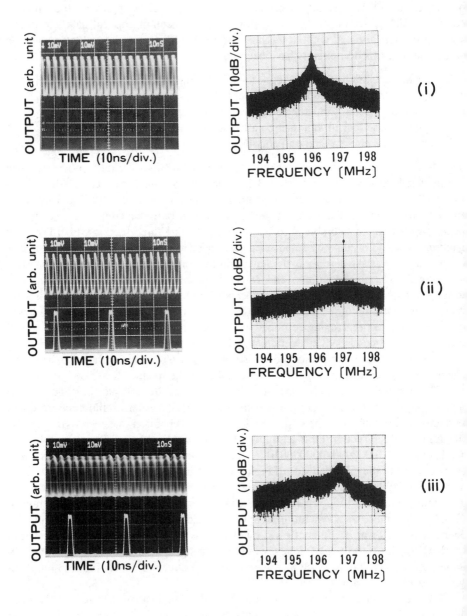

Figure 10.4(a) Injection-locking characteristics observed in both time and spectral regions: (i) free running a 196 MHz without optical injection; (ii) subharmonic injection locking to data stream of . . 0100000001000000010 . . . at 197 MHz; (iii) miss-injection-locking.

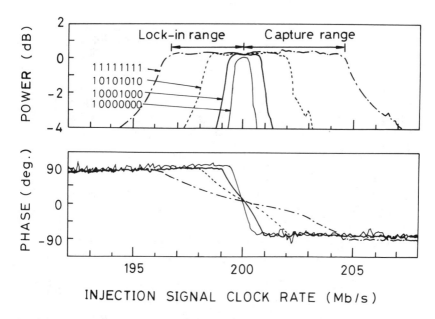

Figure 10.4(b) Injection-locking characteristics for four return-to-zero (RZ) signal patterns of 11111111, 10101010, 10001000, and 10000000 with constant peak power of 210 μW. (©1992 IEEE. From [14].)

patterns of 11111111, 10101010, 10001000, and 10000000 with a constant peak power of 210 μW. The free-running self-pulsation frequency f_0 was set to 200 MHz by adjusting the injection current. The injection signal clock rate was swept from the low rate to the high rate in the experiment. Therefore, the notations of "lock-in range" and "capture range" are used for $f < f_0$ and $f > f_0$, respectively. The results shown indicate that even if the power level and the clock frequency of the input signal are constant, the relative phase of the output pulse varies depending on the signal pattern. This phenomenon causes timing jitter generation in the extracted optical clock if the clock frequency of the input signal is not equal to the free-running pulsation frequency. Consequently, in order to avoid this jitter generation, the free-running pulsation frequency should be set close to the input signal clock frequency.

We can simulate characteristics of the all-optical clock extraction using a self-pulsating LD with rate equations [29]. Figure 10.5 shows the calculated output pulse rate versus injection signal clock rate for RZ signal patterns of 11111111 with various input power levels. At a higher injection signal clock rate than f_0, a wide lock-in range exists, which increases with increasing input power levels. Here, f_0 is the free-running self-pulsation frequency.

The self-pulsation frequency increases with an increase in the injection current. Under multigigahertz operation of self-pulsating LDs, however, the carrier density in the

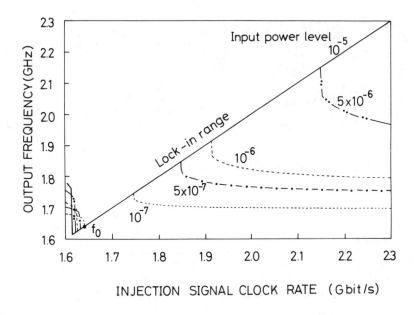

Figure 10.5 Injection-locking characteristics simulated using rate equations. (From [29].)

gain region becomes so high that the reduction in lifetime τ_g reduces the ratio τ_g/τ_a such that the device no longer self-pulsates. This can be overcome somewhat by ensuring that τ_a is also small. To achieve this, Barnsley et al. have selectively doped the absorber regions with zinc. The zinc ions act as centers for nonradiative recombination, thereby reducing the carrier lifetime. The reduction in carrier lifetime was sufficient to obtain strong pulsations at frequencies between about 3 and 5.2 GHz. All-optical clock extraction from 5-Gbps RZ data is demonstrated using a two-section InGaAsP self-pulsating LD whose saturable absorber region was doped with zinc ions [18]. Injection of a 10-μW optical data signal at a wavelength close to one of the Fabry-Perot wavelengths was sufficient to synchronize the self-pulsations to the incoming data stream.

Although a physical mechanism has not yet been clarified, Möhrler et al. observed self-pulsation in an InGaAsP/InP two-section DFB LD without a saturable absorber by adjusting appropriate injection conditions [30]. The DFB LDs have two gain sections. It was demonstrated that--in contrast to Fabry-Perot elements—no selective treatment of one section is required for creating the self-pulsation, and was concluded that the self-pulsation in DFB elements is of a different type than that in Fabry-Perot elements. This new type of self-pulsation has two advantages: first, self-pulsation frequency between several hundreds of megahertz and more than 20 GHz [31] are observed, and second, no selective doping is required to achieve fast recombination in the absorber section. As e al. [22] demonstrated clock extraction from an optically injected $2^{23} - 1$ pseudorandom inary sequence (PRBS) RZ data signal in a two-section ridge waveguide DFB LD with elf-pulsation between 0.4 and 0.7 GHz. A locking range of 50 MHz was measured.

0.2.2 Regenerative Repeaters

Optical retiming was first demonstrated using a 1.5-μm wavelength bistable Fabry-Perot mplifier decision gate at 140 Mbps [32]. Regeneration is achieved by combining the nput signal with an optical clock waveform and coupling them both into an amplifier. The clock waveform consists of a train of pulses at the data rate with a wavelength at which the amplifier is bistable and whose peak power is just below the threshold at which he amplifier jumps to the high-gain state (Figure 10.6(a)). When the input signal is low, slightly amplified clock pulse appears at the output. When the input signal is high, then xtra power is sufficient to exceed the bistable threshold and the amplifier jumps to the igh-gain state where it remains, even if the input signal goes low, until the end of the lock pulse (Figure 10.6(b)). A large clock pulse is seen at the output. At its most sensitive wavelength, it operated at an error rate of 10^{-9}, with an input signal of −40 dBm, and ave 31.5-dB facet-to-facet gain. The input signal could be tuned to any amplifier mode n a 35-nm range with a penalty of less than 5 dB.

An optical retiming regenerator was demonstrated at 200 Mbps using two kinds of onlinear LD functional devices, as shown in Figure 10.7(a) [14,33]. One is an all-optical iming extraction using the self-pulsation LD, and another is an AND gate using a BLD. As shown in Figure 10.7(a), in an optical retiming regenerator, an optical input signal ncoded by a pseudorandom RZ signal at 200 Mbps and launched from a conventional DFB LD ($\lambda = 1.551$ μm) was divided by the first fiber coupler [14]. One part of the input ignal was injected into the self-pulsating three-section DFB LD through a hemispherically nded fiber. The actual injected peak power was estimated to be 230 μW. The free-unning pulsation frequency of the self-pulsating three-section DFB LD was adjusted to 00 MHz by controlling the applied currents. The self-pulsating three-section DFB LD vas injection locked, and optical clock pulses synchronized to the input signal were roduced. Their peak power, wavelength, and the extinction ratio were 2.4 mW, 1.552 m, and 8.7 dB, respectively. In order to compensate for the coupling loss between the hree-section DFB LDs, an EDFA with 4-dB optical gain was used.

The amplified optical clock pulses were combined with the remaining input optical ignal by the next fiber coupler and injected into the bistable three-section DFB LD. njected peak powers for both the input signal and the extracted clock pulses were set to 45 μW, just below the bistable threshold for a 2.5-ns duration pulse. The relative phase f the clock pulse to the input signal was adjusted to −π/2. The output of the bistable hree-section DFB LD was detected with an InGaAs pin photodiode with a bandwidth of GHz.

The experimental results of all-optical retimed regeneration are shown in Figure 0.7(b), where (i), (ii), and (iii) denote input pseudorandom RZ signal at 200 Mbps,

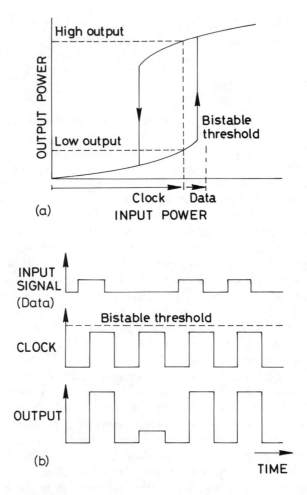

Figure 10.6 Optical retiming using a 1.5-μm bistable Fabry-Perot amplifier decision gate: (a) bistable optical power transfer characteristic; (b) idealized timing diagram. (Reprinted with permission from Chapman and Hall Ltd. From [32].)

combination of the signal and the extracted optical clock, and the retimed and regenerated optical output, respectively. The peak power and the wavelength of the regenerated optical signal were 750 μW and 1.555 μm, respectively. The regenerated output showed the good extinction ratio of >8 dB. Small polarization sensitivities both in the lock-in range and the bistable threshold were confirmed by measurements at various polarization states of the input signal. The wavelength difference between the input and the output signal was about 5 nm, and no substantial increase in the bistable threshold is expected if the

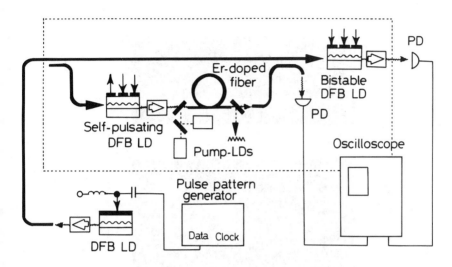

Figure 10.7(a) Experimental setup for demonstrating the all-optical retimed regeneration using three-section DFB LDs.

wavelength of the input signal is within the gain wavelength range of the three-section DFB LD of about 10 nm.

10.2.3 Compensation of Chromatic Dispersion

The use of EDFAs in optical communications gives the potential for long-distance optical links at 1.55 μm without regeneration. The chromatic dispersion of the fiber then becomes the main limitation in high-bit-rate communications. Dispersion compensation techniques feature prominently among proposed solutions, including the use of prechirped signals and dispersion-compensating fiber. A further method of dispersion compensation was proposed by Yariv et al. [34], which relied on optical phase conjugation to give spectral inversion midway along the fiber length. The effects of dispersion accumulated over the first half of the transmission would then be exactly reversed in the second half.

Phase conjugation by NDFWM in an LD amplifier was used to achieve dispersion compensation [35,36]. The NDFWM relies on subpicosecond carrier dynamic effects in the LD amplifier to provide the optical nonlinearity, allowing the potential for use with extremely high bit rates in excess of 100 Gbps (see Section 8.4.3). The use of the LD amplifier increases the efficiency of the phase conjugation. Furthermore, with the use of an LD amplifier, the wavelengths of the optical signals are limited only by the LD amplifier gain bandwidth (typically 50 nm).

Spectrum inversion is essential for applying HNDFWM to an optical-fiber dispersion compensator. In order to confirm spectrum inversion, the modulation sideband spectrum

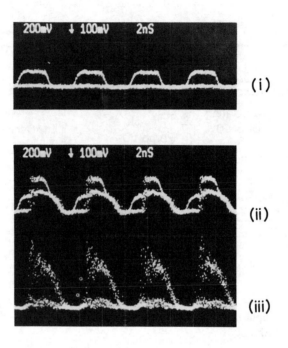

Figure 10.7(b) Results of all-optical retimed regeneration using three-section DFB LDs: (i) input signal; (ii) combination of signal and extracted optical clock; (iii) retimed and regenerated optical output (©1992 IEEE. From [14].)

of the converted signal was measured using a Fabry-Perot compressively strained MQW LD as a conjugator [35]. A 1.5-μm DFB LD monolithically integrated with Franz-Keldysh type optical intensity modulator was used as a probe light source. The input probe light was modulated in a 2-Gbps fixed pattern (1010 . . .) and injected at the second resonance peak. The modulation sideband spectrum of the probe light used was asymmetric due to the residual FM component of the intensity modulator. The converted-signal spectrum was actually the reverse of the probe-light spectrum.

The technique was used by Tatham et al. to remove the dispersion penalty in the 100-km transmission of a directly modulated, highly chirped 2.5-Gbps signal at 1.5 μm [36]. An InGaAsP LD with facets AR-coated to less than 10^{-3} reflectivity was used as nonlinear device. The overall efficiency of the phase conjugation was -12.5 dB.

10.3 WAVELENGTH-DIVISION MULTIPLEXING SYSTEMS

Although the potential transmission bandwidth for single-channel transmission has yet to be realized within installed systems, it is the case that such systems become progressively nore difficult and expensive to implement in the gigabit-per-second range largely due to speed limitations within the terminal electronics. Hence, there is increasing interest in the potential offered by WDM for the achievement of multichannel transmission on a single fiber to provide enhanced bandwidth utilization [37].

WDM involves the transmission of as many different peak wavelength optical signals in parallel on a single optical fiber as possible. Although in spectral terms, optical WDM is analogous to electrical frequency-division multiplexing, it has the distinction that each WDM channel effectively has access to the entire IM fiber bandwidth with current technology on the order of several gigahertz. The technique is illustrated in Figure 10.8, where a single nominal wavelength optical-fiber communication system is shown together with a duplex (i.e., two different nominal wavelength optical signals traveling in opposite directions providing bidirectional transmission), and also a multiplex (i.e., two or more different nominal wavelength optical signals transmitted in the same direction) fiber communication system. It is the latter WDM operation that has generated particular interest within telecommunications.

More recently, interest has grown in the possible application of WDM techniques within optical-fiber networks to provide parallel access to individual pieces of terminal equipment. Such multichannel optical networks have generally adopted a broadcast topol-

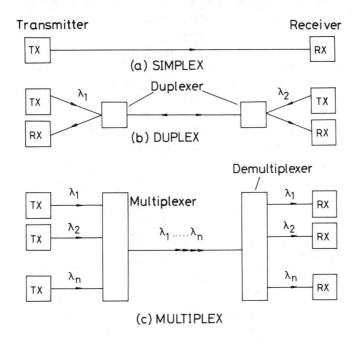

Figure 10.8 Optical-fiber system operating modes: (a) simplex; (b) duplex; (c) multiplex. (From [37]. Reproduced by permission of IEE.)

ogy based on passive star coupler devices as a result of optical-power budget considerations (i.e., the maximum number of node stations increases exponentially with the permitted system loss). Examples of experimental IM WDM star networks are the mixed traffic distribution system investigated by British Telecom [38] and the fixed-assignment LAM-DANET demonstrated by Bell Communications Research [39]. Moreover, WDM is also under investigation within optical-fiber networks employing coherent transmission techniques [40]. Although at an early stage of development, such networks offer the potential for many thousands of finely spaced WDM channels.

High-density (HD) WDM will be implemented in future optical wide-band communication systems. One of the key components of noncoherent HD WDM systems is the tunable narrowband optical filter, which is described in Chapter 9. Recently, the attractive capabilities of the LD as a multifunctional device have been investigated in M^3-NET (M-cubed network) [41]. The LD could be used for light emission, optical filtering or photodetection. An FSK transmission system up to 1.5 Gbps was demonstrated using two-section DFB LDs as both transmitter and receiver. At the receiving end, the two-section DFB LD acted as a tunable filter, FSK discriminator, and photodetector.
M^3-NET is a multiwavelength optical network that stands for multiwavelength, multi-bit-rate, and multifunctional devices when used in HD WDM networks. The broadcasting capability inherent in the M^3-NET design makes it especially suitable for application in video signal distribution. We see the network configuration and experimental results from a three-channel HD WDM M^3-NET testbed demonstrator. A diagram of the basic M^3-NET configuration is shown in Figure 10.9(a).

At the transmitting end, three two-section DFB LDs emitting at different wavelengths $\lambda_1 = 1.5174$ μm, $\lambda_2 = 1.5191$ μm ($\lambda_1 - \lambda_2 = 2.3$ nm, spaced wavelengths), and $\lambda_3 = 1.5193$ μm ($\lambda_3 - \lambda_2 = 0.2$ nm, densely packed wavelengths) were injected into a 3:1 coupler. These LDs had two equal electrodes of 200 μm each and presented a continuous tuning range greater than 1 nm with a minimum spectral linewidth of about 15 MHz. These two-electrode DFB LD transmitters presented a large and flat FM response from 100 kHz to

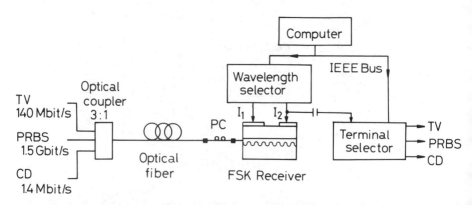

Figure 10.9(a) M^3-NET configuration. (From [41], reproduced by permission of IEE.)

Figure 10.9(b) Fully connected M³-NET. (Courtesy of CNET).

several gigahertz (>3 GHz). The first laser transmitter was modulated by a 140-Mbps digital TV channel, the second was modulated by a high bit rate of 1.5 Gbps NRZ PRBS (higher than the high-definition TV signal bit rate), and the third was modulated by a low bit rate of 1.4 Mbps from a compact disc (CD).

At the receiving end, another identical two-section DFB LD used as a tunable narrowband FM receiver was connected to the transmitter lasers by a few meters of single-mode optical fiber and a polarization controller (PC). A computer controls the injection currents of the receiving laser and the terminal equipment switching unit. A fully connected M^3-NET is shown in Figure 10.9(b).

The M^3-NET configuration is transparent to high and low bit rate transmitted on either spaced or dense wavelengths. A minimum channel spacing of about 11 GHz was measured. Such a network configuration is suitable for video distribution and can be expected for bidirectional broadband network use, with the advantage of simplicity as compared to the heterodyne technique.

Chawki et al. have also demonstrated a new interactive version of a two-channel HD WDM M^3-NET system called M^3I-NET [42]. In this system, the two-section multifunctional DFB LD is used both as an FSK transmitter and as a tunable FSK receiver/ASK transmitter (i.e., an optical transceiver). Furthermore, the two-section DFB LD has separate electrical access ports that are used to reduce the complexity of the electronics. The selection of dense multiplexed wavelengths is achieved for high and low bit rate. The transmitting information (TV or CD) on a selected channel is easily controlled by using the DFB LD as an ASK transmitter. The selection of a TV program or a CD musical part has been successfully achieved by using a simple infrared (IR) TV/CD remote control unit. Such an interactive WDM optical system is suitable for broadband local-area network (LAN) applications.

10.4 OPTICAL EXCHANGING SYSTEMS

Three types of optical switching networks have been proposed using different features of optoelectronics technology such as optical space-division, optical time-division, and optical wavelength-division switches as shown in Figure 10.10 [43]. Combination of these switching systems is also possible. Space-division switching networks are suitable for small-size optical switching systems. Mechanical optical switches can be applied to the space-division switching network, because a high-speed operation is not required. Nonmechanical optical switches have the advantage of ease in integration and reliability, but their loss and crosstalk are greater than those for mechanical optical switches. Time-division switching networks are suitable for large-size optical switching systems and can be easily interfaced with existing time-division optical-fiber transmission systems. However, they require high-speed control circuits. Wavelength-division switching networks are very attractive because of their flexibility.

Optical functional devices required for each switching network are as follows.

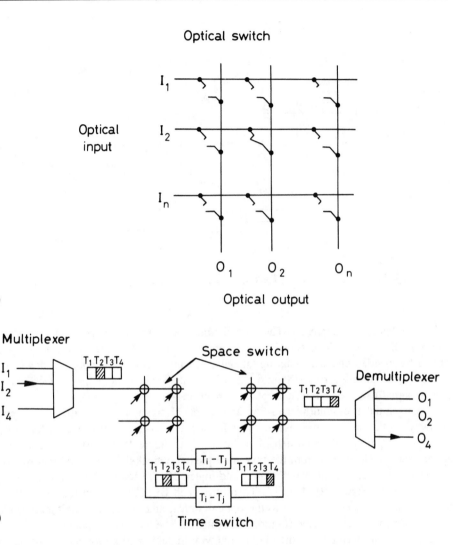

Figure 10.10(a,b) Optical switching networks: (a) optical space-division switch; (b) optical time-division switch (space-time-space construction).

1. Optical switch matrix for space-division switching;
2. Optical memory and optical write/read gate for time-division switching;
3. Tunable wavelength filter and wavelength converter for wavelength-division switching.

10.4.1 Optical Time-Division Switching

An optical time-division switching system was developed using BLDs as optical memories [44]. Figure 10.11 shows a block diagram and a timing chart of the experimental optical

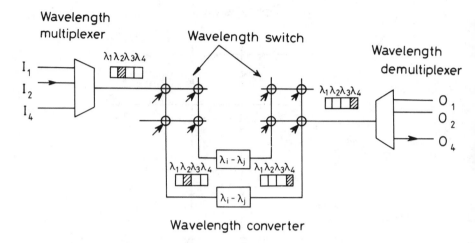

Figure 10.10(c) Optical wavelength-division switch (space-wavelength-space construction). (©1987 IEEE. From [43].)

switching system, respectively. The optical time switch is constructed in a single-stage time switch. A color video signal from a camera is encoded in a 64-Mbps digital pulse stream by a video encoder. The digital pulse stream is combined with the other pulse streams by a multiplexer (MUX) into a 256-Mbps pulse stream and then converted to an optical signal using an electronic-optical converter.

This input time-multiplexed 256-Mbps optical signal is led to the input optical highway. In the optical time-division switch, each BLD is reset every signal frame by a reset current pulse in the first half of each time slot (Figure 10.11(b)(i)). Optical signals are extracted sequentially from the time-multiplexed optical signal on the input highway by the 1 × 4 write optical switch matrix and injected into each BLD in the latter half of each time slot (Figure 10.11(b)(ii)). Therefore, each BLD remains in the OFF state or turns to the ON state, according to the optical signal, and memorizes the binary optical data until the next reset period (Figure 10.11(b)(iii)). The 4 × 1 read optical switch matrix connects the BLD outputs to the output highway in each read period and constructs an output time-multiplexed optical signal, where time switching is accomplished. The time-multiplexed optical signal on the output highway is converted to the electric signal again using an optical-electronic converter and separated into four 64-Mbps digital pulse streams by a DMUX. Finally, this digital pulse stream is decoded into the analog color video signal and applied to four video monitors. It was necessary to inject −15-dBm input power for the BLD set operation. The anticipated optical output value for the BLD was 4 dBm. Therefore, a 19-dB optical gain was obtained by the BLD optical memory.

The speed has been extended to 512 Mbps by using high-speed BLDs with semi-insulating embedded layers [45,46]. Four 128-Mbps full-motion color video signals are multiplexed into a 512-Mbps highway in a bit-interleaved form and sent to a time switch

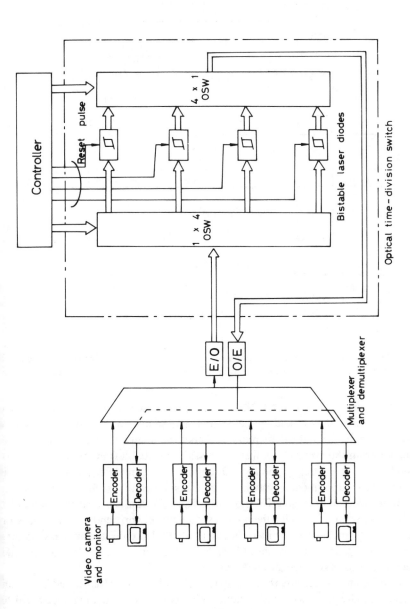

Figure 10.11(a) Experimental optical time-division switch system block diagram. OSW: optical switch matrix, E/O: electronic-optical converter, O/E: optical-electronic converter.

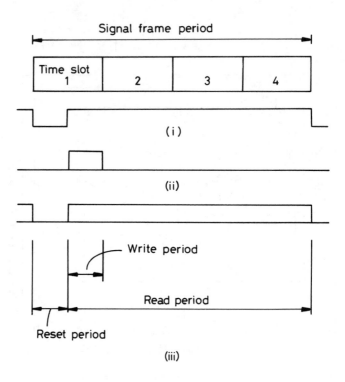

Figure 10.11(b) Timing chart of optical time-division switch: (i) reset pulse; (ii) bistable LD optical input; (iii) bistable LD optical output. (©1986 IEEE. From [44].)

The time switch consists of a 1 × 4 LiNbO₃ write optical switch, four BLDs as optical memories, and a 4 × 1 LiNbO₃ read optical switch. The signals in the input highway are sequentially written in the optical memory and then randomly read out to the output highway according to the controller's instructions. Figure 10.12(a) is the photograph of the BLD module and (b) is that of the LiNbO₃ optical switch module. Figure 10.12(c) is the photograph of the experimental system, which is composed of a rack for the switching system and four video sources and TV monitors. The rack consists of three shelves for the video codecs and control circuits, the optical switching network, and a microprocessor. All switching circuits, including the video codecs, the control circuits, the optical switching network, and a microprocessor, have been installed in the same rack. A BER of less than 10⁻⁹ has been obtained in a practical environment.

If these BLDs are replaced by the high-speed polarization BLDs described in Section 5.2, an optical time-division switching system with much faster switching speed will be realized [47]. Here, we again consider the system in which signals of four channels are multiplexed. The signals of the TE mode in the input highway are sequentially written in the optical memory and then randomly read out to the output highway according to

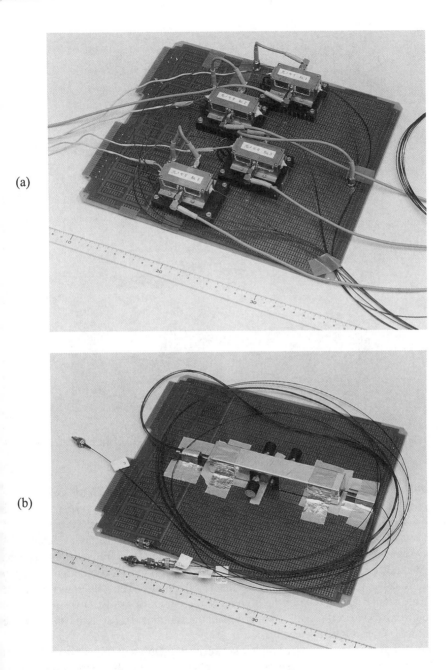

Figure 10.12(a,b) (a) BLD module; (b) LiNbO$_3$ optical switch module.

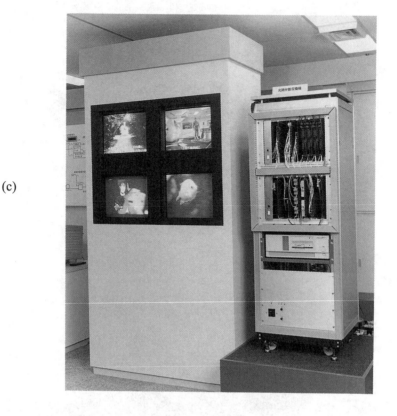

(c)

Figure 10.12(c) Photograph of experimental system for time-division photonic switching. (Courtesy of S. Yamakoshi, Fujitsu Laboratories, 1990.)

the controller's instructions. Each polarization BLD is reset by a reset TM optical pulse. Figure 10.13 shows the calculated timing chart of the optical switching system. Here, we consider a 50-Gbps highway speed system in which the width of each TE set and TM reset pulse is 10 ps. Depending on the optical signal injected into the BLDs, each polarization BLD remains in the OFF state or turns to the ON state. The OFF state corresponds to the TM-mode oscillation and the ON state corresponds to the TE oscillation. For each of the four channels, the "1" states are assumed in the figure. The "0" states of all the channels are similar and shown in the bottom. If the output of the BLD are read out in the period from 100 to 200 ps with an appropriate thresholding, this system can exchange signals at a 50-Gbps highway speed.

Broadband integrated services digital network (B-ISDN) will require over 40 Gbps in the feeder loop, connecting a central office and remote terminals having a drop/insert function [48]. Remote terminals will process large amounts of high-speed data. Optical-signal-processing features will play an important role in this terminal. An optical-signal-

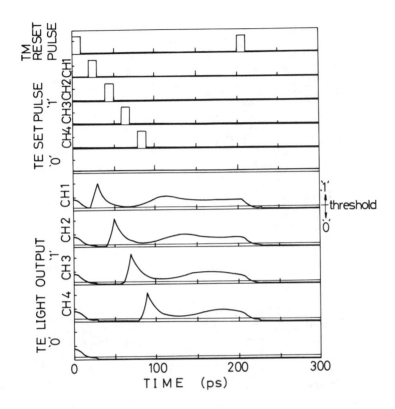

Figure 10.13 Simulation of optical time-division switch using polarization BLDs as memory. (From [47].)

processing-based broadband subscriber loop that uses a ring architecture linking photonic access nodes has been proposed. Each photonic access node has an optical drop/insert function and can synchronize optical frame signals using BLDs acting as optical sampling memories. The principle and configuration of the optical sampling memory and an optical data latch are shown in Figure 10.14. Suppose the optical data stream comes into a BLD. It is set to the lasing state by the optical input beyond the threshold level. To set the BLD for a particular optical data bit, an optical frame pulse is superimposed on the data pulse as shown in the figure. In this way, only the desired data can be latched as the output of the BLD. RZ optical data of 155 Mbps was dropped out of the 622-Mbps data highway with a BER of less than 10^{-9}. An optical insert function has also been constructed using an optical data latch circuit. In order to generate a vacant time slot for new data to be inserted, dropped data is inhibited by the optical inhibit circuit using a BLD in the data sequence. New data from a terminal end is inserted into the vacant time slot by using an optical data latch circuit, which converts the insertion data duty cycle to the correct one for the data sequence.

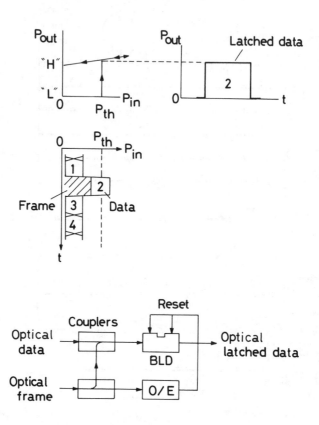

Figure 10.14 Optical sampling memory and optical data latch. (©1989 IEEE. From [48].)

Asynchronous transfer mode (ATM) will become the basis for future networks [49,50]. The ATM packet-switched technique offers a highly flexible means of implementing broadband services. In current proposals, the packet processing is performed electronically. In order to avoid massive multiplexing/demultiplexing costs in multi-gigabit-per-second systems, optical processing could be used to segment the transmitted information into manageable amounts of 10 Gbps where electronic processing can take over. Absorptive nonlinear amplifiers [51–53] have been tested by Barnsley et al. as the devices for optical processing applications in multi-gigabit-per-second TDM packet-switched networks [54]. As shown in Figure 10.15, the source LD was modulated by DATA1, and this signal was injected into the nonlinear optical amplifier (NLOA). The absorber contact was ac coupled to a second data generator that could apply reversed-bias pulses to the absorber section and TDM was accomplished. They demonstrated the use of NLOAs for switching and pulse shaping of information up to bit rates of 500 Mbps, but with rise and fall times of about 100 ps. NLOAs therefore already have suitable switching characteristics for application in

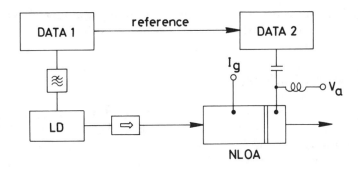

Figure 10.15 Experimental setup for TDM switching using two-segment BLD amplifier. (From [54]. Reproduced by permission of SPIE.)

ATM-type packet networks with 5- to 10-Gbps header speeds. However, for TDM switching, there is a need for higher repetition rate operation.

In view of the evolution of future networks towards B-ISDN, which calls for high transport capacities, the combined use of wavelength routing, optical buffering, and wavelength multiplexing for advanced ATM switching nodes has been proposed as a way to overcome the bottlenecks of existing electronic technologies by exploiting photonics.

Because of the asynchronous nature of the ATM flows, an ATM switch needs cell buffering to solve cell contention. In the above-mentioned proposal, buffering is obtained by optical means; that is, by a fiber-loop memory whose word length corresponds to one ATM cell period and whose memory addresses correspond to a set of optical wavelengths assigned to the cells before they enter the loop [55]. The cell is kept stored (i.e., circulates in the loop) as long as the filter is tuned to the corresponding wavelength, and it is erased by detuning or switching off the filter for a time corresponding to an entire circulation. A following wavelength-routing output stage that is properly synchronized with the loop memory ensures proper routing of packets.

An optical-fiber-loop-memory testbed has been implemented to validate these concepts, and packet-storage experiments with a wavelength-selective element in the loop have been carried out first, in view of the forthcoming introduction of multiwavelength access. The input optical data packets are generated by external modulation of a 1.55-μm DFB LD to avoid excess spectral broadening due to the chirping effect. To overcome round-trip losses and to introduce wavelength selectivity, a switchable optical gain element and a narrowband optical filter, respectively, are inserted into a loop. In principle, the loop length could be equal to the packet duration; however, guard bands are introduced between packets to permit periodic switch-off of the gain elements, which avoids the onset of instabilities in the loop.

Tsuda et al. [56] demonstrated an optical-fiber loop memory with an optical threshold device. A side-light-injection-type BLD described in Section 3.6 was used as the optical threshold device. Stable storage of standard ATM cells (256 bits long) at 500 Mbps has been

observed for 100 consecutive circulations in the fiber loop owing to a digital regeneration b
the threshold device. The highest bit rate was limited by the switching speed of the bistabl
device and was 1.6 Gbps.

10.4.2 Optical Wavelength-Division Switching

Optical wavelength-division switching is expected to facilitate a flexible switching net
work. The main advantages of optical wavelength-division switching are as follows [57]
First, phase synchronization, which is necessary in time-division switching, is not required
The number of devices can be reduced in comparison with space-division switching. I
optical wavelength-division switching, the wavelength of the signal carried through th
optical highway is converted to another wavelength by the wavelength conversion switch
Then one or more signals are selected by the wavelength selection switch to connect wit
the appropriate channel. The signal is reconverted to a different wavelength and set int
the output optical fiber (Figure 10.16).

Figure 10.17(a) shows the arrangement with which we demonstrated the basi
function of optical wavelength-division switching [58,59]. An optical pulse with TI
polarization generated by an LD was injected into a wavelength conversion device throug
an optical isolator. The ON-OFF signal was produced by a mechanical chopper. Th
output wavelength from the wavelength converter was switched by a current pulse superim
posed on a bias current of the wavelength converter and selected by a DFB LD amplifier
The output wavelength from the wavelength converter was monitored by an optica
spectrum analyzer and the selected signal was detected through a Fabry-Perot interferom
eter.

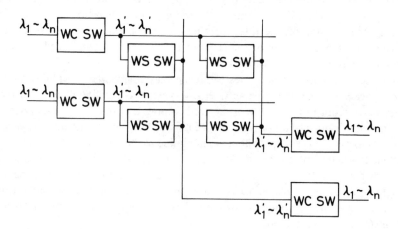

Figure 10.16 Optical wavelength-division switching using wavelength conversion and wavelength selection
switches. WC SW: wavelength conversion switch; WS SW: wavelength selection switch. ([©]1988
IEEE. From [58].)

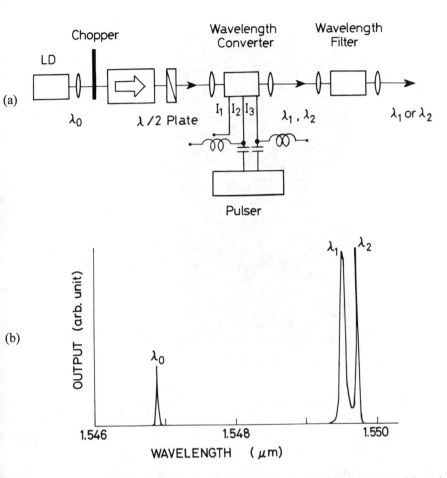

Figure 10.17(a,b) (a) Experimental setup examining the basic function of optical wavelength-division switching (from [59]). (b) Output optical spectrum of the wavelength conversion device. Input wavelength (λ_0) was converted to λ_1 and λ_2. λ_1 and λ_2 are switched with a repetition rate of 10 MHz. (©1988 IEEE. From [58].)

The output optical spectrum of the wavelength conversion device is shown in Figure 10.17(b). The bias currents were set at a current ratio of $I_3/(I_2 + I_3) = 0.45$. Current pulses superimposed on the bias currents alternately changed to a current ratio of 0.4 and 0.5 with a repetition rate of 10 MHz. Accordingly, the input wavelength ($\lambda_0 = 1.5470$ μm) was converted to two output wavelengths separated by 0.2 nm ($\lambda_1 = 1.5495$ μm and $\lambda_2 = 1.5497$ μm). The λ_0 signal was also detected and the power was one-third that of the λ_1 and λ_2 signals in the experiment because part of the optical input pulse passed through the device. This through power can be drastically reduced by using the spatial-filter action of a single-mode fiber.

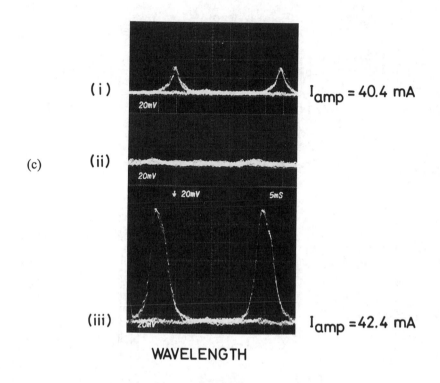

Figure 10.17(c) Optical spectra of the output signal from the DFB LD amplifier. Gain maximum frequencies corresponded to λ_2 and l_1 at (i) I_{amp} = 40.4 mA and (iii) I_{amp} = 42.4 mA, respectively. The amplified signal was not observed at the other bias levels, as shown in (ii) (©1988 IEEE. From [58].)

    Optical spectra of the output signal from the DFB LD amplifier are shown in Figure 10.17(c). The gain maximum frequency corresponded to λ_2 when I_{amp} = 40.4 mA, and a λ_2 signal was selected. When I_{amp} = 42.4 mA, the gain maximum frequency corresponded to λ_1, and a λ_1 signal was selected. The output wavelength of the wavelength conversion device was kept at λ_1 or λ_2 when the photographs shown in Figure 10.17(c) were taken. It was also confirmed that the same result could be obtained when the wavelength was dynamically switched with a repetition rate of 10 MHz. The amplified signal was not observed at other bias levels as shown in Figure 10.17(c)(ii). The lasing threshold current of the amplifier was 44.4 mA. The selection of one signal from the wavelength-converted signal was achieved. However, the output signal was weak at λ_2 because of the decrease in DFB LD amplifier gain with decreasing bias current. This characteristic could be dealt with by using a multisection DFB LD amplifier, which has wide frequency tunability with gain and a constant gain bandwidth.

    A wavelength switch (λ switch), which can interchange wavelengths from an input WDM signal, is a basic component in the WD switching network. There are two kinds f λ switches, as shown in Figure 10.18(a,b) [60]. In the λ switch shown in Figure 10.18(a), an input WDM signal is split and parts thereof are led to individual variable-wavelength filters, each of which extracts a specific wavelength signal from the input signal. The output optical signal from the tunable wavelength filter is then applied to an optically controlled modulator and is intensity modulated onto a preassigned-wavelength light carrier, which is extracted from the WDM wavelength reference light by a fixed-wavelength filter. Through this process, input wavelengths λ_a, λ_b, ... up to λ_z, in Figure 10.18(a), are converted into wavelengths λ_1, λ_2, ... up to λ_n, respectively. Moreover, by controlling multiple tunable filters to select the same wavelength signal, multiple wavelength carriers can be modulated according to the same information. If the multicast function is not necessary, attenuation, caused by optical-signal splitting in the switch shown in Figure 10.18(a), can be avoided by using another type of wavelength filter, which can separate a specific wavelength signal from the WDM signal, as shown in Figure 10.18(b).

Multistage switching networks are necessary to construct large-capacity switching systems. A λ^m switching network [60] was proposed for this purpose. As shown in Figure 10.19, a wavelength DMUX and MUX are mounted in interstage connections. Between individual DMUX output ports and individual MUX input ports, an optical path is provided, thereby giving each λ switch potential connectivity to every next-stage switch. As shown in Figure 10.19, the λ^3 switching network has n input/output links, each of which is n-channel wavelength-division multiplexed. Therefore, line capacity for the λ^3 network is

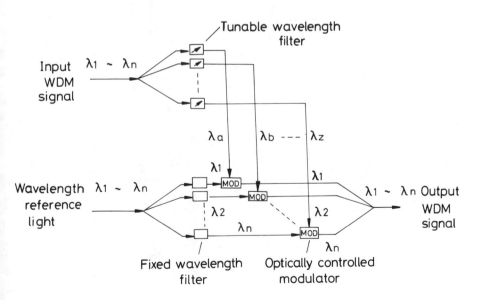

Figure 10.18(a) Wavelength switch: λ switch, which is applicable to a multicast function.

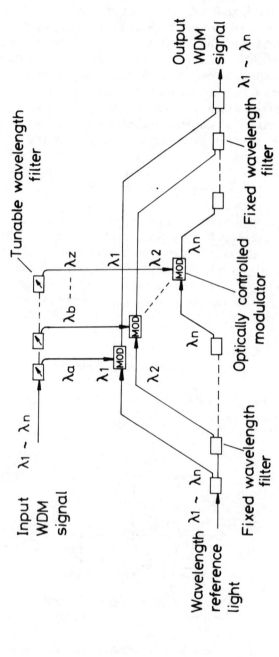

Figure 10.18(b) Attenuation caused by optical-signal splitting in the switch is avoided by using another type of filter. (©1990 IEEE. From [60].)

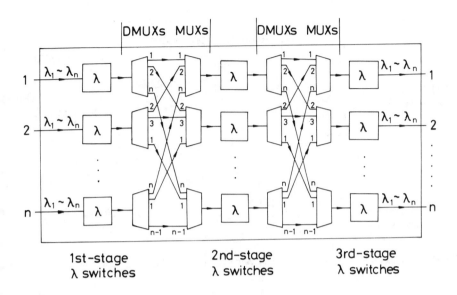

DMUXs MUXs   DMUXs MUXs

1st-stage
λ switches

2nd-stage
λ switches

3rd-stage
λ switches

Figure 10.19 λ^3 switching network. (©1990 IEEE. From [60].)

expressed as n^2. In the case that n is 16, the λ^3 switching networks can achieve 256-line capacity.

An experiment of a WD switching system using InGaAsP/InP phase-shift-controlled DFB LD tunable wavelength filters [61] was performed to test the feasibility of a switching system. Figure 10.20 shows a block diagram of the experimental photonic WD switching system. This system consists of a wavelength MUX, a λ switch, and a wavelength DMUX. The wavelength MUX consists of LD modulators, which intensity modulate light carriers supplied from wavelength reference light sources according to input CH1-CHn signals, and an optical combiner. In the λ switch, an input WDM signal is split and parts are led to individual tunable wavelength filters, each of which extracts a specific wavelength signal. The output signal from the tunable wavelength filter is then converted to an electronic signal by an optical-electronic converter. A preassigned-wavelength light carrier is intensity modulated by an LD modulator according to the electronic signal from the optical-electronic converter. The wavelength DMUX consists of an optical splitter, fixed-wavelength filters, and optical-electronic converters.

The number of WD channels n in this photonic switching network is mainly determined by the tunable-filter performance. The phase-shift-controlled DFB LD tunable filter has the advantages of wide wavelength tuning range and narrow transmission bandwidth with constant gain and bandwidth over the tuning range. Tunable-filter optical gain and spontaneous emission noise characteristics determine the required filter input power. Calculated filter input peak power values, taking the filter spontaneous emission into account, are -31.5 and -38.5 dBm to satisfy the 10^{-10} BER for a 200-Mbps signal, on

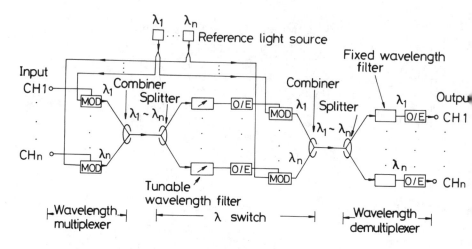

Figure 10.20 Experimental wavelength-division switching system. MOD: modulator; O/E: optical-electronic converter. (©1990 IEEE. From [60].)

the assumption that the tunable filter optical gain is 15 and 75, respectively. Assuming 3-dBm modulator output peak power and 5-dB coupling loss between optical devices and fibers, 21 and 28 dB, respectively, for 15 and 75 optical gains could be assigned to the combiner and splitter maximum loss, while the operation margin is 3.5 dB. Therefore, it is concluded that the n values can reach 8 and 16, respectively, where 0.5-dB single-stage optical coupler excess loss in the combiner and splitter is counted.

With the improvement in filter optical gain and replacement of the optical combiner by a low-loss wavelength MUX, n can be increased. For example, required filter input peak power will be reduced to -43 dBm using a filter with 200 optical gain. As a result, a 32.5-dB loss value can be assigned to the wavelength MUX and optical splitter. Therefore, n can reach 100 by using the wavelength MUX with 10-dB loss and the optical splitter with 22.5-dB loss.

Tunable-filter selectivity and tuning range also bound the n value. The relations between required channel separation $\Delta\lambda$, wavelength tuning range $W = (n - 1)\Delta\lambda$, and the n value are calculated. $\Delta\lambda$ was determined to be the minimum value that satisfies the 1-dB power penalty due to crosstalk, caused by $(n - 1)$ channel WDM signals. Power penalty was determined as the increase of required filter optical input power to satisfy 10^{-10} BER with crosstalk. The error-rate characteristic with crosstalk was calculated using the convolution integral of noise and crosstalk probability density functions. The filter transmission spectrum used for the calculation assume that the passband shape is Lorentzian, whose 10-dB down bandwidth is 1Å, and that the out-band response is flat with X (dB) attenuation from the peak gain. It is concluded that n can reach 8, while $X = 16$ dB and $W = 6$Å. When $X = 20$ dB and $W = 12$Å, the n value can reach 16. With future improvement in the tunable filter, n can reach 100 with $X = 24$ dB and $W = 100$Å.

A wavelength converter based on a semiconductor NLOA was used to switch 155-Mbps data from a 1.3-μm to a 1.55-μm WDM wavelength-routed network [62]. The wavelength-routed optical network offers interconnection between network nodes by using passive wavelength-selective couplers at the nodes and WDM of traffic on the transmission fibers. The central core of each network node consists of a WDM interconnection field. This interconnection field provides the cross-connect links between inputs and outputs. In such hard-wired interconnection, the node reconfiguration time is long. A much more rapid response time is achieved by the inclusion of an optical switch within the interconnection field (see Figure 10.21(a)). Each wavelength on each input fiber can now be switched to any output fiber, thus increasing the flexibility of the network and increasing the utilization factor by offering more protection strategies. Also shown in Figure 10.21(a) is a wavelength switch external to the interconnection field. Networks operating at incompatible wavelengths can be linked to the wavelength-routed network using this wavelength switch. For example, the present 1.3-μm networks could be linked to future 1.55-μm

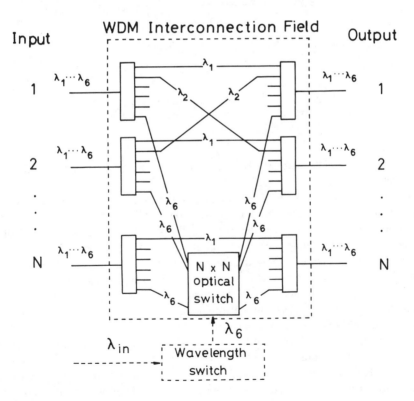

Figure 10.21(a) WDM interconnection field within a wavelength-routed network node. The optical switch allows dynamic reconfiguration of the interconnections and linkages with other wavelength networks using a wavelength switch.

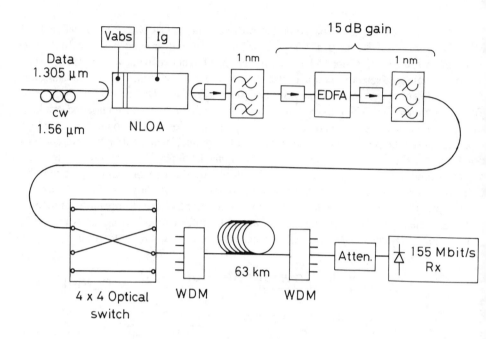

Figure 10.21(b) Schematic diagram of the experimental implementation of (a), where a 1.3-μm-wavelength network is switched into the 1.56-μm-wavelength channel of a wavelength-routed network. ($^{\circ}$1992 IEEE. From [62].)

wavelength-routed WDM networks by the inclusion of a switch capable of transferring 1.3-μm data to 1.55 μm.

Figure 10.21(b) shows the schematic diagram of the experimental system, which is analogous to Figure 10.21(a). The NLOA wavelength conversion device used in these experiments was identical to that described in Section 8.3 [63] and consisted of two independently biased regions. The long (about 450 μm) region was strongly forward-biased to give a gain region, while the shorter (about 35 μm) region was only slightly forward-biased to produce a saturable absorption region. Two input optical signals were injected into the absorber facet, a CW TE signal ($\lambda = 1.56$ μm, mean power about 10 μW) aligned to one of the NLOA Fabry-Perot modes, and a 155-Mbps $2^7 - 1$ pseudorandom optical data signal ($\lambda = 1.305$ μm, mean power about 300 μW) from a directly modulated DFB LD. The data are transferred onto the 1.56-μm-wavelength signal due to a modulation in gain caused by the saturation of the absorber region. The 1.56-μm signal was amplified by about 8 dB (fiber-fiber), while the 1.3-μm signal was attenuated in the NLOA by about 40 dB. In this experiment, a 1-nm optical filter at the NLOA output reduced the spontaneous emission from the NLOA and suppressed the 1.3-μm signal to less than -80 dBm. Filtering of the spontaneous emission from the NLOA resulted in a contrast ratio of about 10:1 for the 1.56-μm data. The converted data was then amplified by about 15 dB by an EDFA to a power level of around 800 μW. Another 1-nm optical filter was used to suppress the level of spontaneous emission from the EDFA. A 4 × 4 digital optical switch (loss = 10 dB) was used to route the converted data into the WDM interconnection field within a wavelength-routed network demonstrator [64]. The signal was then coupled into a 63-km length of step index fiber using a WDM MUX (loss of about 5 dB). After transmission and demultiplexing, the data signal was received using a commercial 155-Mbps receiver. The measured system loss from digital optical switch input to receiver input was 35.4 dB, with a system margin of 5.5 dB. After the NLOA, the converted data showed 10^{-9} BER with a receiver sensitivity penalty of 1.0 dB. This is due to the 10:1 contrast ratio at the NLOA output and residual spontaneous emission passed by the optical filter. After the EDFA, a further 0.5-dB penalty was observed. This is due to the spontaneous emission from the EDFA. The full system BER performance showed an additional penalty of about 0.7 dB. This is due to chromatic dispersion of the spontaneous emission in the data signal.

A 1.5-Gbps transmission experiment using a two-section DFB LD as an optical wavelength converter was also demonstrated [65]. An amplitude modulated (AM) input signal at 1.5171 μm was converted into an FM output signal at 1.5136 μm in the wavelength converter. The performance of the transmission system using an NRZ PRBS was studied. At 10^{-9} BER, the sensitivity was −19 dBm. There was no degradation in the BER for long sequences ($2^{23} - 1$) due to the flat FM response measured at low frequencies.

10.4.3 Frequency-Division Switching Using Nondegenerate Four-Wave Mixing

Bachus et al. carried out a demonstration of a multifrequency switching node using FWM in an LD amplifier and tunable optical filters, where one-way frequency conversion (i.e., $f_1 \rightarrow f_2$, not the simultaneous reverse conversion $f_2 \rightarrow f_1$) was achieved [66]. The carrier frequencies are allocated with 5-GHz spacing and modulated with 34-Mbps NRZ-coded signals in the PSK/FM scheme. The branching filter is a fiber M-Z interferometer with 10-GHz free spectral range (FSR). Signals are converted from f_1 to f_2 in a DFB mixer. The output filter consists of two M-Z interferometers in series with 25- and 5-GHz FSR. All filters are thermally tuned. The BER of the coherent multicarrier switching output signal was measured at 10^{-9}. This configuration could be extended to bidirectional conversion (i.e., $f_1 \rightleftarrows f_2$).

Inoue proposed the configuration composed of two LD amplifiers and two M-Z optical filters [67]. Using the transmittance characteristics of the M-Z filters, a simple configuration is possible for bidirectional frequency conversion. The configuration for frequency exchange is shown in Figure 10.22(a). The inputs and outputs of two LD amplifiers (LDA1, LDA2) are connected with two M-Z filters (MZ1, MZ2). The M-Z filters, with 2 × 2 ports, have sine-shape transmittance as a function of optical frequency, as shown in Figure 10.22(b), where the frequency separation of the transmittance can be set in the fabrication process and the center frequency can be adjusted by a phase shifter

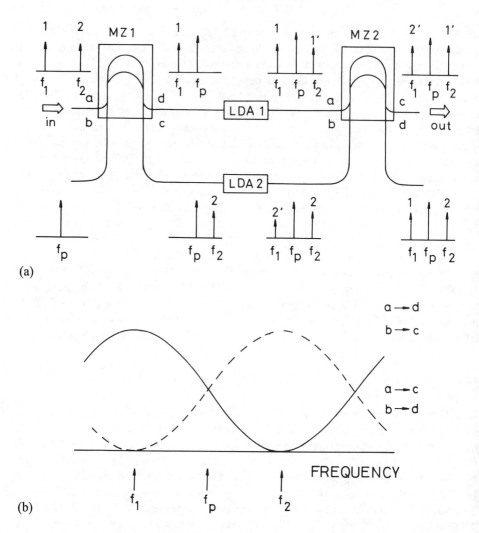

Figure 10.22(a,b) (a) Configuration for frequency exchange using NDFWM in LD amplifiers: LDA1 and LDA2; LD amplifiers, MZ1 and MZ2; and M-Z optical filters. (b) Transmittance of M-Z filters and frequency position of input lights.

contained in the filters. In the configuration shown in Figure 10.22(a), the transmittances of MZ1 and MZ2 are set as illustrated in Figure 10.22(b) versus the input signal frequency. Here, f_1 and f_2 are the optical frequency of the signal lights to be exchanged. A pump light prepared whose frequency f_p is positioned at the middle of the two signal frequencies (i.e., $f_p = (f_1 + f_2)/2$), as shown in Figure 10.22(b).

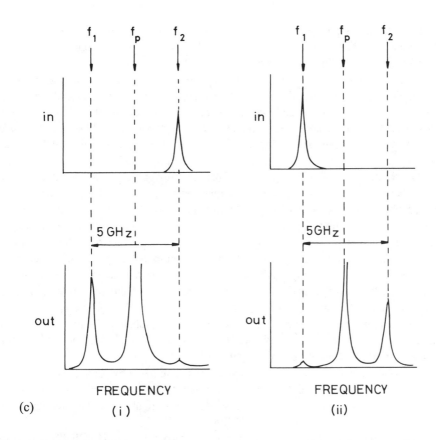

Figure 10.22(c) Experimental results on frequency exchange. Upper part: input spectra at port *a* of MZ1; lower part: output spectra at port *c* of MZ2: (i) input signal frequency is set at f_2; (ii) input signal frequency is set at f_1. (From [67]. Reproduced by permission of IEE.)

The two signal lights $1(f_1)$ and $2(f_2)$ and the pump light (f_p) are launched at ports *a* and *b* of MZ1, respectively. Owing to the transmittance characteristic of MZ1, shown in Figure 10.22(b), frequency pairs of {signal $1(f_1)$, pump (f_p)} and {signal $2(f_2)$, pump (f_p)} appear at ports *d* and *c* of MZ1, and are injected into LDA1 and LDA2, respectively. In LDA1, an FWM process generates additional light of frequency $f_p + (f_p - f_1) = f_2$, which has a modulation signal that is identical to signal $1(f_1)$ when the pump light is unmodulated. Thus, signal $1(f_1)$, pump (f_p), and frequency-converted signal $1(f_2)$ appear at the LDA1 output. At the LDA2 output, on the other hand, signal $2(f_2)$, pump, and frequency-converted signal $2(f_1)$ appear.

Output lights from LDA1 and LDA2 are launched at MZ2. Owing to the transmittance characteristic of MZ2 shown in Figure 10.22(b), signal $1(f_2)$ is transmitted from port a to port c and signal $2(f_1)$ is transmitted from port b to port c, while the original lights, signal $1(f_1)$ and signal $2(f_2)$, are transmitted to port d. That is, frequency-converted lights and the original lights appear at ports c and d of MZ2, respectively. Thus, an optical-frequency exchange operation is obtained by regarding port c of MZ2 as a system output.

An experiment was carried out to demonstrate the operation described above. GaIn-AsP AR-coated LD amplifiers and waveguide M-Z filters were used in the experiment, which were connected by polarization-maintaining fibers. Two DFB LDs ($\lambda = 1.55$ μm) were prepared for the pump and signal light sources. The input signal light frequency was set at f_1 or f_2, where $f_1 - f_2 = 5$ GHz.

The results are shown in Figure 10.22(c). The upper figures are input spectra at port a of MZ1, and the lower figures are output spectra at port c of MZ2. For an input signal f_2, f_1 light appears at output, but f_2 light does not appear (Figure 10.22(c)(i)), whereas f_2 light appears at the output, but f_1 does not appear for input signal f_1 (Figure 10.22(c)(ii)). Thus, bidirectional frequency conversion was demonstrated experimentally. The pump light also appears in the output spectra, which can be eliminated by another optical filter if desired. Conversion efficiency was dependent on the experimental conditions, including loss between fiber and LD amplifier, and the excess loss of the M-Z filters. Intrinsic efficiency is determined by the FWM efficiency in the LD amplifiers. In this experiment, for example, frequency-converted light of -10.9 dBm was obtained at the output of the LD amplifier chip when pump and signal powers were -9.3 dBm and -15.2 dBm, respectively, at the LD amplifier chip input.

The disadvantage of this frequency exchange is that the conversion frequency range is limited to within several gigahertz, owing to the electron carrier lifetime in LD amplifiers. This limitation could be solved by using a second pump light of another frequency and replacing the MZ2 by a four-channel M-Z filter.

To highlight the usefulness of the frequency converter using HNDFWM described in Section 8.4.4, we now discuss the two applications shown in Figure 10.23(a) [68]. In point-to-multipoint switching, an optical data signal at the carrier frequency f_s is converted into one or several optical data signals at the carrier frequencies f_1 to f_N. This is accomplished if one or several probe waves (frequencies f_x, f_x', . . .) are tuned to the frequencies ($f_1 + \Delta f_s$), . . . , ($f_N + \Delta f_s$). Each of these probe waves acts in the same way as the pump wave P_2 shown in Figure 8.18. Δf_s(≤ 2 GHz) is the frequency spacing between the data signal of frequency f_s and the associated pump wave P_p, which is necessary to induce the NDFWM process in the optical amplifier. The pump wave P_p is not explicitly indicated in Figure 10.23(a).

In multipoint-to-point switching, an arbitrary signal out of N optical data signals (carrier frequencies f_1, . . . , f_N) is converted to a selected carrier frequency f_s, which is different from f_1 to f_N. This is accomplished by tuning a pump wave near one of the frequencies f_1 to f_N with a frequency spacing $\Delta f_s \leq 2$ GHz. This pump wave acts in the same way as the pump wave P_1 shown in Figure 8.18. The pump wave P_c associated with the converted data signal at the carrier frequency f_s is again not indicated in Figure

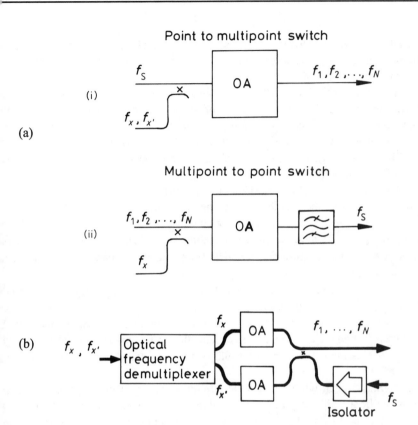

Figure 10.23 (a) Applications for optical frequency-division switching: (i) point-to-multipoint switch; (ii) multipoint-to-point switch. (b) Point-to-multipoint switch with counterpropagating pump waves f_x, f'_x, and f_p (parallel to the signal wave f_s). To enlarge the number of possible channels, a parallel arrangement of two optical amplifiers is used. (Reprinted with permission from Chapman and Hall Ltd. From [68].)

10.23(a). The optical fiber attached to the optical amplifier transmits only frequencies near f_s.

As the two examples show, the desired switching operation is only effective if two pump waves are initiated, one associated with f_s and one associated with one of the frequencies f_1 to f_N. Consequently, the switching operation represents an AND gate. The frequency spacing between $f_1, \ldots, f_N$ and f_s should be >10 GHz. The maximum data rate is restricted by the value Δf_s to <1 Gbps.

Optical filters are an essential component for the two switching applications shown in Figure 10.23(a). At present, the most promising optical filters are periodic filters consisting of two 3-dB directional couplers connected by two waveguides of different lengths. Recently, such a filter was used to separate one out of eight channels that were

5 GHz spaced and intensity modulated by 400 Mbps [69]. Thus, it appears that optical filters will provide room for a large number of channels (N) within the 4,000-GHz bandwidth of the optical amplifier.

Saturation of the optical amplifier involves a serious upper limit for the number of channels. The optical amplifier output power of all signal waves and pump waves is expected not to exceed the saturation output power by very much. If we assume a total optical output power of 10 dBm, we estimate that up to $N = 10$ channels can be used per optical amplifier. However, the number of channels may be increased by the parallel arrangement of several optical amplifiers.

We give an example in Figure 10.23(b) of point-to-multipoint switching. In this example, we make use of the possibility that the two pump waves P_p and P_c may be counterpropagating waves. Therefore, in this example the data signal (of frequency f_s) and its associated pump wave are counterpropagating to the pump waves at frequencies f_x and f'_x through the optical amplifiers. An optical filter is used to separate the frequencies f_x and f'_x into two groups.

The phase noises of the signal laser, local laser, and two pump lasers are summed by the NDFWM process and can seriously degrade the quality of the received signal. For this reason, narrowed linewidth (<100 kHz) external cavity lasers or subcarrier modulation schemes are used. When we use a phase noise cancellation method for optical-frequency conversion, it enables the use of DFB LDs and is not restricted in modulation schemes [70].

The experimental configuration of optical-frequency conversion is shown in Figure 10.24. The signal light is mixed with two pump lights (called *pump* and *converter* below) and is injected into the LD amplifier. The frequency-converted lights (n,n') appearing on both sides of the converter spectrum and the amplified converter light (c) come out from the optical bandpass filter as output. The LD amplifier has a 16-dB (TE) fiber-to-fiber gain and 3.5-dBm saturation output power. Signal, pump, and converter powers at the LD amplifier fiber input are −10.75, −4.96, and −7.66 dBm, respectively. Polarizations of all of them are adjusted to TE by manual polarization controllers. To obtain high conversion efficiency, the LD amplifier is used in the highly saturated region. The total conversion efficiency measured by a scanning Fabry-Perot spectrometer is −3.0 dB.

The phase noise cancellation circuit located in the receiver extracts both received frequency-converted light (n) and amplified converter light (c) components using a band-pass filter (BPF-1), squares them, and extracts their mutual multiplication component as output using BPF-2. Both lights are equally subjected to the phase noise of the converter and local lights, so they can be canceled by multiplication.

The spectrum linewidth of the frequency-converted signal is reduced from 46 to 5.5 MHz after phase noise cancellation, while the linewidths of the signal, pump, converter, and local lasers are 2.1, 2.8, 25, and 20 MHz, respectively. The resultant linewidth almost equals the sum of the linewidths of the signal and pump lasers. It is clearly shown that the phase noises of the converter and local lasers are canceled.

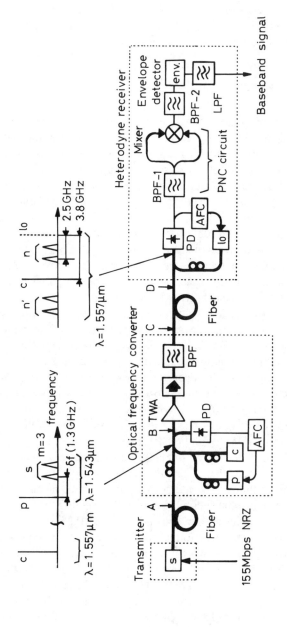

Figure 10.24 Experimental configuration of optical-frequency conversion. s: signal laser; p: pump laser; c: converter laser; lo: local laser; n, n': frequency-converted replica of signal light. (©1993 IEEE. From [70].)

A 155-Mbps FSK transmission experiment with 1,750-GHz (+14 nm) frequency conversion was demonstrated. Using the phase noise cancellation circuit, the floor at BER = 10^{-6} vanishes and a clear eye pattern is obtained. The received power penalty is about 1.7 dB at BER = 10^{-9}, considered mainly due to imperfection of the phase noise cancellation circuit and/or limited signal bandwidth of the LD amplifier.

10.4.4 Free-Space Switching

The main objective of free-space technology is to exploit the spatial bandwidth (pin-outs or connections) available in the optical domain [71]. This has allowed researchers to look for connection-intensive switching fabrics, as opposed to maximizing the available bandwidth in a limited number of connections (temporal bandwidth). The devices used in these free-space structures could be either large 2D arrays of optical logic gates, such as symmetric self-electro-optic effect devices, vertical surface transmission electrophotonic devices, and nonlinear interference filters, or 2D optoelectronic integrated circuits (2D-OEICs) or *smart pixels*. 2D-OEICs in this case are devices composed of electronics (transistors) for information processing and photonics (e.g., detectors, modulators, lasers, and light-emitting diodes) for transporting the information between integrated circuits. The optics used to interconnect free-space devices could be composed of either holographic elements or bulk optical elements such as lenses and mirrors.

A high-performance optically addressable optical crossbar switch has been implemented based on Fabry-Perot nonlinear LD amplifiers [72]. The principle of such a crossbar switch is shown in Figure 10.25(a). The operation of the switch can be represented by a vector-matrix multiplication with the input and output as a vector and the 2D array as a matrix. In this approach, the Fabry-Perot resonance peak of each of the switch elements in the same row is adjusted to be close to that of the corresponding input channel. Since the transmission state (1 or 0) of each of the switch elements can be set independently, the maximum aggregate throughput rate grows as n^2. As a result of this, any interconnect pattern is possible, from WDM to full broadcasting. In addition, the low switching power and high optical gain of the nonlinear LD amplifiers translate to potentially large switch size. An important advantage that distinguishes this crossbar switch from the previous ones is that it has the capability to be addressed optically based on an AND gate operation for each switching element, which permits fast, parallel reconfiguration of the interconnect pattern. (The addressing optics are not shown in Figure 10.25(a).)

In order to investigate the performance characteristics of the optically addressed crossbar switch, a 1×3 crossbar switch by using three discrete LD amplifiers has been set up (Figure 10.25(b)) [72]. Three GaAs LDs biased just below the threshold serve as a 1×3 crossbar switch. The Fabry-Perot peaks of these LD amplifiers are aligned to be close to each other. An optical ASK data signal is generated by modulating a single-wavelength (λ_d) CW laser beam from a GaAs LD operated well above threshold using a LiNbO$_3$ M-Z modulator with a 1.4-Gbps NRZ $2^{15} - 1$ bit PRBS. This optical data beam

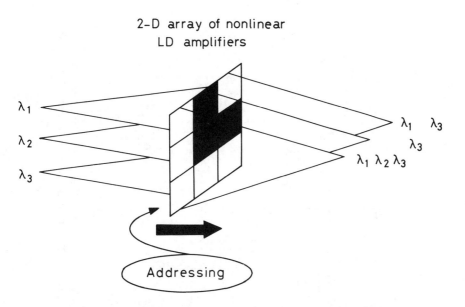

Figure 10.25(a) Proposed 3 × 3 optical crossbar switch based on nonlinear LD amplifiers.

s then split into three and injected to the three LD amplifiers from the front facets. An optical addressing signal is taken from a tunable external cavity laser (λ_a). Because the LD amplifier has the same switching characteristics in both transmission and reflection, the addressing signal is injected into the LD amplifiers from the rear facets to simplify the coupling. Both wavelengths, λ_a and λ_d, are monitored by a scanning Fabry-Perot cavity. At the output, a grating is used to remove λ_a from the data signal at λ_d.

The scheme for optically controlled data transmission is based on the switching power dependence on the detuning of the incoming signal off the Fabry-Perot resonance. λ_d and λ_a are tuned to be close to two Fabry-Perot transmission peaks of the LD amplifier three mode spacings away in this experiment and one mode spacing corresponds to 124 GHz), with a detuning from the corresponding Fabry-Perot peak $\delta\lambda_a$ (0.1Å) smaller than $\delta\lambda_a$ (0.35Å). With this arrangement, the output data signal at λ_d is normally low, since the input power is too weak to switch the LD amplifier at the detuning $\delta\lambda_d$. If a second beam, the control signal, is injected at a smaller detuning, the Fabry-Perot resonance will shift to the vicinity of the control signal. As a result of this, the data signal will switch and exit the device with amplification. Thus, the LD amplifier acts as a normally off optical switch with gain. Pan and Dagenais demonstrated a per-channel data bandwidth of 1.4 Gbps, instrument limited. To measure the sensitivity of the crossbar switch, BER measurements were performed on the output of LD amplifier 1 (NL-DLA-1) at 1 and 1.4 Gbps. A BER of 10^{-9} is obtained with an input signal power of −26.4 and −24.6 dBm, respectively.

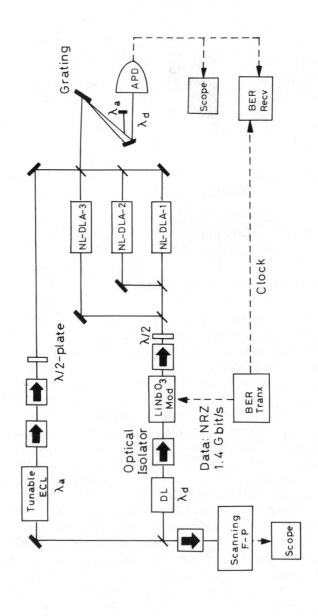

Figure 10.25(b) Experimental setup to demonstrate a discrete 1 × 3 optical crossbar switch. ECL: external cavity laser; LiNbO₃ Mod: LiNbO₃ M-Z electro-optic modulator; BER Tranx: BER transmitter; NL-LDA: nonlinear LD amplifier; BER RCVR: BER test receiver. (From [72]. Reproduced by permission of AIP.)

REFERENCES

[1] Midwinter, J. E., and Y. L. Guo, *Optoelectronics and Lightwave Technology*, Chichester, England: John Wiley & Sons Ltd, 1992.

[2] For example, Basov, N. G., W. H. Culver, and B. Shal, *Application of Lasers to Computers, Laser Handbook 1*, F. T. Arecchi and E. O. Schulz-DuBois, eds., Amsterdam, 1972, pp. 1649–1693.

[3] Shapiro, S. L., ed, Ultrashort Light Pulses, Berlin: Springer-Verlag, 1984.

[4] Kaiser, W., ed., *Ultrashort Laser Pulses*, Berlin: Springer-Verlag, 1988.

[5] Tucker, R. S., G. Eisenstein, and S. K. Korotky, "Optical Time-Division Multiplexing for Very High Bit-Rate Transmission," *J. Lightwave Technology*, Vol. 6, No. 11, November 1988, pp. 1737–1749.

[6] Hasegawa, A., and F. Tappert, "Transmission of Stationary Nonlinear Optical Pulses in Dispersive Dielectric Fibers. I. Anomalous Dispersion," *Appl. Phys. Lett.*, Vol. 23, No. 3, August 1973, pp. 142–144.

[7] Mollenauer, L. F., R. H. Stolen, and J. P. Gordon, "Experimental Observation of Picosecond Pulse Narrowing and Solitons in Optical Fibers," *Phys. Rev. Lett.*, Vol. 45, No. 13, September 1980, pp. 1095–1098.

[8] For example, Taylor J. R., ed, *Optical Solitons-Theory and Experiment*, Cambridge, England: Cambridge University Press, 1992.

[9] Mollenauer, L. F., B. M. Nyman, M. J. Neubelt, G. Raybon, and S. G. Evangelides, "Demonstration of Soliton Transmission at 2.4 Gbit/s Over 12000 km," *Electron. Lett.*, Vol. 27, No. 2, January 1991, pp. 178–179.

[10] Hartman, D. H., "DigitaL High Speed Interconnects; A Study of the Optical Alternative," *Opt. Eng.*, Vol. 25, No. 10, October 1986, pp. 1086–1102.

[11] Delfyett, P. J., D. H. Hartman, and S. Z. Ahmad, "Optical Clock Distribution Using a Mode-Locked Semiconductor Laser Diode System," *J. Lightwave Technology*, Vol. LT-9, No. 12, December 1991, pp. 1646–1649.

[12] Von der Linde, D., "Characterization of Noise in Continuosly Operating Mode Locking Lasers," *Appl. Phys. B*, Vol. 39, 1986, pp. 201–217.

[13] Jinno, M., and T. Matsumoto, "All-Optical Timing Extraction Using a 1.5 μm Self Pulsating Multielectrode DFB LD," *Electron. Lett.*, Vol. 24, No. 23, November 1988, pp. 1426–1427.

[14] Jinno, M., and T. Matsumoto, "Nonlinear Operations of 1.55 μm Wavelength Multielectrode Distributed-Feedback Laser Diodes and Their Applications for Optical Signal Processing," *J. Lightwave Technology*, Vol. LT-10, No. 4, April 1992, pp. 448–457.

[15] Barnsley, P. E., and H. J. Wickes, "All-Optical Clock Recovery From 2.5 Gbit/s NRZ Data Using Self-Pulsating 1.58 μm Laser Diode," *Electron. Lett.*, Vol. 28, No. 1, January 1992, pp. 4–6.

[16] Barnsley, P. E., G. E. Wickens, H. J. Wickens, and D. M. Spirit, "A 4 × 5 Gb/s Transmission System With All-Optical Clock Recovery," *IEEE Photonics Tech. Lett.*, Vol. 4, No. 1, January 1992, pp. 83–86.

[17] Takayama, K., and K. Habara, "3.2 GHz Operation of All-Optical Synchronisation Circuit," *Electron. Lett.*, Vol. 25, No. 25, December 1989, pp. 1739–1741.

[18] Barnsley, P. E., H. J. Wickes, G. E. Wickes, and D. M. Spirit, "All-Optical Clock Recovery From 5 Gb/s RZ Data Using a Self-Pulsating 1.56 μm Laser Diode," *IEEE Photonics Tech. Lett.*, Vol. 3, No. 10, October 1991, pp. 942–945.

[19] Farrell, G., P. Phelan, and J. Hegarty, "All-Optical Synchronisation With Frequency Division Using Self-Pulsating Laser Diode," *Electron. Lett.*, Vol. 28, No. 8, April 1992, pp. 738–739.

[20] Farrell, G., P. Phelan, and J. Hegarty, "All-Optical Clock Distribution With Synchronous Frequency Division and Multiplication," *Electron. Lett.*, Vol. 28, No. 15, July 1992, pp. 1387–1389.

[21] Farrell, G., P. Phelan, J. Hegarty, and J. A. Shields, "All-Optical Timing Extraction With Frequency Division Using a Twin-Section Laser Diode," *IEEE Photonics Tech. Lett.*, Vol. 5, No. 6, June 1993, pp. 718–721.

[22] As, D. J., R. Eggemann, U. Feiste, M. Möhrle, E. Patzak, and K. Weich, "Clock Recovery Based on a

New Type of Self-Pulsation in a 1.5 μm Two-Section InGaAsP-InP DFB Laser,'' *Electron. Lett.*, Vol. 29, No. 2, January 1993, pp. 141–142.

[23] Smith, K., and J. K. Lucek, ''All-Optical Clock Recovery Using a Mode-Locked Laser,'' *Electron. Lett.*, Vol. 28, No. 19, September 1992, pp. 1814–1816.

[24] Ellis, A. D., K. Smith, and D. M. Patrick, ''All Optical Clock Recovery at Bit Rates up to 40 Gbit/s,'' *Electron. Lett.*, Vol. 29, No. 15, July 1993, pp. 1323–1324.

[25] Barnsley, P. E., ''NRZ Format All-Optical Clock Extraction at 3.2 Gbit/s Using Two-Contact Semiconductor Devices,'' *Electron. Lett.*, Vol. 28, No. 13, June 1992, pp. 1253–1255.

[26] Barnsley, P. E., and P. J. Fiddyment, ''Clock Extraction Using Saturable Absorption in a Semiconductor Nonlinear Optical Amplifier,'' *IEEE Photonics Tech. Lett.*, Vol. 3, No. 9, September 1991, pp. 832–834.

[27] Kawanishi, S., and M. Saruwatari, ''10 GHz Timing Extraction From Randomly Modulated Optical Pulses Using Phase-Locked Loop With Travelling-Wave Laser-Diode Optical Amplifier Using Optical Gain Modulation,'' *Electron. Lett.*, Vol. 28, No. 5, February 1992, pp. 510–511.

[28] Jinno, M., T. Matsumoto, and M. Koga, ''All-Optical Timing Extraction Using an Optical Tank Circuit,'' *IEEE Photonics Tech. Lett.*, Vol. 2, No. 3, March 1990, pp. 203–204.

[29] Hidayat, I. S., and H. Kawaguchi, ''Analysis of All-Optical Clock Extraction Using Self-Pulsating Laser Diode,'' *Technical Report of IEICE*, OQE 93–165, 1994-2, pp. 1–6.

[30] Möhrle, M., U. Feiste, J. Hörer, R. Molt, and B. Sartorius, ''Gigahertz Self-Pulsation in 1.5 μm Wavelength Multisection DFB Lasers,'' *IEEE Photonics Tech. Lett.*, Vol. 4, No. 9, September 1992, pp. 976–978.

[31] Sartorius, B., U. Feiste, J. Hörer, M. Möhrle, R. Molt, and M. Rosenzweig, ''Self-Pulsation at More Than 20 GHz in InGaAsP/InP DFB Lasers,'' *Conf. Dig. of 13th IEEE Int. Semiconductor Laser Conf.*, 1992, pp. 104–105.

[32] Webb, R. P., ''Error-Rate Measurements on an All-Optically Regenerated Signal,'' *Optical and Quantum Electronics*, Vol. 19, 1987, pp. S57–S60.

[33] Jinno, M., and T. Matsumoto, ''Optical Retiming Regenerator Using 1.5 μm Wavelength Multielectrode DFB LDs,'' *Electron. Lett.*, Vol. 25, No. 20, September 1989, pp. 1332–1333.

[34] Yariv, A., D. Fekete, and D. M. Pepper, ''Compensation for Channel Dispersion by Nonlinear Optical Phase Conjugation,'' *Opt. Lett.*, Vol. 4, No. 2, February 1979, pp. 52–54.

[35] Murata, S., A. Tomita, J. Shimizu, and A. Suzuki, ''THz Optical-Frequency Conversion of 1 Gbit/s-Signals Using Highly Nondegenerate Four-Wave Mixing in an InGaAsP Semiconductor Laser,'' *IEEE Photonics Tech. Lett.*, Vol. 3, No. 11, November 1991, pp. 1021–1023.

[36] Tatham, M. C., G. Sherlock, and L. D. Westbrook, ''Compensation Fibre Chromatic Dispersion by Optical Phase Conjugation in a Semiconductor Laser Amplifier,'' *Electron. Lett.*, Vol. 29, No. 21, October 1993, pp. 1851–1852.

[37] Senior, J. M., and S. D. Cusworth, ''Devices for Wavelength Multiplexing and Demultiplexing,'' *IEE Proc. Pt. J*, Vol. 136, No. 3, June 1989, pp. 183–202.

[38] Payne, D. B., and J. R. Stern, ''Transparent Single-Mode Fiber Optical Networks,'' *IEEE J. Lightwave Technology*, Vol. LT-4, No. 7, July 1986, pp. 864–869.

[39] Kobrinski, H., R. M. Bulley, M. S. Goodman, M. P. Vecchi, and C. A. Brackett, ''Demonstration of High Capacity in the LAMBDANET Architecture: A Multiwavelength Optical Network,'' *Electron. Lett.*, Vol. 23, No. 16, July 1987, pp. 824–826.

[40] Glance, B., K. J. Pollock, P. J. Fitzgerald, C. A. Burrus, B. L. Kasper, J. Stone, and J. Stulz, ''Densely Spaced WDM Coherent Star Network With Optical Frequency Stabilization,'' *Optical Fiber Commun. Conf.*, OFC' 88, 1988, pp. 93–94.

[41] Chawki, M. J., R. Auffret, E. Le Coquil, L. Berthou, J. Le Rouzic, and L. Demeure, ''M³-NET: Demonstration of HD WDM Optical Network Using Two-Electrode DFB-LD Filter as Tunable Narrowband FM Receiver,'' *Electron. Lett.*, Vol. 28, No. 2, January 1992, pp. 147–149.

[42] Chawki, M. J., R. Auffret, L. Berthou, and E. Le Coquil, ''M³I NET: Demonstration of an HD-WDM Optical Information-Distribution Network With Interactive Audio/Video-Channel Using a Two-Electrode

DFB LD as a Multifunction Optical Transceiver," *Electron. Lett.*, Vol. 29, No. 12, June 1993, pp. 1085–1086.

[43] Yasui, T., and H. Goto, "Overview of Optical Switching Technologies in Japan," *IEEE Communications Magazine*, Vol. 25, No. 5, May 1987, pp. 10–15.

[44] Suzuki, S., T. Terakado, K. Komatsu, K. Nagashima, A. Suzuki, and M. Kondo, "An Experiment on High-Speed Optical Time-Division Switching," *J. Lightwave Technology*, Vol. LT-4, No. 7, July 1986, pp. 894–899.

[45] Yamakoshi, S., "Application of Bistable Laser Diodes," in Gallium *Aresenide and Related Compounds*, IOP Publising Ltd, 1990, pp. 33–42.

[46] Shimoe, T., S. Kuroyanagi, K. Murakami, H. Rokugawa, N. Mekada, and T. Odagawa, "An Experimental 512 Mbps Time-Division Photonic Switching System," *Photonic Switching '89*, FC2.

[47] Kawaguchi, H., "Polarization Bistable Laser Diodes" *Int. J. Nonlinear Optical Physics*, Vol. 2, No. 3, 1993, pp. 367–389.

[48] Fujimoto, N., H. Rokugawa, K. Yamaguchi, S. Masuda, and S. Yamakoshi, "Photonic Highway: Broad-Band Ring Subscriber Loops Using Optical Signal Processing," *J. Lightwave Technology*, Vol. LT-7, No. 11, November 1989, pp. 1798–1805.

[49] Matsunaga, T., K. Yukimatsu, and H. Ishikawa, "Photonic ATM Switching Technology," No. 30, IEEE Tokyo Section, Denshi Tokyo, 1991, pp. 138–142.

[50] Suzuki, S., and K. Kasahara, "Electro-Photonic Devices for Gigabit Networks," *The IEEE Magazine of Lightwave Telecommunication Systems*, Vol. 3, No. 3, August 1992, pp. 36–40.

[51] Marshall, I. W., M. J. O'Mahony, D. M. Cooper, P. J. Fiddyment, J. C. Regnault, and W. J. Devlin, "Gain Characteristics of a 1.5 μm Nonlinear Split Contact Laser Amplifier," *Appl. Phys. Lett.*, Vol. 53, No. 17, October 1988, pp. 1577–1579.

[52] Adams, M. J., "Theory of Two-Section Laser Amplifiers," *Optical and Quantum Electronics*, Vol. 21, 1989, pp. S15–S31.

[53] Barnsley, P. E., I. W. Marshall, H. J. Wickes, P. J. Fiddyment, J. C. Regnault, and W. J. Devin, "Absorptive and Dispersive Switching in a Three Region InGaAsP Semiconductor Laser Amplifier at 1.57 μm," *J. Modern Optics*, Vol. 37, No. 4, 1990, pp. 575–583.

[54] Barnsley, P. E., I. W. Marshall, P. J. Fiddyment, and M. J. Robertson, "Absorptive Nonlinear Semiconductor Amplifiers for Fast Optical Switching," *SPIE Symposium, Optically Activated Switching*, OE/Boston '90, Vol. 1378, pp. 116–126.

[55] Calzavara, M., P. Gambini, M. Puleo, B. Bostica, P. Cinato, and E. Vezzoni, "Optical-Fiber-Loop Memory for Multiwavelength Packet Buffering in ATM Switching Application," *Tech. Dig. of OFC'93*, TuE3, pp. 19.

[56] Tsuda, H., K. Nonaka, K. Hirabayashi, and T. Kurokawa," Digitally Regenerating Optical Loop Memory With an Optical Threshold Device," *IECE Japan*, 1993, B-917 (in Japanese).

[57] Yasui, N., and K. Kikuchi, "Photonic Switching/Network Architectural Possibilities," *Photonic Switching*, T. K. Gustafson and P. W. Smith, eds., New York: Springer, p. 24.

[58] Kawaguchi, H., K. Magari, H. Yasaka, M. Fukuda, and K. Oe, "Tunable Optical-Wavelength Conversion Using an Optical Triggerable Multielectrode Distributed Feedback Laser Diode," *IEEE J. Quantum Electronics*, Vol. QE-24 1988, pp. 2153–2159.

[59] Kawaguchi, H., "Bistable Laser Diodes and Their Applications for Photonic Switching," *Int. J. Optoelectronics*, Vol. 7, No. 3, 1992, pp. 301–348.

[60] Suzuki, S., M. Nishio, T. Numai, M. Fujiwara, M. Itch, S. Murata, and N. Shimosaka, "A Photonic Wavelength-Division Switching System Using Tunable Laser Diode Filters, " *J. Lightwave Technology*, Vol. 8, No. 5, May 1990, pp. 660–666.

[61] Numai, T., S. Murata, and I. Mito, "1.5 μm Tunable Wavelength Filter Using a Phase-Shift-Controlled Distributed Feedback Laser Diode With a Wide Tuning Range and a High Constant Gain," *Appl. Phys. Lett.*, Vol. 54 No. 19, May 1989, pp. 1859–1860.

[62] Barnsley, P. E., and P. J. Chidgey, "All-Optical Wavelength Switching From 1.3 μm to a 1.55 μm WDM

Wavelength Routed Network: System Results," *IEEE Photonics Tech. Lett.*, Vol. 4, No. 1, January 1992, pp. 91–94.

[63] Barnsley, P. E., and P. J. Fiddyment, "Wavelength Conversion From 1.3 μm to 1.5 μm Using Split Contact Optical Amplifier," *IEEE Photonics Tech. Lett.*, Vol. 3, No. 3, March 1991, pp. 256–259.

[64] Chidgey, P. J., I. Hawker, G. R. Hill, and H. J. Westlake, "Role of Reconfigurable Wavelength-Multiplexed Networks and Links for Future Optical Network," *Proc. Photonic Switching '91*, Salt Lake City, Utah, March 1991, paper FD1.

[65] Pottier, P., M. J. Chawki, R. Auffret, G. Claveau, A. Tromeur, "1.5 Gbit/s Transmission System Using All Optical Wavelength Converter Based on Tunable Two-Electrode DFB Laser," *Electron. Lett.*, Vol. 27, No. 23, November 1991, pp. 2183–2185.

[66] Bachus, E.-J., R. P. Braun, Ch. Caspar, H.-M. Foisel, K. Heimes, N. Keil, B. Strebel, J. Vathke, and M. Weickhmann, "Coherent Optical Multicarrier Switching Node," *OFC'89*, 1989, PD 13-1.

[67] Inoue, K., "Optical Frequency Exchange Utilising LD Amplifiers and Mach-Zehnder Filters," *Electron. Lett.*, Vol. 25, No. 10, May 1989, pp. 630–632.

[68] Großkopf, G., L. Küller, R. Ludwig, R. Schnabel, and H. G. Weber, "Semiconductor Laser Optical Amplifiers in Switching and Distribution Networks," *Optical Quantum Electronics*, Vol. 21, 1989, pp. S59–S74.

[69] Toba, H., K. Oda, K. Nosu, N. Takato, and H. Miyazawa, "5GHz-Spaced, Eight-Channel Optical FDM Transmission Experiment Using Guided-Wave Tunable Demultiplexer," *Electron. Lett.*, Vol. 24, No. 2, January 1988, pp. 78–80.

[70] Kikuchi, N., and S. Sasaki, "Noise Analysis for Optical Frequency Conversion Using Nearly Degenerate Four Wave Mixing in Semiconductor Optical Amplifier," *J. Lightwave Technology*, Vol. 11, No. 5/6, May/June 1993, pp. 819–828.

[71] Hinton, H. S., "Photonics in Switching," *IEEE LTS*, August 1992, pp. 26–35.

[72] Pan, Z., and M. Dagenais, "Subnanosecond Optically Addressable Generalized Optical Crossbar Switch With an Aggregate Throughput Rate of 4.2 Gbit/s," *Appl. Phys. Lett.*, Vol. 62, No. 18, May 1993, pp. 2185–2187.

Chapter 11
Future Prospects

1.1 IMPROVEMENT OF LASER DIODE PERFORMANCE BY NEW TECHNOLOGIES

Photonic functional devices based on LD nonlinearities will have greater performance and richer functions as semiconductor laser technologies improve. Here, we will outline two new research fields, which are rapidly progressing: (1) quantum wire and quantum box (quantum dot) and (2) microcavity.

1.1.1 Quantum Wire and Quantum Box

Considering the huge success of physics systems and devices based on 2D heterostructures, it is a natural trend to continue to diminish systems' dimensionality and recent years have witnessed a huge effort toward reducing the dimensionality of systems to one (1D) and zero (0D) dimensions. Arakawa et al. [1] proposed the concept of quantum-wire LDs or quantum-box LDs with, respectively, a 1D and a 0D electronic system.

Figure 11.1(a) shows simple illustrations of the active layer in MQW, multiquantum-wire, and multiquantum-box lasers. By making such multidimensional microstructures, the freedom of the carrier motion is reduced to one or zero. The density of states of electrons in these structures is expressed as

$$\rho_c^{\text{wire}}(\epsilon) = \left(\frac{m_c}{2\hbar^2\pi^2}\right)^{1/2} \sum_{l,m} \frac{1}{\sqrt{\epsilon - \epsilon_l - \epsilon_m}}$$

for the quantum-wire laser   (11.1)

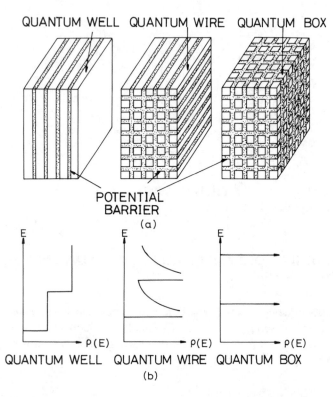

Figure 11.1 (a) Illustration of the active layer with a MQW structure, a multi-quantum-wire structure, and a multiquantum-box structure; (b) density of states of electrons in these structures. (©1986 IEEE. From [1].)

$$\rho_c^{box}(\epsilon) = \sum_{l,m,k} \delta(\epsilon - \epsilon_l - \epsilon_m - \epsilon_k)$$

for the quantum-box laser   (11.2)

where ϵ_l, ϵ_m, and ϵ_k are the quantized energy levels of a quantum-wire laser and a quantum-box laser. As shown in Figure 11.1(b), the density of states has a more peaked structure with the decrease of the dimensionality. This leads to a change in the gain profile, a reduction of threshold current density.

Figure 11.2 shows the maximum gain as a function of current density calculated for GaAs/Ga$_{0.8}$Al$_{0.2}$As bulk (conventional DH) and QWs with various quantization dimensions [2]. The following assumptions have been made: the thickness for bulk crystal is 0.15 μm; for the quantum wire, wires are arrayed with the separation length equal to the wire width; and for the quantum dots, cubic boxes are arrayed in the structure. The layer

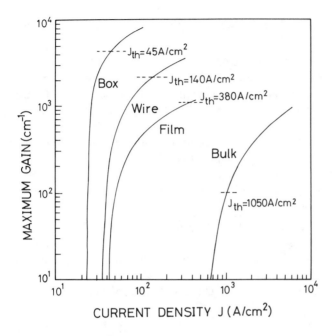

Figure 11.2 Maximum gain as a function of injection current density calculated for GaAs/Ga$_{0.8}$Al$_{0.2}$As quantum box, quantum wire, quantum film, and bulk crystal (conventional DH). Dashed line on each curve is the level of gain required for the laser threshold. Dimensions of active region are as follows. Box: $10 \times 10 \times 10$ nm; wire: 10×10 nm; film: 10 nm; bulk: 0.15 μm. ($^\circ$1986 IEEE. From [2].)

number n is one for all cases. The increases of gain with the quantization dimension is observed in the figure.

The threshold gains calculated are shown in Figure 11.2 by the dashed lines for bulk and all the QWs. The threshold current densities are obtained for GaAs/GaAlAs to be 1,050 A/cm^2 for bulk, 380 A/cm^2 for quantum film, 140 A/cm^2 for quantum wire, and 45 A/cm^2 for quantum box. These threshold current densities can be reduced further by optimizing the well thicknesses or dot dimensions.

11.1.2 Surface-Emitting Lasers

For many applications requiring a 2D laser array or monolithic integration of lasers with electronic components, it is desirable to have the laser output normal to the surface of the wafer. Such lasers are known as *surface-emitting lasers* [3]. There is a class of surface-emitting lasers that have their optical cavity normal to the surface of the wafer. These devices are known as vertical-cavity surface-emitting lasers (VCSELs) in order to distinguish them from other surface emitters such as the grating-coupled surface-emitting lasers.

As we can see from the scheme of the VCSEL, it has many advantages, as follows [4].

1. A 2D laser array could be monolithically fabricated.
2. The initial laser test could be made before separating into chips.
3. Single-mode longitudinal operation is expected because of its large mode spacing
4. Other electric devices can be monolithically integrated on the same substrate two-dimensionally around the laser. Also, it is possible to stack vertically multi-thin-film functional optical devices on the surface-emitting laser, such as an optical switch, modulator, optical amplifier.
5. A narrow circular beam is achievable.
6. A heteromultilayer can be used as a reflecting mirror in a way similar to a DBR laser or multi-active-layer DFB laser.

Room-temperature CW operation was achieved in GaAlAs/GaAs (about 890 nm) [5] and GaInAs/GaAs (about 980 nm) systems [6], and very recently in a GaInAsP/InP (about 1.37 μm) system [7]. Low (submilliampere) threshold currents are now possible, and device efficiency has increased dramatically. Coldren et al. observed on 10-μm square devices CW output power of more than 3 mW at room temperature and 0.4 mW at 100°C in a GaInAs/GaAs system [8].

11.1.3 Microcavity Laser Diodes

Lately, a lot of interest has been focused on the spontaneous emission properties of materials in microcavities [9]. These structures are resonators having at least one dimension size on the order of a wavelength. In the framework of the Fermi golden rule, the effect of optical confinement in one or more dimensions is understood as a rearrangement of the usual free-space density of photon states. The density of photon states (mode density) at some frequencies will be increased, whereas at others it will be decreased. Furthermore, this increase or decrease will be accompanied by a spatial redistribution of mode density. Thus, if a photon-emitting medium is introduced into such a cavity, its spontaneous emission rate and spatial emission intensity distribution will be altered, depending on the cavity-mode density at the emission frequency. In the last decade, much work has been done in this research field, which is called *cavity quantum electrodynamics* (cavity QED) particularly as a means of studying the interaction of matter with vacuum field fluctuations.

Altering the spontaneous emission, however, is also interesting from the point of view of device application. Of particular interest is the concept of a thresholdless laser proposed by Kobayashi et al. [10]. Recent successful demonstration of controlling spontaneous emission and nearly thresholdless laser operation using condensed materials hold technological promise for constructing LDs with ultralow power consumption. It should be noted that after the first success in the current injection VCSEL, marked progress has been seen in constructing high-performance VCSELs. Further technological progress in these VCSELs will be naturally combined with the cavity QED approach.

Figure 11.3(a) schematically represents the operation principle of a thresholdless laser. Suppose that a light-emitting material has a single narrow emission band with an extremely high quantum efficiency. In the mode point of view, the excited atoms are mostly coupled with free-space modes in a conventional large-sized cavity, even though there is only one cavity mode within the emission bandwidth. That is to say that most of the spontaneous emission radiates out from the side of a conventional laser cavity. In that situation, the cavity-mode photon number can only increase rapidly above the threshold due to stimulated emission. Thus, the phase transition (threshold) appears in the cavity-mode output. On the other hand, in the ideal microcavity, all the emitted photons couple into the single-cavity resonance mode. Therefore, by increasing pumping, the emission process gradually changes from spontaneous to stimulated emission without a phase transition (threshold) in the input-output curve.

Steady-state solutions for an ideal four-level laser are shown in Figure 11.3(b) with logarithmic scales. β is the parameter in this calculation; when the spontaneous emission coupling β is quite small, clear thresholds are observed in the input-output curves. (Note that in conventional semiconductor laser devices, β ranges from 10^{-6} to 10^{-5} per cavity mode.) It can be seen, however, that the threshold becomes unclear as β increases, and it disappears in the input-output curve at $\beta = 1$. Although it may not be meaningful to distinguish spontaneous emission and stimulated emission if there is no threshold and all

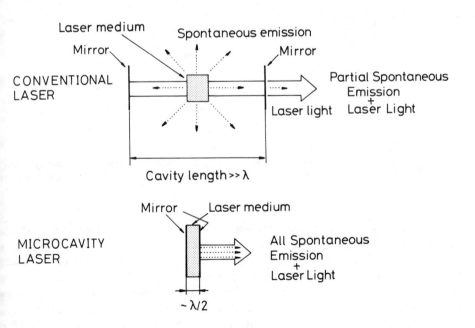

Figure 11.3(a) Operating principle of microcavity laser. (Reprinted with permission from Chapman & Hall Ltd. From [9].)

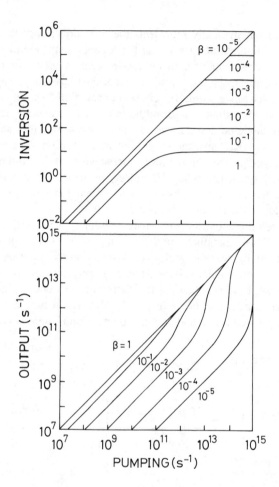

Figure 11.3(b) Calculated light output S_{out} and population inversion n versus pumping p of microcavity four level lasers in logarithmic scales. (Reprinted with permission from Chapman & Hall Ltd. From [9].)

the photon emission processes are controlled by the cavity, for convenience, we designate the emission rate proportional to s in the equation as stimulated emission. If we pay attention to the behavior of the population inversion, the difference between the spontaneous emission dominant region and the laser oscillation region is separated in the figure even for $\beta = 1$. With increasing pumping, n linearly increases within the spontaneous emission dominant region. On the other hand, in the laser oscillation region, n is clamped at the lasing threshold level. Thus, in that sense a fuzzy threshold still exists, although it does not appear in the input-output curves. The threshold in the population inversion n and also in the pumping rate (i.e., input power) decreases with increasing β. This is because of decrease in the mode volume.

A microcavity structure has the potential for fast switching as well as lessening the amount of power it uses. It has been recently pointed out that, with microcavity lasers, the dynamic response speed is improved [11]. If this technology is successfully applied to nonlinear LD devices, their dynamic response speed will be much improved. For example, the BLD that includes both an active and a saturable absorber layer in its microcavity will have very fast switching speed [12].

Let us consider the case that an active layer and a saturable absorber are successively grown, and a microcavity is fabricated using a multilayer DBR mirror structure. We extend the rate equations of microcavity lasers [11] to the microcavity BLD with a saturable absorber [12]. In the case of a closed cavity resonator, where β is close to unity, we cannot obtain bistability because the absorber in the cavity saturates through pumping owing to spontaneous emission. However, by choosing the open resonator with the appropriate β values, we can obtain bistability and dramatically improve the switching speed of the bistable operation.

A number of researchers have studied cavity QED effects and spontaneous emission suppression or enhancement in 3D microresonators. The utility of 1D resonators for quantum optics is limited in that radiation in the other two dimensions is not confined. The hemispherical geometry is promising since it provides transversal as well as longitudinal mode confinement on a wavelength scale [13].

A photonic bandgap crystal may overcome the limitations of resonators. The electronic energy band structure in semiconductors is well understood. An electron's momentum is modulated by a lattice of ions, yielding a gap of forbidden electronic energies irrespective of the electron's momentum. Yablonovitch [14–16] proposed that a periodic electromagnetic modulation created by a lattice composed of dielectric "atoms" could yield a gap of forbidden photon energies irrespective of a photon's momentum. Such a dielectric "crystal" has been dubbed a *photonic bandgap* (PBG) structure. Inhibition of optical spontaneous emission from a dye has already been observed in a PBG-like structure made of a monodispersed colloidal suspension, composed of negatively charged polystyrene spheres crystallized into a face-centered cubic structure [17]. Light localization is also possible [18].

11.2 CONSUMED POWER AND SWITCHING SPEED OF BISTABLE LASER DIODES

One of the most important remaining problems with photonic functional devices is the reduction of the optical switching energy and the electrical biasing power. The switching energy strongly depends on how close the bias current is to the laser threshold in an absorptive BLD. In a dispersive BLD, it depends on the initial detuning and the bias current. Figure 11.4(a) shows the progress in bistable switching energy from 1981 to the present. The minimum switching power reported so far is 200 nW and 1 μW for absorptive and dispersive bistability, respectively. The minimum switching energies are 700 and 500 aJ, respectively.

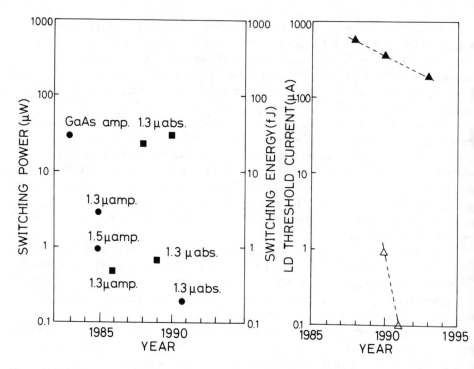

Figure 11.4(a) Progress in performances of BLDs and LDs. Switching power (●) and energy (■) of BLDs as well as threshold current of LDs (▲ experimental, △ theoretical).

A reduction in the electrical biasing power can be realized by reducing the threshold current of LDs. LDs with threshold currents of less than 1 mA have already been reported by some research groups. For example, strained V-groove lasers with a threshold current as low as 203 μA have been demonstrated at room temperature [19]. An airpost microcavity surface-emitting laser with a threshold current as low as 190 μA has been demonstrated under pulsed operation at room temperature [20]. It is believed that by using advanced technologies such as quantum-confined effects, LDs with a much lower threshold current (1 μA threshold LDs) will be achieved [14]. Yamanishi and Yamamoto have proposed a semiconductor surface-emitting laser structure in which all spontaneous emission is coupled into a single lasing mode by means of a quantum microcavity, and discrete electron-hole pair emission is made free of absorption by means of a dc-biased quantum dot. They estimated that the threshold current of such an LD can be reduced to below 100 nA [21]. The minimum LD threshold currents versus year are also shown in Figure 11.4(a).

Another important problem is that of improving the switching speed and repetition rate. As described in Chapter 3, the repetition rate of the flip-flop operation has been improved to 5 Gbps using a BLD with reset electrical pulses (Figure 11.4(b)). In dispersive

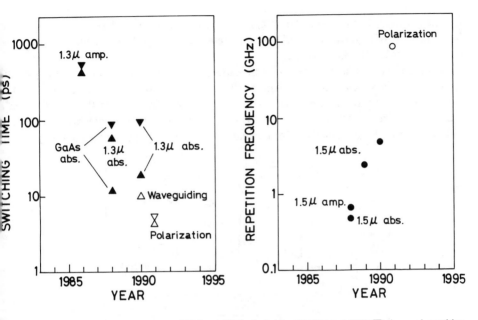

Figure 11.4(b) Progress in performances of BLDs and LDs. Switching ON (▲) and OFF (▼) time and repetition frequency (●) of BLDs. Δ , ∇, and O are theoretical values.

BLDs, about 1 Gbps seems to be the maximum repetition rate so far. In two-mode BLDs such as polarization BLDs, however, extremely high repetition rate (>50 Gbps) can be expected with the help of fast optical trigger pulses.

11.3 ULTRAFAST SWITCHING BASED ON NONLINEAR EFFECTS WITH FAST RESPONSE TIME

Up to now, we have mainly considered the loss (or gain) change and the refractive index change due to the interband carrier generation (or recombination) in LDs as nonlinear mechanisms. The typical lifetime of these mechanisms is on the order of nanoseconds if there are no additional accelerating mechanisms such as a reversed-bias voltage or a strong stimulated emission. In LDs, however, there are ultrafast nonlinear mechanisms due to the nonequilibrium carrier distribution in the bands, such as carrier-heating or spectral-hole-burning effects, as described in Section 2.3.3. Relaxation is characterized with the intraband relaxation time on the order of 10^{-12} to 10^{-13} sec.

The gain and refractive index dynamics were studied in the three regimes of LD operation: gain, transparency, and absorption [22]. These three regimes are characterized by whether the gain for the probe pulses is positive, zero, or negative, respectively. Each regime can be accessed by tuning the wavelength of the probe pulses or by changing the

LD bias current. For the data shown here, the probe pulses' (dye laser) wavelength was held fixed at 836 nm and the LD bias current was changed. The device used in the experiment is a CSP AlGaAs LD.

Figure 11.5(a) shows the probe transmission signal as a function of pump-probe delay. The pulses were 440 fs in duration. The LD bias current is varied as indicated in the figure. In all three regimes we observe a negative gain transient that recovers on a picosecond time scale. The level that the gain recovers to is determined by the pump-induced carrier density changes. For instance, in the gain regime, where the pump stimu-

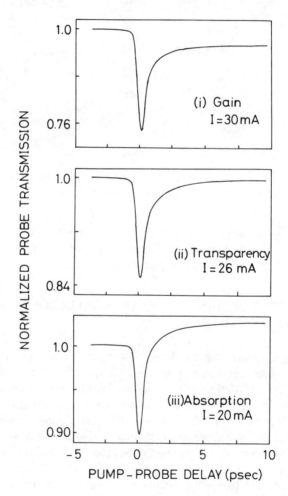

Figure 11.5(a) Probe transmission as a function of probe delay relative to the pump in the (i) gain, (ii) transparency, and (iii) absorption regimes. The pump-probe wavelength was 836 nm, and the pulse duration was 440 fs. The diode laser bias currents are as indicated.

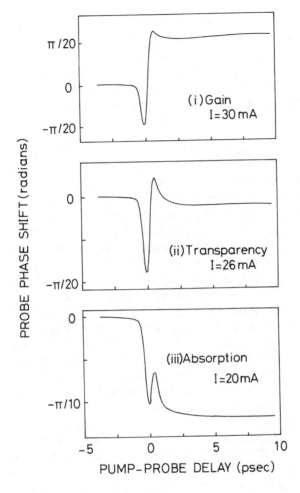

Figure 11.5(b) Probe phase shift as a function of probe delay relative to the pump in the (i) gain, (ii) transparency, and (iii) absorption regimes. The pump-probe wavelength was 836 nm, and the pulse duration was 440 fs. The diode laser bias currents are as indicated. (From [22].)

lates photon emission, the carrier density is reduced, and so the gain recovers to a lower value. These long-term gain changes persist on the order of a nanosecond (the carrier lifetime). Free-carrier heating, in which the electron and hole distributions are heated through free-carrier absorption of pump photons, is the mechanism to explain the picosecond gain transient. Heating the carrier distributions results in a gain decrease that relaxes on a picosecond time scale as the carriers cool back to the lattice temperature.

Figure 11.5(b) shows the measured probe phase shift as a function of pump-probe delay. A positive probe phase shift corresponds to an increase in the LD refractive index.

Note the negative refractive index transient that appears in all three regimes of LD operation. This index change appears to be instantaneous in so far as it is not resolved by the 440-fs pulses used here. This refractive index transient is consistent with measurements of below bandgap refractive index changes in AlGaAs waveguides and is attributed to a rapid electronic or virtual process. Hultgren and Ippen also observe a positive refractive index change that settles to a quasiequilibrium value on a picosecond time scale. They attribute this positive refractive index change to free-carrier heating, the same process that is responsible for the negative gain transient discussed above. By relating a gain change to an index change through the Kramers-Kronig relation, it can be shown that, for photon energies near the bandgap energy, an increase in carrier temperature results in an increase in the refractive index. Thus, the gain decrease in Figure 11.5(a) and the refractive index increase in Figure 11.5(b) are consistent with the free-carrier heating model. As with the probe transmission measurements, the long-term behavior of the refractive index is determined by the pump-induced carrier density change. A given carrier density change results in a refractive index change that is opposite in sign at these wavelengths. Thus, in the gain regime, for example, where the pump reduces the carrier density through stimulated emission, the long-term refractive index change is positive.

To exploit such effects for switching, an interferometric structure was constructed to produce an intensity switch. Davies et al. [23] have built devices based on the nonlinear directional coupler. Directional couplers were constructed based on LD amplifiers with an InGaAsP ($\lambda = 1.55 \ \mu$m) active region 0.15 μm thick. Wet chemical etching was used to define ridge waveguides with an effective index step of about 0.02 for TE-polarized light. The couplers consisted of 2-μm waveguides separated by 2 μm, and were designed to only support zero and first-order supermodes. Contact windows on top of the waveguides allowed current injection through each ridge. The top electrical contact was continuous across the device, so that the injected current was shared equally between the two waveguides. Single-ridge devices 3 μm wide for assessment purposes were also produced.

The coupling coefficients for the couplers were found by measuring the splitting of the Fabry-Perot resonances in the transmission spectra when the devices were operated just below their lasing threshold. The coupling length for these devices was thus deduced to be 1.4 mm for TE-polarized input light. Couplers of this length were cleaved and AR-coated, and were found to show crosstalks of better than 10 dB (i.e., for light input into guide 1, at the device output more than 90% of the transmitted light is in guide 2).

Experimental evaluation of the nonlinear operation of these devices was undertaken with a mode-locked color-center laser, which could be tuned between 1.47 and 1.56 μm. The input polarization was maintained as TE for all the experiments to be described.

When tested with the mode-locked color-center laser giving 14-ps pulses, the coupler showed nonlinear transmission characteristics. Nonlinear input-output plots were seen above and below transparency, with the magnitude of the effect being reduced as the current was increased. The normalized transmission in the bar and cross states at a transparency of 120 mA is shown in Figure 11.6. The average power into the device is shown in the figure: an average power of 10 mW corresponds to a peak power of 8.9W

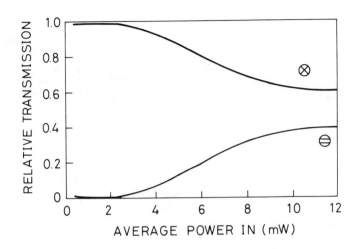

Figure 11.6 Relative transmission of directional coupler showing nonlinear operation with 14-ps input pulses. 10 mW average power corresponds to 8.9W peak power launched into device. (From [23].)

launched into the device. As can be seen, the outputs do not cross over (although they do at lower currents, where band-filling effects will also contribute to the switching). The poor switching behavior is ascribed to the high internal loss and agrees well with theory for a lossy coupler with loss and nonlinear index as calculated from the measurements described on a device having one waveguide. The performance shown in Figure 11.6 is also degraded from the CW characteristics because it represents an averaged measurement that includes unswitched light from the low-intensity pulse wings. Time-resolved measurement of this device is required to fully characterize the switching speed. However, from the form of the self-phase modulation measurements, and from earlier results [22,24,25], we can expect the switch-on and switch-off times to be below 10 ps.

11.4 TWO-DIMENSIONAL ARRAY

To date, BLDs have been constructed using a single LD chip. However, using surface-emitting LD technology, we will also be able to make 2D arrays of photonic functional devices such as a BLD array. When the threshold current is reduced to 1 μA, a 1,000 $\times$ 1,000 array can be operated with 1W of power, since the operating voltage is about 1V. Von Lehmen et al. have reported the fabrication of an 8 $\times$ 8 independently addressable InGaAs/GaAs VCSEL array and a 32 $\times$ 32 matrix-addressable VCSEL array [26].

Three types of surface-emitting BLDs have been proposed so far, as shown in Figure 11.7 [27]. Vertical-cavity-laser-type optical logic and gate devices were proposed by Nitta et al (Figure 11.7(a)) [28]. These devices include a saturable absorber in the vertical laser cavity. Pan et al. [29] have reported observation of bistable polarization switching in a

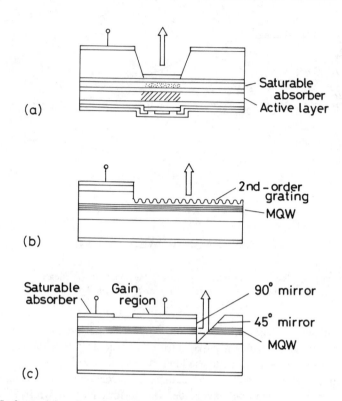

Figure 11.7 Surface-emitting absorptive BLDs: (a) vertical-cavity BLD; (b) DBR BLD with a second-order grating; (c) BLD with an integrated beam deflector. (From [27].)

VCSEL under optical injection. This polarization switching is achieved through injection locking (see Section 4.2), where both the wavelength and the polarization of the vertical-cavity laser are locked to the injected optical signal. Although the switching speed of this polarization bistability has not yet been clarified, polarization self-modulation at frequencies up to 6 GHz has been observed in a VCSEL [30]. This self-oscillation is observed when a portion of the laser output power is injected back into the laser after having rotated its polarization by 90 degrees with respect to the initial laser polarization state. A surface-emitting BLD using output coupling from a second-order grating of a DBR (Figure 11.7(b)) [31] and the laser with an integrated beam deflector (Figure 11.7(c)) [32] have also been demonstrated. However, the threshold currents of these devices are more than 50 mA at present. To construct 2D arrays, the threshold currents have to be reduced.

These technologies will open up the possibility of using such devices as key components for optical computing in which parallel processing is used. An optical crossbar switch and optical associative memory using BLD arrays have also been proposed [33].

REFERENCES

[1] Arakawa, Y., and A. Yariv, "Quantum Well Lasers--Gain, Spectra, Dynamics," IEEE J. Quantum Electronics, Vol. QE-22, No. 9, September 1986, pp. 1887–1899.

[2] Asada, M., Y. Miyamoto, and Y. Suematsu, "Gain and the Threshold of Three-Dimensional Quantum-Box Lasers," IEEE J. Quantum Electronics, Vol. QE-22, No. 9, September 1986, pp. 1915–1921.

[3] Evans, G. A., and J. M. Hammer, ed., Surface Emitting Semiconductor Laser and Arrays, Boston: Academic Press, 1993.

[4] Iga, K., and S. Uchiyama, "GaInAsP/InP Surface-Emitting Laser Diode," Optical and Quantum Electronics, Vol. 18, 1986, pp. 403–422.

[5] Koyama, F., F. Kinoshita, and K. Iga, "Room Temperature CW Operation of GaAs Vertical Cavity Surface Emitting Laser," Trans. of the IEICE, Vol. E71, No. 11, November 1988, pp. 1089–1090.

[6] Lee, Y. H., J. L. Jewell, A. Scherer, S. L. McCall, J. P. Harbison, and L. T. Florez, "Room-Temparature Continuous-Wave Vertical-Cavity Single-Quantum-Well Microlaser Diodes," Electron. Lett., Vol. 25, No. 20, September 1989, pp. 1377–1378.

[7] Baba, T., Y. Yogo, K. Suzuki, F. Koyama, and K. Iga, "First Room Temperature CW Operation of GaInAsP/InP Surface Emitting Laser," IEICE Trans. Electron., Vol. E76-C, No. 9, September 1993, pp. 1423–1424.

[8] Coldren, L. A., R. S. Geels, S. W. Corzine, and J. W. Scott, "Efficient Vertical-Cavity Lasers," Optical and Quantum Electronics, Vol. 24, 1992, pp. S105–S119.

[9] Yokoyama, H., K. Nishi, T. Anan, Y. Nambu, S. D. Brorson, E. P. Ippen, and M. Suzuki, "Controlling Spontaneous Emission and Threshold-less Laser Oscillation With Optical Microcavities," Optical and Quantum Electronics, Vol. 24, 1992, pp. S245–S272.

[10] Kobayashi, T., T. Segawa, A. Morimoto, and T. Sueta, "A New Type of Lasers, Light Emitters, and Optical Functional Devices, in Which the Spontaneous Emission Is Controlled," Tech. Dig. of 43th Fall Meeting of Japanese Appl. Phys. Society, September 1982, paper 29a-B-6 (in Japanese).

[11] Yokoyama, H., and S. D. Brorson, "Rate Equation Analysis of Microcavity Lasers," J. Appl. Phys., Vol. 66, No. 10, November 1989, pp. 4801–4805.

[12] Kawaguchi, H., "Optical Nonlinearities in Semiconductor Lasers and Their Applications for Functional Devices," Int. J. Nonlinear Optical Phys., Vol. 1, No. 1, January 1992, pp. 203–221.

[13] Matinaga, F. M., A. Karlsson, S. Machida, Y. Yamamoto, T. Suzuki, Y. Kadota, and M. Ikeda, "Low-Threshold Operation of Hemispherical Microcavity Single-Quantum-Well Lasers at 4 K," Appl. Phys. Lett., Vol. 62, No. 5, February 1993, pp. 443–445.

[14] Yablonovitch, E., "Inhibited Spontaneous Emission in Solid-State Physics and Electronics," Phys. Rev. Lett., Vol. 58, No. 20, May 1987, pp. 2059–2062.

[15] Yablonovitch, E., and T. J. Gmitter, "Photonic Band Structure: The Face-Centered-Cubic Case," Phys. Rev. Lett., Vol. 63, No. 18, October 1989, pp. 1950–1953.

[16] Yablonovitch, E., "Photonic Band-Gap Structures," J. Opt. Soc. Am. B, Vol. 10, No. 2, February 1993, pp. 283–295.

[17] Martorell, J., and N. M. Lawandy, "Observation of Inhibited Spontaneous Emission in a Periodic Dielectric Structure," Phys. Rev. Lett., Vol. 65, No. 15, October 1990, pp. 1877–1880.

[18] John, S., and J. Wang, "Quantum Electrodynamics Near a Photonic Band Gap: Photon Bound States and Dressed Atoms," Phys. Rev. Lett.,Vol. 64, No. 12, May 1990, pp. 2418–2421.

[19] Tiwari, S., G. D. Pettit, K. R. Milkove, R. J. Davis, J. M. Woodall, and F. Legoues," 203 μA Threshold Current Strained V-Groove Laser," IEDM'92, pp. 859–862.

[20] Kawakami, T., T. Numai, T. Yoshikawa, M. Sugimoto, H. Yokoyama, Y. Sugimoto, K. Kasahara, and K. Asakawa, "The Airpost Microcavity Surface Emitting Laser (III): Sub 500 μA Laser Oscillation at Room Temperature," Extended Abstracts, 54th Autumn Meeting, The Japan Society of Applied Physics, September 1993, paper 28a-H-3 (in Japanese).

[21] Yamanishi, M., and Y. Yamamoto, "An Ultimately Low-Threshold Semiconductor Laser With Separate Quantum Confinements of Single Field Mode and Single Electron-Hole Pair," Japanese J. Appl. Phys., Vol. 30, No. 1A, January 1991, pp. L60–L63.

[22] Hultgren, C. T., and E. P. Ippen, "Ultrafast Refractive Index Dynamics in AlGaAs Diode Laser Amplifiers," Appl. Phys. Lett., Vol. 59, No. 6, August 1991, pp. 635–637.

[23] Davies, D. A. O., M. A. Fisher, D. J. Elton, S. D. Perrin, M. J. Adams, G. T. Kennedy, R. S. Grant, P. D. Roberts, and W. Sibbett, "Nonlinear Switching in InGaAsP Laser Amplifier Directional Couplers Biased at Transparency," Electron. Lett., Vol. 29, No. 19, September 1993, pp. 1710–1711.

[24] Grant, R. S., and W. Sibbett., "Observations of Ultrafast Nonlinear Refraction in an InGaAsP Optical Amplifier," Appl. Phys. Lett., Vol. 58, No. 11, March 1991, pp. 1119–1121.

[25] Fisher, M. A., H. Wickes, G. T. Kennedy, R. S. Grant, and W. Sibbett, "Ultrafast Nonlinear Refraction in an Active MQW Waveguide," Electron. Lett., Vol. 29, No. 13, June 1993, pp. 1185–1186.

[26] Von Lehmen, A., C. Chang-Hasnain, J. Wullert, L. Carrion, N. Stoffel, L. Florez, and J. Harbison, "Independently Addressable InGaAs/GaAs Vertical-Cavity Surface-Emitting Laser Arrays," Electron. Lett., Vol. 27, No. 7, March 1991, pp. 583–585.

[27] Kawaguchi, H., "Bistable Laser Diodes and Their Applications for Photonic Switching," Int. J. Optoelectronics, Vol. 7, No. 3, 1992, pp. 301–348.

[28] Nitta, J., Y. Koizumi, and K. Iga, "GaAs/AlGaAs Surface-Emitting-Laser-Type Optical Logic and Gate Devices," CLEO'86 Dig. of Technical Papers, San Francisco, FO4, p. 9.

[29] Pan, G. Z., S. Jiang, M. Dagenais, R. A. Morgan, K. Kojima, M. T. Asom, R. E. Leibenguth, G. D. Guth, and M. W. Focht, "Optical Injection Induced Polarization Bistability in Vertical-Cavity Surface-Emitting Lasers," Appl. Phys. Lett., Vol. 63, No. 22, November 1993, pp. 2999–3001.

[30] Jiang, S., Z. Pan, M. Dagenais, R. A. Morgan, and K. Kojima, "High-Frequency Polarization Self-Modulation in Vertical-Cavity Surface-Emitting Lasers," Appl. Phys. Lett., Vol. 63, No. 26, December 1993, pp. 3545–3547.

[31] Kojima, K., K. Kyuma, S. Nada, J. Ohta, and K. Hamanaka, "Ultrafast Switching Characteristics of a Bistable Surface-Emitting Multiple Quantum Well Distributed Bragg Reflector Laser," Appl. Phys. Lett., Vol. 52, No. 12, March 1988, pp. 942–944.

[32] Sugimoto, M., N. Hamao, N. Takado, M. Ueno, H. Iwata, M. Uchida, K. Onabe, K. Asakawa, and T. Yuasa, "Surface Emitting Bistable Multiquantum Well Lasers With a 45° Dry Etched-Mirror," Extended Abstracts of the 19th Conf. on Solid State Devices and Materials, Tokyo, p. 523.

[33] Sharfin, W. F., and M. Dagenais, "The Role of Nonlinear Diode Laser Amplifiers in Optical Processors and Interconnects," Optical and Quantum Electronics, Vol. 19, 1987, pp. S47–S56.

List of Acronyms

Each scientific field has its own jargon, and the field of laser diodes is no exception. Although an attempt was made to avoid the extensive use of acronyms, many still appear throughout the book. Each acronym is defined the first time it appears in the text. As a further help, the appendix lists all acronyms in alphabetical order.

AM	amplitude modulation
AO	acousto-optic
APD	avalanche photodiode
AR coating	antireflection coating
ASE	amplified spontaneous emission
ASK	amplitude-shift keying
ATM	asynchronous transfer mode
BER	bit error rate
BH	buried heterostructure
B-ISDN	broadband integrated services digital network
BLD	bistable laser diode
BPF	bandpass filter
C^3 laser	cleaved-coupled-cavity laser
CD	compact disc
CPM	colliding-pulse mode locking
CSP LD	channeled-substrate-planar stripe laser diode
CW	continuous wave
2D	two-dimensional
3D	three-dimensional
DBR	distributed-Bragg reflector
DFB	distributed feedback
DFWM	degenerate four-wave mixing
DH	double heterostructure
DMUX	demultiplexer

DPSK	differential phase-shift keying
EDFA	erbium-doped fiber amplifier
EL	electroluminescence
FDM	frequency-division multiplexing
FK effect	Franz-Keldysh effect
FM	frequency modulation
FSK	frequency-shift keying
FSR	free spectral range
FWHM	full width at half maximum
FWM	four-wave mixing
HD WDM	high-density wavelength-division multiplexing
hh	heavy hole
HNDFWM	highly nondegenerate four-wave mixing
IM	intensity modulation
LAN	local-area network
LC-FPI	liquid crystal Fabry-Perot interferometer
LD	laser diode
lh	light hole
LOC	large optical cavity
LPE	liquid-phase epitaxy
MOVPE	metalorganic vapor phase epitaxy
MQW	multiple quantum well
MUX	multiplexer
M-Z	Mach-Zehnder
NDFWM	nearly degenerate four-wave mixing
NLOA	nonlinear optical amplifier
NRZ	nonreturn to zero
OEIC	optoelectronic integrated circuits
PBG	photonic bandgap
PBRS	pseudorandom binary sequence
PBS	polarization beam splitter
PC	polarization controller
PSK	phase-shift keying
QCFK effect	quantum-confined Franz-Keldysh effect
QCSA	quantum-confined Stark effect
QED	quantum electrodynamics
QW	quantum well
RZ	return to zero

SEM	scanning electron microscope
SMRR	side-mode rejection ratio
TDM	time-division multiplexing
TTG	tunable twin guide
TWA	traveling wave type LD amplifier
VCSEL	vertical-cavity surface-emitting laser
WDM	wavelength division multiplexing
XCBLD	cross-coupled bistable laser diode

About the Author

Hitoshi Kawaguchi was born in Japan on May 29, 1949. He received a B.E. degree in electronics from Niigata University, Niigata, Japan, in 1972, and M.E. and Ph.D. degrees in electronics from Tohoku University, Sendai, Japan, in 1975 and 1982, respectively.

In 1975, he joined NTT Electrical Communication Laboratories where he worked on semiconductor lasers, optical bistability and instability, and semiconductor optical switches. Since 1988, he has been a faculty member of Yamagata University in Yonezawa. He spent one year (1990–1991) at Professor John Carroll's research group at Cambridge University, in Cambridge, England, as a U.K. SERC Visiting Fellow. He is presently a professor of the Optoelectronics Research Group at Yamagata University. His research interests are in the areas of optoelectronics, photonic devices, and nonlinear materials.

Index

The Artech House Optoelectronics Library

Brian Culshaw, Alan Rogers, and Henry Taylor, *Series Editors*

Principles of Modern Optical Systems, Volumes I and II, I. Andonovic and
  D. Uttamchandani, editors

Reliability and Degradation of LEDs and Semiconductor Lasers, Mitsuo Fukuda

Semiconductors for Solar Cells, Hans Joachim Möller

Single-Mode Optical Fiber Measurements: Characterization and Sensing, Giovanni
  Cancellieri

For further information on these and other Artech House titles, contact:

Artech House
685 Canton Street
Norwood, MA 02062
617-769-9750
Fax: 617-769-6334
Telex: 951-659
email: artech@world.std.com

Artech House
Portland House, Stag Place
London SW1E 5XA England
+44 (0) 71-973-8077
Fax: +44 (0) 71-630-0166
Telex: 951-659